Lecture Notes in Networks and Systems

Volume 484

The series "Lecture Notes in Networks and Systems" publishes the latest developments in Networks and Systems—quickly, informally and with high quality. Original research reported in proceedings and post-proceedings represents the core of LNNS.

Volumes published in LNNS embrace all aspects and subfields of, as well as new challenges in, Networks and Systems.

The series contains proceedings and edited volumes in systems and networks, spanning the areas of Cyber-Physical Systems, Autonomous Systems, Sensor Networks, Control Systems, Energy Systems, Automotive Systems, Biological Systems, Vehicular Networking and Connected Vehicles, Aerospace Systems, Automation, Manufacturing, Smart Grids, Nonlinear Systems, Power Systems, Robotics, Social Systems, Economic Systems and other. Of particular value to both the contributors and the readership are the short publication timeframe and the world-wide distribution and exposure which enable both a wide and rapid dissemination of research output.

The series covers the theory, applications, and perspectives on the state of the art and future developments relevant to systems and networks, decision making, control, complex processes and related areas, as embedded in the fields of interdisciplinary and applied sciences, engineering, computer science, physics, economics, social, and life sciences, as well as the paradigms and methodologies behind them.

Indexed by SCOPUS, INSPEC, WTI Frankfurt eG, zbMATH, SCImago.

All books published in the series are submitted for consideration in Web of Science.

For proposals from Asia please contact Aninda Bose (aninda.bose@springer.com).

More information about this series at https://link.springer.com/bookseries/15179

Wojciech Zamojski · Jacek Mazurkiewicz ·
Jarosław Sugier · Tomasz Walkowiak ·
Janusz Kacprzyk
Editors

New Advances in Dependability of Networks and Systems

Proceedings of the Seventeenth International Conference on Dependability of Computer Systems DepCoS-RELCOMEX, June 27 – July 1, 2022, Wrocław, Poland

Springer

Editors
Wojciech Zamojski
Wrocław University of Science and Technology
Wrocław, Poland

Jacek Mazurkiewicz
Wrocław University of Science and Technology
Wrocław, Poland

Jarosław Sugier
Wrocław University of Science and Technology
Wrocław, Poland

Tomasz Walkowiak
Wrocław University of Science and Technology
Wrocław, Poland

Janusz Kacprzyk
Polish Academy of Sciences
Systems Research Institute
Warsaw, Poland

ISSN 2367-3370 ISSN 2367-3389 (electronic)
Lecture Notes in Networks and Systems
ISBN 978-3-031-06745-7 ISBN 978-3-031-06746-4 (eBook)
https://doi.org/10.1007/978-3-031-06746-4

This Springer imprint is published by the registered company Springer Nature Switzerland AG
The registered company address is: Gewerbestrasse 11, 6330 Cham, Switzerland

Preface

In this volume, we would like to present proceedings of the Seventeenth International Conference on Dependability of Computer Systems DepCoS-RELCOMEX which is scheduled to be held online in Wrocław, Poland, from June 27 to July 1, 2022.

It is the third time when the conference will be organized with only remote access of participants. Two years ago, during preparation of the proceedings in February/March 2020, we did not predict that then-local pandemic would spread so fast also in Europe and the event, organized yearly since 2006, would have to be held exclusively online. Despite technical difficulties and limited experience in this area, the remote sessions successfully accomplished conference goals, although limited possibilities of online contacts could not match benefits of a real-life meeting in Brunów Palace—our traditional venue. With these positive experiences, it was easier to decide that the 2021 edition will be held also online, and now, despite of what we firmly believed in and hoped for a year ago, the difficult situation forces us again to cancel a traditional meeting and organize the conference exclusively with remote access. But we still have faith in better times coming and in gathering all of us in a real meeting where we will be able to freely discuss future of computer systems and of our planet, like we used to do every year since the first DepCoS meeting in 2006.

The conference is now organized by the Department of Computer Engineering at the Faculty of Information and Communication Technology, Wrocław University of Science and Technology, but its roots go back to the heritage of two other cycles of events: RELCOMEX (1977–1989) and Microcomputer School (1985–1995) which were organized by Institute of Engineering Cybernetics (predecessor of the Department) under the leadership of prof. Wojciech Zamojski, now also the DepCoS chairman. These proceedings are the first ones published in the series "Lecture Notes in Networks and Systems". Previous DepCoS volumes were printed, chronologically, first by the IEEE Computer Society (2006-2009), then by Wrocław University of Science and Technology Publishing House (2010-2012) and recently by Springer Nature in the "Advances in Intelligent Systems and Computing" volumes no. 97 (2011), 170 (2012), 224 (2013), 286 (2014), 365

(2015), 479 (2016), 582 (2017), 761 (2018), 987 (2019), 1173 (2020) and 1389 (2021). Springer Nature is one of the largest and most prestigious scientific publishers, with the LNNS titles being submitted for indexing in CORE Computing Research and Education database, Web of Science, SCOPUS, INSPEC, DBLP, and other indexing services.

DepCoS-RELCOMEX scope has always been focused on diverse issues which are constantly arising in performability and dependability analysis of contemporary computer systems and networks. Being probably the most complex technical systems ever engineered by man (and also—the most dynamically evolving ones), their organization cannot be any longer interpreted only as a structure built on the base of technical resources but their evaluation must take into account a unique blend of interacting people, networks, and a large number of users dispersed geographically and producing an unimaginable number of applications. Ever-growing number of research methods being continuously developed for such analyses apply the newest results of artificial and computational intelligence. Selection of papers in these proceedings illustrate broad variety of multi-disciplinary topics which are considered in contemporary dependability explorations; an increasing role of the latest methods based on machine/deep learning and neural networks in their analysis is also worth noticing.

Concluding this preface, we would like to thank everyone who participated in organization of the conference and in preparation of this volume: authors, members of the program and the organizing committees, and all who helped in this difficult time. Especially, we would like to gratefully acknowledge work of the reviewers whose opinions and comments were invaluable in selecting and enhancing the submissions. Our thanks go this year to: Ilona Bluemke, Dariusz Caban, DeJiu Chen, Frank Coolen, Manuel Gil Perez, Zbigniew Gomółka, Alexander Grakovski, Ireneusz Jóźwiak, Igor Kabashkin, Vyacheslav Kharchenko, Urszula Kużelewska, Alexey Lastovetsky, Jan Magott, Jacek Mazurkiewicz, Marek Młyńczak, Przemysław Śliwinski, Czesław Smutnicki, Robert Sobolewski, Janusz Sosnowski, Jarosław Sugier, Victor Toporkov, Tomasz Walkowiak, Min Xie, Irina Yatskiv (Jackiva), and Wojciech Zamojski. Their work, not mentioned anywhere else in this book, deserves to be highlighted and recognized in this introduction.

Finally, we would like to thank all authors who selected DepCoS-RELOCMEX as the platform to publish and discuss their research results. We believe that included papers will contribute to progress in design, analysis, and engineering of dependable computer systems and networks and will be an interesting source material for scientists, researchers, engineers, and students working in these areas.

Wojciech Zamojski
Jacek Mazurkiewicz
Jarosław Sugier
Tomasz Walkowiak
Janusz Kacprzyk

Organization

Seventeenth International Conference on Dependability of Computer Systems DepCoS-RELCOMEX

Wrocław, Poland, June 27—July 1, 2022

Program Committee

Wojciech Zamojski (Chairman)	Wrocław University of Science and Technology, Poland
Ali Al-Dahoud	Al-Zaytoonah University, Amman, Jordan
Andrzej Białas	Research Network ŁUKASIEWICZ—Institute of Innovative Technologies EMAG, Katowice, Poland
Ilona Bluemke	Warsaw University of Technology, Poland
Wojciech Bożejko	Wrocław University of Science and Technology, Poland
Eugene Brezhniev	National Aerospace University "KhAI," Kharkov, Ukraine
Dariusz Caban	Wrocław University of Science and Technology, Poland
De-Jiu Chen	KTH Royal Institute of Technology, Stockholm, Sweden
Jacek Cichoń	Wrocław University of Science and Technology, Poland
Frank Coolen	Durham University, UK
Mieczysław Drabowski	Cracow University of Technology, Poland
Francesco Flammini	University of Linnaeus, Sweden
Manuel Gill Perez	University of Murcia, Spain
Franciszek Grabski	Gdynia Maritime University, Gdynia, Poland
Aleksander Grakowskis	Transport and Telecommunication Institute, Riga, Latvia

Organizing Committee

Chair

Wojciech Zamojski	Wrocław University of Science and Technology, Poland

Members

Jacek Mazurkiewicz	Wrocław University of Science and Technology, Poland
Jarosław Sugier	Wrocław University of Science and Technology, Poland
Tomasz Walkowiak	Wrocław University of Science and Technology, Poland
Tomasz Zamojski	Wrocław University of Science and Technology, Poland
Mirosława Nurek	Wrocław University of Science and Technology, Poland

Contents

Robustness on Diverse Data Disturbance Levels of Tabu Search for a Single Machine Scheduling

Wojciech Bożejko[1], Paweł Rajba[2(✉)], and Mieczysław Wodecki[3]

[1] Department of Automatics, Mechatronics and Control Systems, Faculty of Electronics, Wrocław University of Technology, Janiszewskiego 11-17, 50-372 Wrocław, Poland
wojciech.bozejko@pwr.wroc.pl

[2] Institute of Computer Science, University of Wrocław, Joliot-Curie 15, 50-383 Wrocław, Poland
pawel@cs.uni.wroc.pl

[3] Telecommunications and Teleinformatics Department, Faculty of Electronics, Wrocław University of Science and Technology, Wybrzeże Wyspiańskiego 27, 50-370 Wrocław, Poland
mieczyslaw.wodecki@pwr.wroc.pl

Abstract. Uncertainty is deeply entrenched in many optimization problems including scheduling problems where different parameters such as processing times, setup times, release dates or due dates are not known at the time of determining the solution. Therefore we can observe more and more research on creating robust methods which can remain stable when dealing with the real data during the actual algorithm execution. In many such methods there is the focus to design an algorithm for a specific type of data distribution and data disturbance level.

In this paper we investigate a single machine scheduling problem and data modeled with the normal distribution with different standard deviation levels; for comparison we include also data with the uniform distribution within a specific range. We analyze how much tailoring an algorithm for a given disturbance level gives better results. The conducted computational experiments show that the considered robust optimization method provides better results when configured for given disturbed data distributions, however different variants in a slightly different way.

Keywords: Single machine scheduling · Robust optimization · Uncertain parameters · Normal distribution · Data analysis · Tabu search

1 Introduction

Uncertainty is a natural property of many optimization problems in different domains like production, manufacturing, delivering goods, supply chain and a lot

W. Zamojski et al. (Eds.): DepCoS-RELCOMEX 2022, LNNS 484, pp. 1–10, 2022.
https://doi.org/10.1007/978-3-031-06746-4_1

more and it depends on many factors, for instance in the transportation domain delivery on time may be impacted by weather conditions, cars' breakdowns, traffic jams, driver's condition and others. It is also important to understand specific problem's properties to be able to design more efficient and robust algorithms and decide on the best approach for uncertainty modeling which usually is realized by probabilistic distributions, fuzzy numbers or bound form where values come from a specific range.

Solving problem in a deterministic way is based on the fact that all parameters are well defined at the phase of finding a solution for a given instance problem. However, in case data is uncertain or we experience data disturbance during the process execution, applying the deterministic approach may result in the increased actual execution cost (and by that losing the optimality) or even losing the acceptability (feasibility) of solutions.

Even if majority of research in optimization is focused on deterministic problems, uncertainty has been also investigated for many decades. Basics of stochastic scheduling one can find in Pinedo [9] and more extensive reviews dedicated to methods solving scheduling problems in stochastic models are presented in Cai et al. [6], Dean [7], Soroush [13], Urgo and Vancza [14], Vondrák [15], and Zhang et al. [16]. This approach with focus on a specific probabilistic distribution was investigated in Bożejko et al. [1–5], Rajba et al. [10,11] and Rajba [12] where effective methods were proposed for single machine scheduling problem where parameters are modelled with random variables with the normal distribution. In [1,2] and [10] additionally Erlang distribution was investigated and those papers cover $\sum w_i U_i$ and $\sum w_i T_i$ problem variants. On top of that in [3] and [4] techniques to shorten the computational time (i.e. elimination criteria and random blocks) were introduced keeping the robustness of the determined solutions on an expected good level.

In this paper we investigate a tabu search method for a single machine scheduling problem tailored for uncertain data with the normal distribution as described in [1]. The considered test data is generated using the normal distribution with different deviation levels based on the recognized in the field deterministic data set as a reference baseline. As the approach assumes deterministic data as an input and then test data is generated based on it, we refer to this test data as a *disturbed data* even though in some real scenarios we wouldn't refer to the data as "disturbed" due to the fact that uncertainty and variety of values may be viewed as a nature of the considered phenomenon. Nevertheless, this approach and terminology difference doesn't prevent to find the results useful as in case the real scenario data distribution matches the investigated variants, the provided conclusion can be applied with success. Having the test data, we execute algorithm configured for a specific disturbance level several times for each data set with data disturbed with the normal distribution with a given disturbance level. As an additional data set we included also data with the uniform distribution within a specific range. All the executions are to verify how much tailoring an algorithm for a given disturbance level gives better results when targeting data set of that specific disturbance level. The conducted computational experiments show that the considered robust optimization method provides better results when configured for given disturbed data distributions, however different variants in a slightly different way.

The rest of the paper is structured as follows: in Sect. 2 we describe a classic deterministic version of the problem, then in Sect. 3 we introduce a randomized variant of the one. In Sect. 4 we present disturbed data sets and the analysis approach, and in Sect. 5 the main results and a summary of computational experiments are described. Finally, in Sect. 6 conclusions and future directions close the paper.

2 Deterministic Scheduling Problem

Let $\mathcal{J} = \{1, 2, \ldots, n\}$ be a set of jobs where for each $i \in \mathcal{J}$ we define p_i as a *processing time*, d_i as a *due date* and w_i as a cost for a delay. All jobs shall be executed on a single machine under the following main conditions: (1) at any given moment at most one job can be executed and (2) all jobs must be executed without preemption.

Let Π be the set of all permutations of the set $\mathcal{J}$. For each permutation $\pi \in \Pi$ we define

$$C_{\pi(i)} = \sum_{j=1}^{i} p_{\pi(j)}$$

as the completion time of a job $\pi(i)$.

Then we introduce the delay indicator as

$$U_{\pi(i)} = \begin{cases} 0 & \text{for} \quad C_{\pi(i)} \leqslant d_{\pi(i)}, \\ 1 & \text{for} \quad C_{\pi(i)} > d_{\pi(i)}, \end{cases}$$

and the cost function for the permutation π as

$$\sum_{i=1}^{n} w_{\pi(i)} U_{\pi(i)}. \tag{1}$$

Finally, the goal is to find a permutation $\pi^* \in \Pi$ which minimizes

$$W(\pi^*) = \min_{\pi \in \Pi} \left(\sum_{i=1}^{n} w_{\pi(i)} U_{\pi(i)} \right).$$

3 Probabilistic Model

In this section we introduce a probabilistic version of the problem and we investigate two variants: (a) uncertain processing times and (b) uncertain due dates.

In order to simplify the further considerations we assume w.l.o.g. that at any moment the considered solution is the natural permutation, i.e. $\pi = (1, 2, \ldots, n)$.

3.1 Random Processing Times

Random processing times are represented by random variables with the normal distribution $\tilde{p}_i \sim N(p_i, c \cdot p_i))$ ($i \in \mathcal{J}$, c determine the disturbance level and will be defined later) while due dates d_i and weights w_i are deterministic. Then, completion times $\tilde{C}_i$ are random variables

$$\tilde{C}_i \sim N\left(p_1 + p_2 \ldots + p_i, c \cdot \sqrt{p_1^2 + \ldots + p_i^2}\right). \tag{2}$$

and the delay's indicators are random variables

$$\tilde{U}_i = \begin{cases} 0 & \text{for} \quad \tilde{C}_i \leqslant d_i, \\ 1 & \text{for} \quad \tilde{C}_i > d_i. \end{cases} \tag{3}$$

For each permutation $\pi \in \Pi$ the cost in the random model is defined as a random variable:

$$\widetilde{W}(\pi) = \sum_{i=1}^{n} w_i \tilde{U}_i. \tag{4}$$

3.2 Random Due Dates

Random due dates are represented by random variables with the normal distribution $\tilde{d}_i \sim N(d_i, c \cdot d_i)$, $i \in \mathcal{J}$ while processing times p_i and weights w_i are deterministic. Completion times are the same as in the processing times variant and the delay's indicators are random variables

$$\tilde{U}_i = \begin{cases} 0 & \text{dla} \quad C_i \leqslant \tilde{d}_i, \\ 1 & \text{dla} \quad C_i > \tilde{d}_i. \end{cases} \tag{5}$$

Again, cost functions are the same as for the variant with random processing times.

3.3 Comparison Functions

As in the Tabu Search implementation we need a possibility to compare 2 candidate solution, we introduce the following comparison function for both variants:

$$W(\pi) = \sum_{i=1}^{n} w_i E(\tilde{U}_i). \tag{6}$$

It is easy to observe that for the variant with random processing times $\tilde{p}_i$ we have:

$$E(\widetilde{U}_{\pi(i)}) = P(\widetilde{C}_{\pi(i)} > d_{\pi(i)}) = 1 - F_{\widetilde{C}_{\pi(i)}}(d_{\pi(i)}),$$

and for the variant with random due dates $\tilde{d}_i$ we have:

$$E(\widetilde{U}_{\pi(i)}) = P(C_{\pi(i)} > \widetilde{d}_{\pi(i)}) = F_{\widetilde{d}_{\pi(i)}}(C_{\pi(i)}),$$

where $F_X(x)$ is a cumulative distribution function (CDF) for the random variable X. Finally we obtain the formula

$$W(\pi) = \sum_{i=1}^{n} w_i(1 - F_{\tilde{C}_{\pi(i)}}(d_{\pi(i)})), \tag{7}$$

for random processing times and the formula

$$W(\pi) = \sum_{i=1}^{n} w_i F_{\tilde{d}_{\pi(i)}}(C_{\pi(i)}), \tag{8}$$

for random due dates.

4 Disturbed Data and Its Analysis

The considered method was applied also for other target functions and was considered in several papers [1–5,10,12]. One of the key assumptions was the fact that data comes for a certain distribution with a specific parameters. In majority of considered cases we investigated data generated from normal distribution with a specific mean μ and standard deviation σ being a fraction of μ. In all considered variants on average the proposed solution offered better results than the method based on deterministic approach.

In this paper we investigate how much the model and based on it Tabu Search implementation is vulnerable on deviations from the assumed distribution parameters.

Baseline test instances come from OR-Library [8] where there are 125 examples for n = 40, 50 and 100 (in total 375 examples). All the disturbed data has been generated targeting a specific problem variant, i.e. 100 disturbed data instances have been generated assuming:

- problem parameter which is uncertain, i.e. either processing times p_i or due dates d_i,
- disturbance level c introduced in Sect. 3.1 and 3.2 which takes the following values $c \in \{0.02, 0.04, 0.06, 0.08, 1.0, 1.5, 2.0, 2.5, 3.0\}$. On top of that we also generated data with the uniform distribution where each disturbed value comes from the range $[0.8x, 1.2x]$ where x is the considered uncertain variable (i.e. either p_i or d_i).

In our analysis we take a data set for a specific disturbance context, i.e. problem variant (uncertain p_i or d_i) and disturbance level (normal distribution with c value or uniform distribution) and then execute tabu search algorithm 9 times each time configured for a different disturbance value c. As a result we get a table with raw data of the following structure (Table 1):

The goal is to determine for a considered method if and how much configuring algorithm for a specific method make a difference in obtained results. For instance, having a data disturbed a little like for $c = 0.02$, if there is difference

Table 1. Raw result data for each n, distribution, instance number and disturbed item number

N	Factor	Task no.	Dist. item	$AP_{.02}$	$AP_{.04}$	$AP_{.06}$	$AP_{.08}$	$AP_{.1}$	$AP_{.15}$	$AP_{.2}$	$AP_{.25}$	$AP_{.3}$
40	0.02	0	0	v_{11}	v_{12}	v_{13}	v_{14}	v_{15}	v_{16}	v_{17}	v_{18}	v_{19}
40	0.02	0	1	v_{21}	v_{22}	v_{23}	v_{24}	v_{25}	v_{26}	v_{27}	v_{28}	v_{29}
40	...	...	...	...	...	...	...	...	...	...	...	...

between executing algorithm configured for $c = 0.02$ and algorithm configured for $c = 0.3$. In our analysis we check all those combinations and in this paper we do that on an aggregated level, i.e. we consider all values for given n and disturbance level as a comparison baseline. To confirm the statistically significant difference between obtained results for different variants we perform ANOVA with Bonferroni post-hoc tests.

5 Computational Experiments

All the tests are executed on the data described in Sect. 4 and using standard tabu search method implementation with small adjustments related to the way how the two candidate solutions are compared. The algorithm has been configured with the following parameters: **(a)** $\pi = (1, 2, \ldots, n)$ is an initial permutation, **(b)** n is the length of tabu list and **(c)** n is the number of algorithm iterations, where n is the tasks number. Due to the limited space in this paper we skip the algorithm description and refer to [1] for more details.

5.1 Results

On Figs. 1, 2, 3, 4, 5, 6 we can observe the visual representation of the obtained results. On each figure we can see 10 groups for each disturbance data set (i.e. for each disturbance level c and uniform distribution). Within each group we can see 9 values representing execution of the algorithm configured for all 9 different disturbance levels, i.e. all c values.

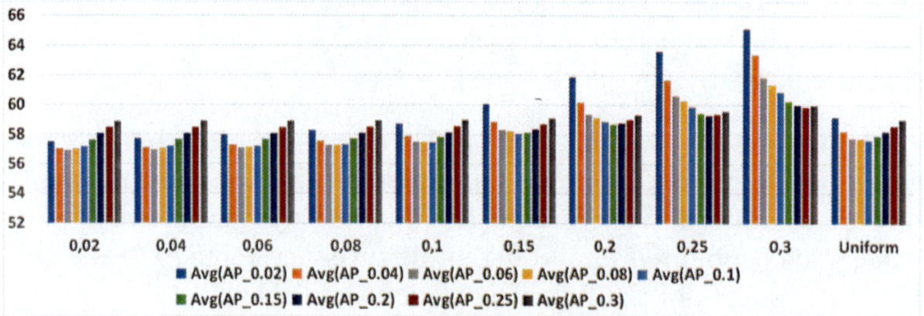

Fig. 1. Average values for each disturbance data group, random p_i, $n = 40$

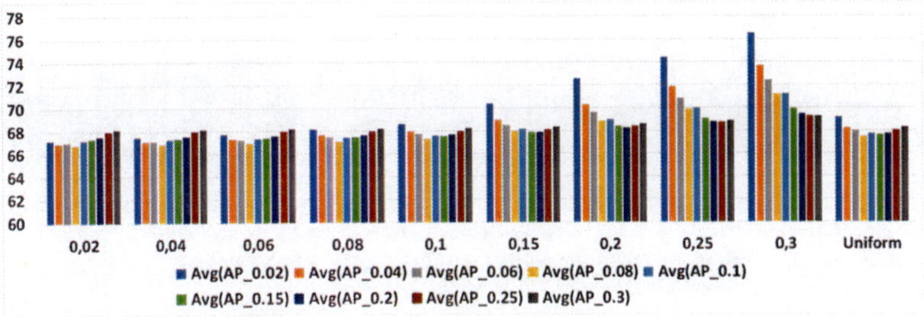

Fig. 2. Average values for each disturbance data group, random p_i, $n = 50$

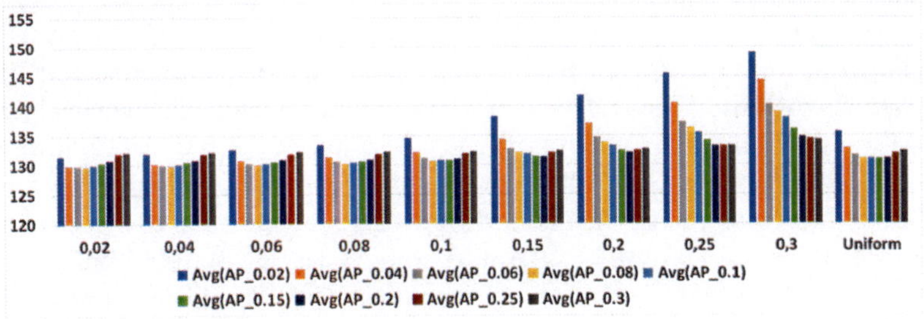

Fig. 3. Average values for each disturbance data group, random p_i, $n = 100$

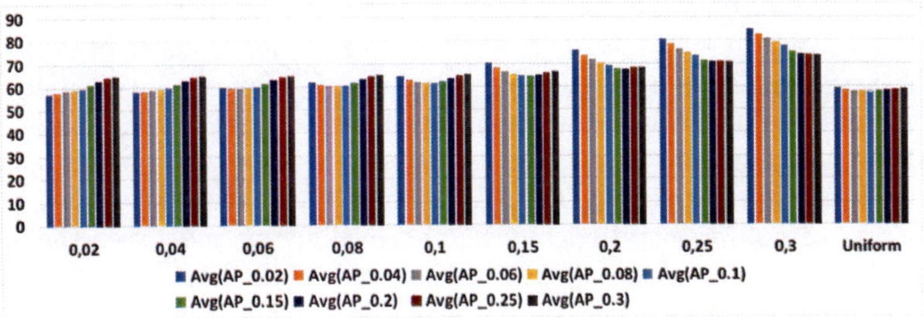

Fig. 4. Average values for each disturbance data group, random d_i, $n = 40$

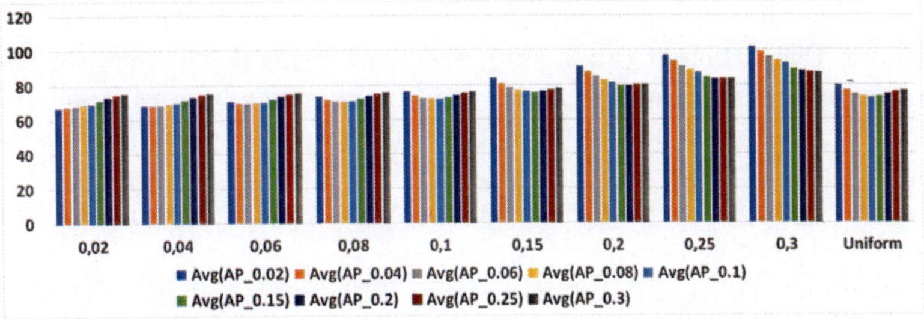

Fig. 5. Average values for each disturbance data group, random d_i, $n = 50$

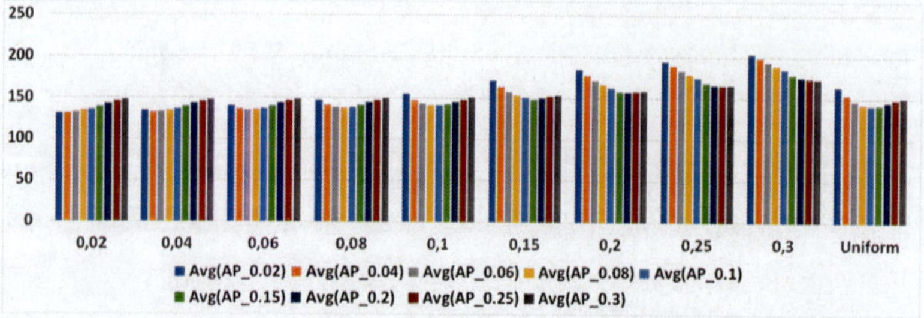

Fig. 6. Average values for each disturbance data group, random d_i, $n = 100$

We can immediately observe a trend that algorithm configured for a specific disturbance level on average gives better results than the algorithm configured for other disturbance level. Even this rule is not strict, the general trend is definitely visible, however much stronger for random d_i than random p_i.

For random p_i it is interesting to see that for smaller disturbances (0.02 – –1.0) distribution is rather stable and that the best choice is to select algorithm configured for $c = 0.06$ ($n = 40$) or $c = 0.08$ ($n = 50, 100$), but differences are negligible. However, for bigger disturbances (>1.0) tailored configuration is getting important. For the uniform data the best choice is to select algorithm configured for $c = 0.1$ ($n = 40$) or $c = 0.08$ ($n = 50, 100$).

For random d_i we can observe the trend that algorithm configured for a specific disturbance level on average gives better results than the algorithm configured for other disturbance level works almost strict. Almost because in just several cases (i.e. 3 for $n = 40$, 1 for $n = 50$, and 0 for $n = 100$) adjacent configuration works better, however those differences are negligible. So, for random d_i tailored configuration for a given disturbance level makes a lot of sense.

Finally, we apply one-way ANOVA with Bonferroni post-hoc test to check whether we can observe statistically significant differences between the obtained values where we compare groups defined by random parameter (p_i vs. d_i), n and data disturbance level.

Results presented in Table 2 are structured as follows: each row defines random parameter and n while in the column we can find respective disturbed data set. In each data cell we can find a number which represents how many of comparisons between different algorithms execution provided statistically significant differences. Even though this results provide a general overview we can make the following observations: (1) there is a big difference between results for random p_i and random d_i, for random d_i (2a) distributions of values correspond to the presented diagrams, and (2b) almost 1/2 of the overall comparisons resulted in statistically significant. These results gives a strong recommendation for the further investigation.

Table 2. Statistically significant differences summary between all considered tabu search disturbance configurations executed on all considered disturbed data sets

Random	n	Normal distribution									u.d.	avg
		0.02	0.04	0.06	0.08	0.1	0.15	0.2	0.25	0.3		
p_i	40	0	0	0	0	0	0	7	9	12	0	3,1
	50	0	0	0	0	0	0	7	11	15	0	3,7
	100	0	0	0	0	0	6	7	11	14	0	4,2
d_i	40	22	20	18	16	13	12	21	26	27	0	19,4
	50	20	20	17	15	12	14	23	28	28	16	19,7
	100	20	19	16	18	16	19	25	28	27	20	20,9

6 Conclusions

In this paper we analyzed the results of solving a single machine scheduling problem with a tabu search method configured for a specific disturbance level several times for each data set with data disturbed with the normal distribution with a given disturbance level. As an additional reference data set we included also data disturbed with the uniform distribution within a specific range. The conducted computational experiments show that the closer the considered robust optimization method is configured to the disturbance level of the data set, the better results it provides. Even though the overall trend has been observed in all considered problem variants, the actual results depend on the disturbance levels and the considered uncertain parameter what has been presented on the respective diagrams. On top of that a summary of statistically significant differences between different results group has been also provided.

As the results of the analysis are very promising and interesting we plan to continue the investigation by performing more deep analysis of the already obtained results as well as by checking other problems and target functions. Finally, it seems to be interesting to include in the analysis other ways of approaching uncertainty and make appropriate comparisons.

References

1. Bożejko, W., Rajba, P., Wodecki, M.: Stable scheduling with random processing times. In: Klempous, R., Nikodem, J., Jacak, W., Chaczko, Z. (eds.) Advanced Methods and Applications in Computational Intelligence, pp. 61–77. Springer, Heidelberg (2014). https://doi.org/10.1007/978-3-319-01436-4_4
2. Bożejko, W., Rajba, P., Wodecki, M.: Stable scheduling of single machine with probabilistic parameters. Bull. Polish Acad. Sci. Tech. Sci. **65**(2), 219–231 (2017)
3. Bożejko, W., Rajba, P., Wodecki, M.: Robustness of the uncertain single machine total weighted tardiness problem with elimination criteria applied. In: Zamojski, W., Mazurkiewicz, J., Sugier, J., Walkowiak, T., Kacprzyk, J. (eds.) DepCoS-RELCOMEX 2018. AISC, vol. 761, pp. 94–103. Springer, Cham (2019). https://doi.org/10.1007/978-3-319-91446-6_10

4. Bożejko, W., Rajba, P., Wodecki, M.: Robust single machine scheduling with random blocks in an uncertain environment. In: Krzhizhanovskaya, V.V., et al. (eds.) ICCS 2020. LNCS, vol. 12143, pp. 529–538. Springer, Cham (2020). https://doi.org/10.1007/978-3-030-50436-6_39
5. Bożejko, W., Rajba, P., Uchroński, M., Wodecki, M.: A job shop scheduling problem with due dates under conditions of uncertainty. In: Paszynski, M., Kranzlmüller, D., Krzhizhanovskaya, V.V., Dongarra, J.J., Sloot, P.M.A. (eds.) ICCS 2021. LNCS, vol. 12742, pp. 198–205. Springer, Cham (2021). https://doi.org/10.1007/978-3-030-77961-0_17
6. Cai, X., Wu, X., Zhou, X.: Optimal Stochastic Scheduling, vol. 4. Springer, New York (2014). https://doi.org/10.1007/978-1-4899-7405-1
7. Dean, B.C.: Approximation algorithms for stochastic scheduling problems. Doctoral dissertation, Massachusetts Institute of Technology (2005)
8. OR-Library. http://www.brunel.ac.uk/~mastjjb/jeb/info.html. Accessed 11 May 2020
9. Pinedo, M.L.: Scheduling: Theory, Algorithms, and Systems. Springer, Boston (2016). https://doi.org/10.1007/978-1-4614-2361-4
10. Rajba, P., Wodecki, M.: Stability of scheduling with random processing times on one machine. Applicationes Mathematicae **2**(39), 169–183 (2012)
11. Rajba, P., Wodecki, M.: Sampling method for the flow shop with uncertain parameters. In: Saeed, K., Homenda, W., Chaki, R. (eds.) CISIM 2017. LNCS, vol. 10244, pp. 580–591. Springer, Cham (2017). https://doi.org/10.1007/978-3-319-59105-6_50
12. Rajba, P.: Sampling method for the robust single machine scheduling with uncertain parameters. In: Paszynski, M., Kranzlmüller, D., Krzhizhanovskaya, V.V., Dongarra, J.J., Sloot, P.M.A. (eds.) ICCS 2021. LNCS, vol. 12747, pp. 594–607. Springer, Cham (2021). https://doi.org/10.1007/978-3-030-77980-1_45
13. Soroush, H.M.: Scheduling stochastic jobs on a single machine to minimize weighted number of tardy jobs. Kuwait J. Sci. **40**(1), 123–147 (2013)
14. Urgo, M., Váncza, J.: A branch-and-bound approach for the single machine maximum lateness stochastic scheduling problem to minimize the value-at-risk. Flex. Serv. Manuf. J. **31**, 472–496 (2019)
15. Vondrák, J.: Probabilistic methods in combinatorial and stochastic optimization. Doctoral dissertation, Massachusetts Institute of Technology (2005)
16. Zhang, L., Lin, Y., Xiao, Y., Zhang, X.: Stochastic single-machine scheduling with random resource arrival times. Int. J. Mach. Learn. Cybern. **9**(7), 1101–1107 (2018)

Using Evolutionary Algorithm in On-line Deployment

Wiktor B. Daszczuk[1(✉)], Rafał Biedrzycki[1], and Piotr Wilkin[2]

[1] Institute of Computer Science, Warsaw University of Technology, Nowowiejska Str. 15/19, 00-665 Warsaw, Poland
{wiktor.daszczuk,rafal.biedrzycki}@pw.edu.pl
[2] Syndatis Ltd., Puławska Str. 12a/10b, 02-566 Warsaw, Poland
piotr.wilkin@syndatis.com

Abstract. There is a problem with arranging many elements in the window in many applications. It is more difficult when the elements are dynamically created or selected from a particular set. There is then a risk of obstruction by objects, which reduces the legibility of the window. Such a problem arises when generating dynamic help for items in a window in the workflow management system designed by Syndatis laboratories. Artificial intelligence algorithms can be used for this purpose, but the obstacle is their long calculations, which in the described case should be completed in less than a second. We proposed to use the evolutionary algorithm with a limited number of generations, which gives satisfactory results.

Keywords: Evolutionary algorithm · User interface · Dynamic graphic layout

1 Introduction

Syndatis BPM4 workflow management system [1] is designed in Syndatis laboratories to develop commercial workflow applications. Among several other units, an extended context help subsystem supports the users in the navigation between the windows and graphical components (controls) put inside the windows. The help consists of a set of balloons that comment on the meaning and operation of individual controls. Many components are put dynamically; therefore the help balloons need to be arranged online in a most readable configuration.

The controls have arbitrary positions and sizes. Every control is equipped with a short text, whose size is independent of the control size. The texts are displayed in balloons. Each balloon has a text field and a handle pointing to the right-bottom corner of the control (Fig. 1a). The scrollbar is added if the balloon is too small to display the entire text (Fig. 1b). However, this reduces the readability because important information can be hidden from the user. The user can resize or hide the balloon using appropriate small icons (Fig. 1c and 1d). The hidden balloon is displayed as a balloon icon (Fig. 1e).

The rules of balloon placement are slightly different in the real system Syndatis BPM4, for which the algorithm was developed. The actual window is presented in Fig. 2. However, the differences are not significant for the presentation of our solution.

W. Zamojski et al. (Eds.): DepCoS-RELCOMEX 2022, LNNS 484, pp. 11–21, 2022.
https://doi.org/10.1007/978-3-031-06746-4_2

The algorithm applied online should arrange the balloons in the most readable layout, which is not an easy task considering contradicting partial goals for the balloons to be mostly close to the controls, not to cover themselves nor the controls to use scrollbars inside.

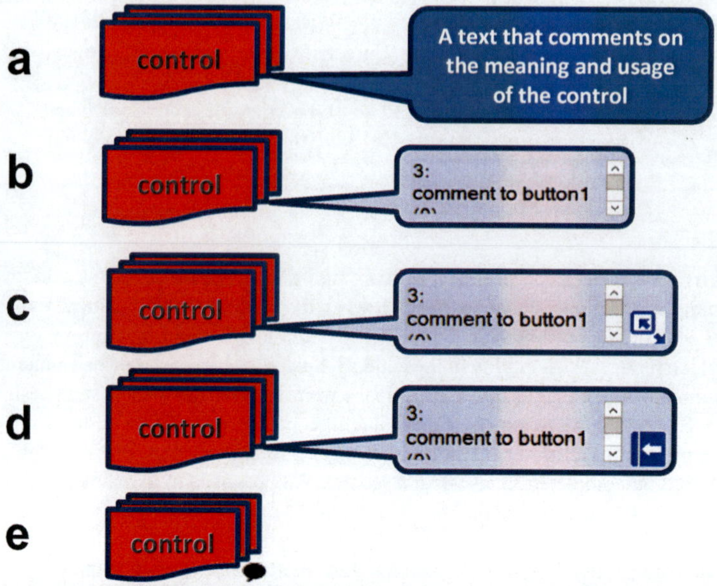

Fig. 1. Appearance of balloons: **a**-the idea, **b**-shrinking, **c**-resizing, **d**-hiding, **e**-hidden

Since the problem has many input parameters and should fulfill several goals, some contradicting each other, we decided to use an Evolutionary Algorithm (EA). EAs [2, 3] are widely used to optimize various multidimensional problems. As EAs were inspired by natural evolution, they use a biological naming convention. An individual contains an encoded solution to the problem. At each iteration algorithm, the process population consists of many individuals. The optimization starts from a randomly initialized population. At each iteration, individuals produce their offspring using selection, crossover, and mutation. Such a procedure, applied to many generations, successfully finds solutions to many practical problems [4, 5]. Usually, such optimization takes much time. Therefore, it is used in offline search, where it is reasonable to wait quite long for a solution possibly closest to the global optimum.

In our case, the help balloons need to be arranged dynamically, so the online search is more applicable. The aim is not to finda global optimum but to quickly find a "good enough" solution. Thus, we decided to apply EA with a small number of generations, which should allow us to display the help balloons in less than one second after the windows occur on the screen. For this reason, additional heuristics are applied.

The paper is organized as follows: Sect. 2 presents related work on applying artificial intelligence methods to user interface (UI) layout. Section 3 describes the specific needs to design dynamic UI in Syndatis BPM4 system. In Sect. 4, the setup of the algorithm

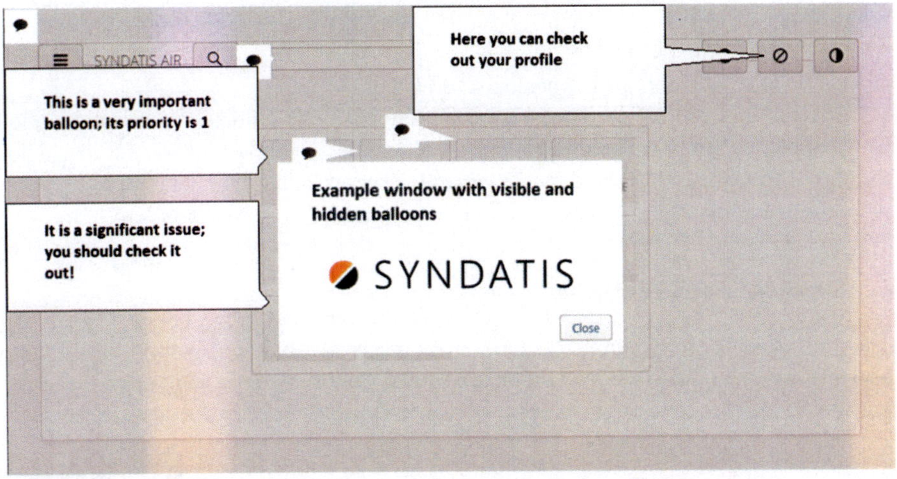

Fig. 2. The example of a set of visible and hidden balloons (rectangles with handle and ●, respectively) in Syndatis BPM4 workflow management system

is defined. Section 5 presents the results for various values of the algorithm parameters. Section 6 concludes the research.

2 Related Work

EAs are mainly used for offline search where we could wait hours or even longer for the results. There are much fewer applications in online or real-time search, where acceptable results should be available in a short time. There are so many applications of EAs that even "much less" is a large number. In [6], authors survey applications of evolutionary algorithms in control systems engineering. In [7], it is shown that the Particle Swarm Optimization and the Evolutionary Algorithm are the best for online drive train efficiency optimization in electric vehicles. The authors concluded that online efficiency optimization in electric vehicles is possible concerning computing time and success probability.

The most obvious way to speed up the algorithm is to use parallel implementation. In [8], a parallel evolutionary algorithm is used for real-time path planning for unmanned aerial vehicles (UAVs). Another possibility to speed up when the objective function is computationally expensive is to use surrogate models of the objective function [9]. In [10], it is shown that domain-specific EAs are the most efficient for model predictive control and real-time decision support.

Evolutionary algorithms are also used in UI design and optimization. In [11], a page layout in the web-based application is evolved. The layout is specified in a textual form using XML. The authors defined several criteria of a good album and composed an objective function on that basis.

Several approaches require user input to build an objective function. One of them is called the interactive genetic algorithm (IGA) [12]. The IGA was applied to brochure design [13], poster design [14], web page design [15], and other types of UI. As the requirement of direct assessment is problematic, there are also attempts to replace it with indirect feedback. In [16], authors propose to use eye-tracking to obtain users' subjective feedback and evolve personalized UI.

As there are applications of evolutionary algorithms to UI design and optimization, we were not able to find a solution to our problem. The papers usually do not reveal how the algorithms were configured, but in general, it seems that a small number of iterations were used, and the algorithms were strongly tailored to the solved problem.

3 Problem Formulation

The goal of the optimization is to minimize undesirable balloon features. The criteria used to assess the quality of a solution are shown symbolically in Fig. 3.

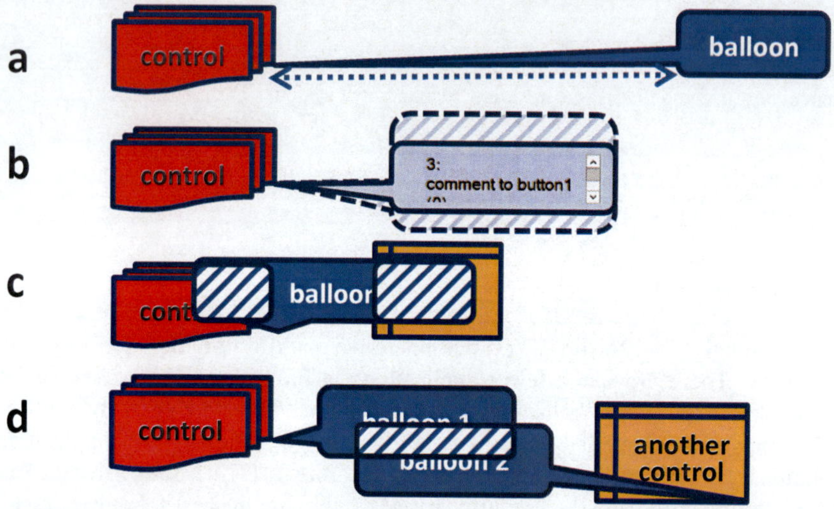

Fig. 3. Measures of undesirable balloon features: **a**-distance, **b**-size (the shrunken part is dashed), **c**-covering (balloon areas covering the controls are dashed), **d**-overlapping (the overlapping area is dashed)

The criteria are as follows:

- The *distance* criterion (Fig. 3a). We measure how close the balloon is to the control (Fig. 4) as the length of a segment connecting the corners of the balloon and the control (balloons A, C, E, G) or the edges (balloons B, D, F, H). If the balloon covers the control (Fig. 4, dashed balloon), the distance is assumed 0. To compare the distances with other features, we normalize them to the diagonal of the window.
- The *size* criterion (Fig. 3b). We measure the size of the actual balloon (solid edge) being the part of the hypothetical balloon displaying the entire text (dashed edge).
- The criterion of *covering* the controls (Fig. 3c). We measure the part of the control size that is covered (dashed part). A balloon can cover multiple controls so that this value can be >1. Therefore, we divide the result by the number of controls.

- The criterion of *overlapping* balloons (Fig. 3d). We measure the part of the other balloon size that is overlapping (dashed part). A balloon can overlap multiple balloons so that this value can be >1. Therefore, we divide the result by the number of balloons. Additionally, if a given balloon overlaps another one, the latter balloon overlaps the former one as well (Fig. 3d). So, the result is divided by 2.

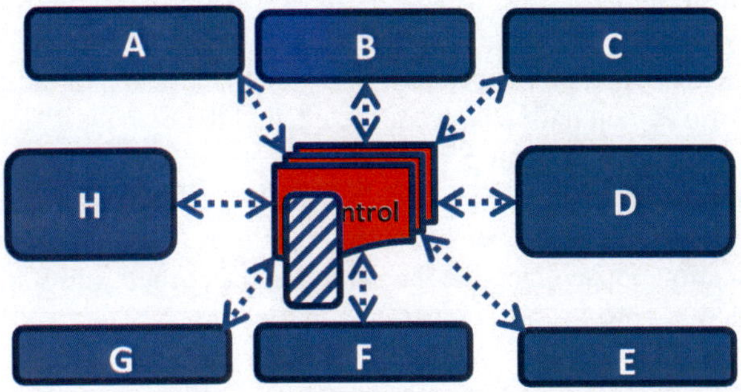

Fig. 4. Distance from a balloon in various positions to the control. Dashed balloon – distance 0

To sum up, we have four criteria of value in the range [0,1]. It is not easy to combine them, and we describe the construction of the objective function in the next section.

4 Setup of the Algorithm

An individual's genome (the chromosome) is represented by a vector of integer coordinates and sizes of all balloons (horizontal, vertical, width, height), which gives search space dimensionality equal to 4 * number of balloons. The minimized objective function is based on values of penalties calculated for four criteria defined in Sect. 3. Using a simple sum of the penalties is not a good idea. A balloon close to the control but overlaps with the other control will have the same quality as one that is too far but does not cover the control. On the other hand, the third balloon with intermediate distance and partial covering is much better than the former two, but it can have an equal or even greater sum of penalties. Thus multiplying penalties associated with the criteria is better, but the product of values between 0 and 1 gets even smaller. Thus, we decided to apply a formula: (1 + distance) * (1 + size) * (1 + covering) * (1 + overlapping). Additionally, we square the penalties to exclude terrible solutions.

As users can favor some criteria over others, we give them such freedom using the weights of the compounds: w_{dist}, w_{size}, w_{cov} and w_{ov}. The final formula for a balloon evaluation is:

$$q_b = (1 + w_{dist} * \text{distance}^2) * (1 + w_{size} * \text{size}^2) * (1 + w_{cov} + \text{covering}^2) * (1 + w_{ov} * \text{overlapping}^2).$$

To calculate the value of an objective function for an individual, we sum the values calculated for each balloon.

In the evolution loop, the pairs of parents are drawn randomly, and for a given pair, we take the value of a given offspring gene (size or position of the given balloon) randomly from one of the parents. Then we apply the mutation to 40% of balloons. A balloon is drawn randomly; then we draw which coordinates are modified: position or size. The selected balloon feature (shift or resize) is mutated by adding a random value from the normal distribution $\mathbb{N}(0, \sigma)$ to the mutated value. The value σ depends on the upper boundary b of value under mutation, i.e., $\sigma = b * s$, where s is a user-defined parameter called mutation strength.

After crossover and mutation, all balloons that protrude outside the window are shifted to the nearest position inside the window. In the last step of the algorithm's loop, parents and offspring populations are connected, and the worst individuals from the set are removed to maintain constant population size.

It is usual that the window is well-designed and the controls have some space around them. Therefore, we take a number of balloon positions (first 12 individuals) fixedly in the initial population, i.e., connecting the control at the corners and at the edges, aligned to the perpendicular edges. This is illustrated in Fig. 5a.

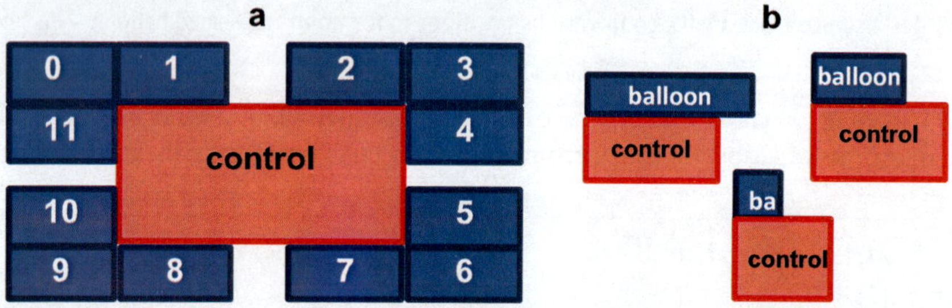

Fig. 5. Heuristic initial placement of a balloon in first 12 individuals: **a**-position of the balloon relative to a control, numbers in balloon positions: number of an individual, **b**-example controls (red) and initial balloon positions (blue) in the individual number 1.

5 Results

We use the population of 100 individuals and 100 generations. The numbers are small, but they allow for displaying a window with 20 controls in 2 s (Intel Core i7-4700HQ CPU @ 2.40 GHz, Windows 10–64). The number of 20 controls is quite typical for business systems in Syndatis BPM4 commercial applications.

Table 1. The results of running the algorithm (mean, standard deviation, best, worst and median) for different s values (0.1 – 0.4) and various cases (random 2–6 to 2–21 lines of text in a balloon). The best mean values are marked in bold, on grey background.

s		max number of lines in a balloon			
		6	11	16	21
0.1	Mean	**74**	485	1317	1929
	Std dev	57	239	584	1263
	Best	0	1	192	408
	Worst	215	995	2073	5741
	Median	56	527	1315	2309
0.2	Mean	83	400	1232	2213
	Std dev	72	213	558	917
	Best	1	42	104	606
	Worst	275	697	1862	3450
	Median	64	415	1362	2294
0.3	Mean	102	**247**	**789**	**1261**
	Std dev	76	83	219	326
	Best	0	4	223	276
	Worst	288	295	926	1432
	Median	82	295	926	1432
0.4	Mean	**74**	392	1353	2574
	Std dev	55	227	636	1277
	Best	0	7	9	154
	Worst	144	898	1299	4476
	Median	64	389	624	2432

The influence of the mutation strength (s) on the achieved objective function values is presented in Table 1. All results are given using the unit 0.0001. The results in the table are obtained for 50 independent runs of the algorithm for every pair (average balloon size, s). Note that a new set of balloon texts is drawn for every run, which increases the scatter of the input parameters. For 20 balloons, the only hope is that we can significantly reduce or collapse them, as they cover a large part of the window. However, such analysis, as well as aesthetic and ergonomic aspects of a window layout, are beyond the scope of the paper.

The table shows that the best value of s is 0.3.

The quality of the results of the algorithm measured using objective function was shown in Table 1. However, we are also interested in organoleptic rating because the entire optimization should improve the comfort of the human. Thus, we will show the results by examples in Fig. 6. A window with 5 controls of different sizes is used (label, button, panel, checkBox, and textBox), and balloon texts of random size (2 to 8 lines of text). We used weights: $w_{dist} = 0.5$, $w_{size} = 0.5$, $w_{cov} = 2$ and $w_{ov} = 0.5$. The coverage criterion is most important because we do not want the controls hidden under the balloons. However, a designer can apply other weigths. The figure shows various

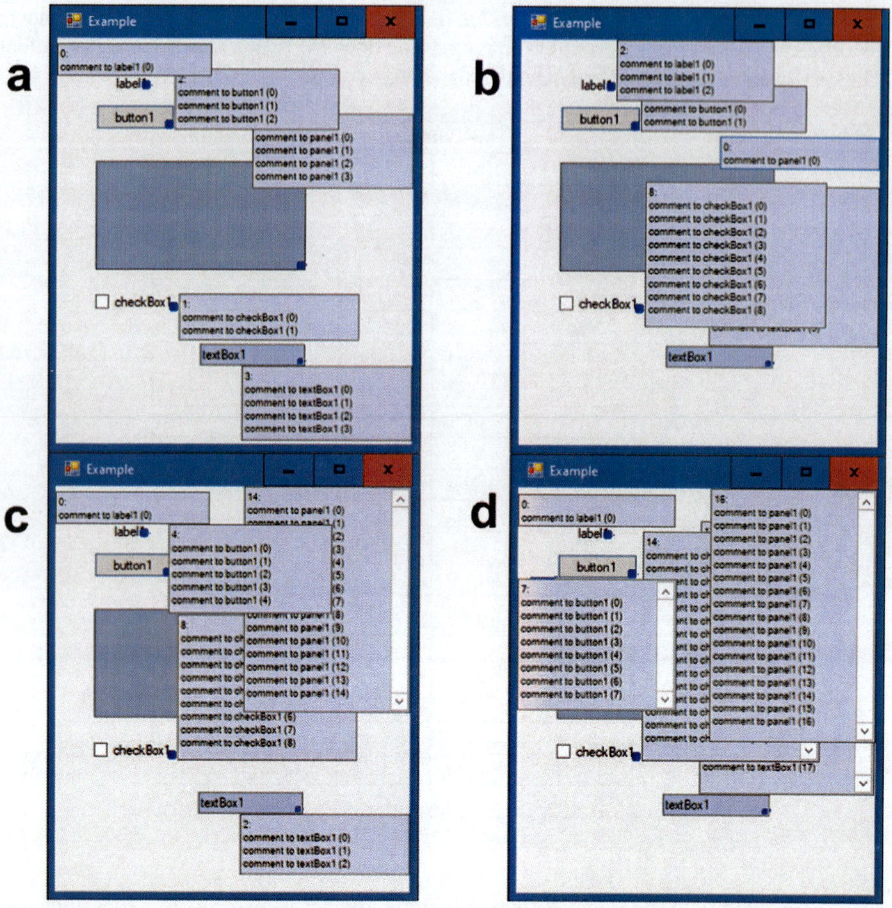

Fig. 6. Results of the algorithm for various sizes of balloons – **a**: 2–6, **b**: 2–11, **c**: 2–16, **d**: 2–21

balloon sizes, for random 2–6 text lines, 2–11 lines, 2–16 lines, and 2–21 lines in Fig. 6a–d, respectively. We see that if the texts are large (2–16 or 2–21 lines, Fig. 6c,d), little can be done to arrange the balloons readily. However, in Fig. 6c (2–16 lines), only one balloon is reduced, and all controls are visible (the panel in large part). In Fig. 6d (2–21 lines), one control is almost fully covered (panel) and four balloons are reduced.

To show that all criteria are important, in Fig. 7, we provide results of omitting a single criterion in the objective function (in the order: distance, size, covering, overlapping). We can observe balloons too far, too much reduced, covering the controls, and too much overlapping (although there is room for them to be moved).

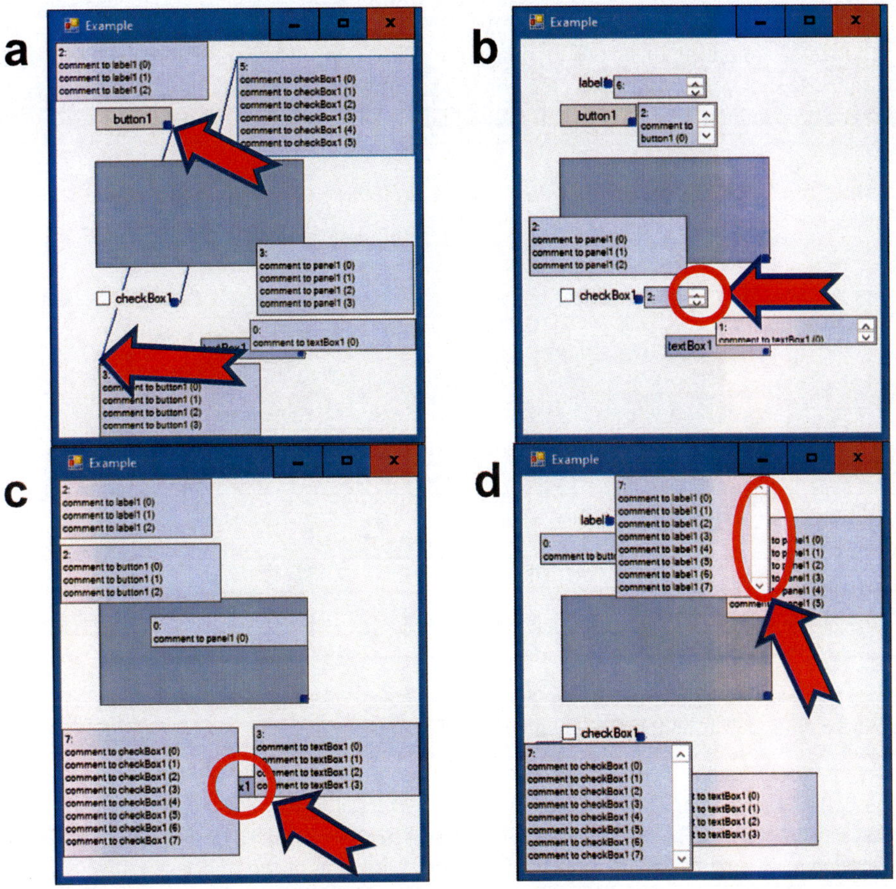

Fig. 7. Results of the algorithm with given criterion disabled: **a**-distance, **b**-size, **c**-covering, **d**-overlapping

6 Conclusions and Future Work

We presented the results of the online execution of the evolutionary algorithm with a small number of generations. In our opinion, the results are satisfactory for the cases in which the total size of all balloons fits in the free space of the window. For more complex cases, little can be done, but the balloons and most of the controls are at least visible. Because in many cases the user screen layout is repetitive, it would sometimes be wise to run the algorithm in two versions: fast online version, to show the hints to the user on the screen as quickly as possible, and an optimal version running in the background with more generations/individuals. The assumption is that, while the scheme may differ between individual users (different browser = different viewport size, different scaling parameters), the shape of a given widow displayed in given circumstances is constant or differs slightly.

We also plan to apply more improvements to the algorithm. For example, the likelihood of participating in the production of the offspring for better parents can be increased. Figure 8 shows few possible changes in the probability distribution of parents being drawn to produce next-generation individuals.

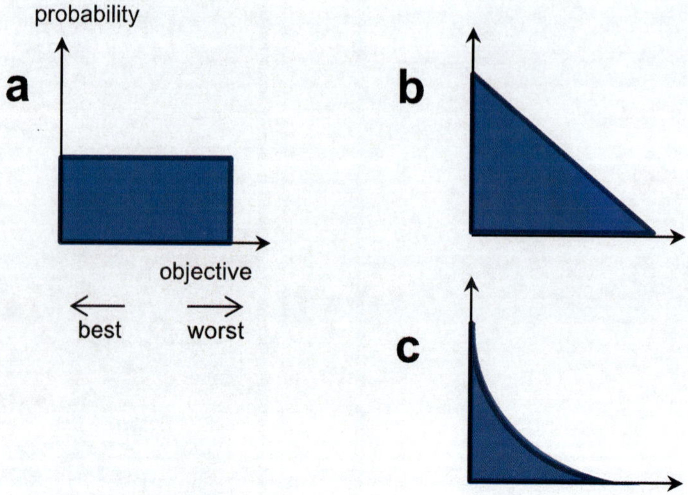

Fig. 8. Possible distributions of probability of drawing best and worst individuals to become parents: **a**-used distribution, **b**, **c**-possible enhancements

Acknowledgment. The research presented in this paper is co-financed by the European Regional Development Fund under the Regional Operational Program of the Lubelskie Voivodeship for 2014–2020 (RPLU.01.02.00-IP.01–06-001/15). Project No. RPLU.01.02.00–06-0048/16.

References

1. Daszczuk, W.B., Rybiński, H., Wilkin, P.: Using domain specific languages and domain ontology in workflow design in syndatis BPM4 environment. In: Zamojski, W., Mazurkiewicz, J., Sugier, J., Walkowiak, T., Kacprzyk, J. (eds.) DepCoS-RELCOMEX 2020. AISC, vol. 1173, pp. 143–154. Springer, Cham (2020). https://doi.org/10.1007/978-3-030-48256-5_15
2. Goldberg, D.E.: Real-coded genetic algorithms, virtual alphabets, and blocking. Complex Syst. **5**, 139–167 (1991). https://content.wolfram.com/uploads/sites/13/2018/02/05-2-2.pdf
3. Michalewicz, Z.: Genetic Algorithms + Data Structures = Evolution Programs. Springer, Berlin (1996). https://doi.org/10.1007/978-3-662-03315-9
4. Slowik, A., Kwasnicka, H.: Evolutionary algorithms and their applications to engineering problems. Neural Comput. Appl. **32**(16), 12363–12379 (2020). https://doi.org/10.1007/s00521-020-04832-8
5. Biedrzycki, R., Kwiatkowski, K., Cichosz, P.: Compressor schedule optimization for a refrigerated warehouse using metaheuristic algorithms. In: 2021 IEEE Congress on Evolutionary Computation (CEC), pp. 201–208. IEEE (2021). https://doi.org/10.1109/CEC45853.2021.9504924

6. Fleming, P.J., Purshouse, R.C.: Evolutionary algorithms in control systems engineering: a survey. Control Eng. Pract. **10**, 1223–1241 (2002). https://doi.org/10.1016/S0967-0661(02)00081-3
7. Apitzsch, T., Klöffer, C., Jochem, P., Doppelbauer, M., Fichtner, W.: Metaheuristics for online drive train efficiency optimization in electric vehicles (2016). https://doi.org/10.5445/IR/1000063608.
8. Jia, D., Vagners, J.: Parallel evolutionary algorithms for UAV path planning. In: AIAA 1st Intelligent Systems Technical Conference, Chicago, Illinois, 20–22 Sept 2004, pp. 1–12 (2004). https://doi.org/10.2514/6.2004-6230
9. Nguyen, T.-H., Vu, A.-T.: Speeding up composite differential evolution for structural optimization using neural networks. J. Inf. Telecommun. 1–20 (2021). https://doi.org/10.1080/24751839.2021.1946740
10. Zimmer, A., Schmidt, A., Ostfeld, A., Minsker, B.: Evolutionary algorithm enhancement for model predictive control and real-time decision support. Environ. Model. Softw. **69**, 330–341 (2015). https://doi.org/10.1016/j.envsoft.2015.03.005
11. Geigel, J., Loui, A.C.P.: Automatic page layout using genetic algorithms for electronic albuming. In: Proceedings of SPIE, Internet Imaging II, vol. 4311, pp. 79–90 (2000). https://doi.org/10.1117/12.411879
12. Kim, H.-S., Cho, S.-B.: Application of interactive genetic algorithm to fashion design. Eng. Appl. Artif. Intell. **13**, 635–644 (2000). https://doi.org/10.1016/S0952-1976(00)00045-2
13. Quiroz, J.C., Banerjee, A., Louis, S.J., Dascalu, S.M.: Document design with interactive evolution. In: Damiani, E., Jeong, J., Howlett, R.J., Jain, L.C. (eds.) New Directions in Intelligent Interactive Multimedia Systems and Services - 2, pp. 309–319. Springer, Heidelberg (2009). https://doi.org/10.1007/978-3-642-02937-0_28
14. Kitamura, S., Kanoh, H.: Developing support system for making posters with interactive evolutionary computation. In: Fourth International Symposium on Computational Intelligence and Design, Hangzhou, China, 28–30 October 2011, pp. 48–51. IEEE (2011). https://doi.org/10.1109/ISCID.2011.21
15. Monmarché, N., Nocent, G., Venturini, G., Santini, P.: On generating HTML style sheets with an interactive genetic algorithm based on gene frequencies. In: Fonlupt, C., Hao, J.-K., Lutton, E., Schoenauer, M., Ronald, E. (eds.) AE 1999. LNCS, vol. 1829, pp. 99–110. Springer, Heidelberg (2000). https://doi.org/10.1007/10721187_7
16. Cheng, S., Dey, A.K.: I see, you design: user interface intelligent design system with eye tracking and interactive genetic algorithm. CCF Trans. Perv. Comput. Interact. **1**(3), 224–236 (2019). https://doi.org/10.1007/s42486-019-00019-w

Measures of Outlierness in High-Dimensional Data Under Correlation of Features – With Application for Open-Set Classification

Szymon Datko(✉), Henryk Maciejewski, and Tomasz Walkowiak

Department of Computer Engineering, Wrocław University of Science and Technology, Wrocław, Poland
szymon.datko@pwr.edu.pl

Abstract. This work deals with the problem of open-set classification in high-dimensional data. This task is realized using outlier detection methods which rely on outlierness measures appropriate for high-dimensional data. We numerically evaluate which of the outlierness measures (outlierness factors) proposed in literature, such as the local density-based LOF, angle-based ABOF, etc., is most suitable in terms of sensitivity and specificity of detection of outstanding observations in the specific task of open-set recognition. The analysis is done for varying dimensionality of data, varying distance between the outliers and the typical data, as well as varying correlation between features.

Keywords: Open-set classification · High-dimensional data · Statistical hypothesis testing · Nonparametric tests

1 Introduction - Problem Formulation

One of limitations in commonly used classification algorithms is related to the closed-set classification procedure. Closed-set classifiers associate unknown objects to one of the trained classes, even if the object is not related (not similar enough) to any of the classes known by the model. In many practical applications this should be avoided: classifiers should label such an object as unrecognized rather than spuriously assign it to an unrelated class. It is especially important in any of safety-critical applications of machine learning [1]. The models which are capable of this are known as open-set classifiers. The open-set classification becomes vital in the tasks where it is infeasible to provide training examples for all possible classes/categories which the model is likely to see in the future. Examples of such tasks include classification of texts in terms of subject, authorship attribution, or image recognition. This problem was recently discussed in a number of publications, e.g. [2,6–8,11,12].

One of approaches to open-set classification proposed in [13] consists in performing the closed-set classification and then quantifying the dissimilarity (outlierness) of the recognized object and the class to which the closed-set model

W. Zamojski et al. (Eds.): DepCoS-RELCOMEX 2022, LNNS 484, pp. 22–31, 2022.
https://doi.org/10.1007/978-3-031-06746-4_3

assigns the object. If the object in the latter step is recognized as an outlier (outstanding observation) with respect to this category, then the object is labelled as unrecognized. Hence, in this approach, the open-set classification relies on outlier detection techniques. A challenging aspect of this method is related to high-dimensionality of feature vectors used in many applications (e.g. text mining, image recognition, etc.). Hence efficient procedures for outlier detection in high dimensional data are essential.

Although vast literature is available on the problem of outlier detection (and on related problems of anomaly or novelty detection), the detection of outliers/anomalies in high dimensional data still remains a challenging task [14]. Many different approaches to outlier detection proposed in the literature [4,5,10,14] can be categorized as: probabilistic methods which estimate the density of the normal class; distance-based methods (based on nearest neighbours or clustering algorithms); or methods which attempt to find the boundaries around the normal data. However, most of these methods are not applicable in high-dimensional data. Most numerical evaluations of these methods are commonly done using datasets with up to 10 dimensions (see e.g. [5]).

Recently Kriegel et al. [9,14] proposed an angle-based method (ABOF - angle based outlierness factor) which was shown by numerical simulations to outperform a popular LOF (local density-based outlierness factor, proposed in [3]). Such methods as ABOF or LOF are based on a measure of dissimilarity (or *outlierness*) of a point in high dimensional data calculated with respect to a cluster (set of points). Low or high value of the measure (e.g., low value of ABOF, or high value of LOF) are supposed to indicate an outlier with respect to the cluster.

In this work we evaluate different popular outlierness measures (or outlierness factors, OF), including LOF and ABOF. The purpose is to suggest the preferred method for the open-set classification procedure proposed in [13]. We perform a numerical study to estimate sensitivity of detection of outliers as a function of dimensionality of data, distance between the points and the reference cluster, as well as the correlation between features.

2 Outlier Detection for Open-Set Classification

Here we provide the idea of the open-set classification procedure proposed in [13]. Different outlierness factors, OF, (such as the ones evaluated in this work) are used in step 2 of the procedure. We denote as $OF(v, K)$ the value of the OF measure calculated for a point $v \in R^p$ with respect to the cluster (set of points) K in R^p.

1. Step 1 of the procedure consists in training a standard closed-set classifier (e.g., SVM or MLP), $g : R^p \to C$. In order to perform open-set classification of an observation $w \in R^p$, we first classify the observation using g, $g(w) = c$, (we refer to $c \in C$ as the winning class).

2. Step 2 of the procedure consists in quantifying if w is an outlier with respect to the set of points X_c which represents the training data vectors for the winning class c. In [13] the following procedure is proposed to achieve this:
   - Based on the set of values $(OF(v, X_c) : v \in X_c)$, its median $Q2$ and the first and third quartiles $Q1$, $Q3$ are calculated, with the inter-quartile range $IQR = Q3 - Q1$.
   - Outlierness measure for the observation w is calculated with respect to X_c, $OF(w, X_c)$.
   - The observation w is declared as an outlier with respect to X_c if $Q1 - OF(w, X_c) > r \cdot IQR$ or $OF(w, X_c) - Q3 > r \cdot IQR$.
   - If w is declared as an outlier, the open-set classifier labels it as unrecognized, otherwise assigns w to class c.

The value of the r parameter, commonly used in standard statistical procedures to detect outliers or extreme observations in univariate distributions, is typically between 1.5 and 3.

3 Numerical Evaluation of Outlierness Factors

The purpose of the numerical study is to compare performance of different OF measures for known characteristics of the reference cluster K and the second (testing) cluster L assumed to represent outlier observations with respect to the reference cluster. In the simulation study we observe how different measures OF tend to quantify outlierness of observations from L, as a function of dimensionality of data, intercluster shift between L and K, and correlation between feature (components of the observations quantified as outliers).

3.1 Outlierness Measures Analyzed

In our research we involved 5 algorithms. Two out of these are popular, commonly used and well described in the literature techniques applied in the task of outlier detection. Three simpler measures we provide for comparison.

Proposed by Breunig et al. [3], Local Outlier Factor, abbreviated as LOF, is an outlierness measure claimed to be effective even in high-dimensional spaces due to its design. It works by computing the so called local density of points, calculated as the ratio between average distance from a given point to its neighbors and average distance form these neighbors to their (neighbors) nearest neighbors. If this ratio is high, this means that the point is relatively further from other cluster's points than expected (typical distance between points), meaning it is an outlier.

The Angle-Based Outlier Factor, described by Kriegel et al. [9], is another commonly-used technique for detecting abnormal data (meaning here: not similar to previously known observations). It is claimed to be superior for applications in high-dimensional spaces, as it does not rely on euclidean metrics. Instead it utilizes the cosine function – angles calculated between a given point and all possible pairs of points from the target data cluster. Finally a variance of all angles is computed - if this value is big, a wide spectrum on angles is observed, so the given point lies within a cluster; otherwise it is an outlier.

For comparison, we chose the two simple Euclidean metrics, denoted in this paper as $Lp1$ and $Lp2$. These corresponds to the classical distances calculated in Lebesgue spaces of order p (Lp; in our case $p = 1$ and $p = 2$; another common notation is $L1$ and $L2$ metric, correspondingly). Distance in such spaces is calculated as a root of order p from the sum of each vector elements to the power of p. We calculate the average distance of given point to the target cluster. If such distance is big, it means the point is distant from the cluster and therefore can be marked as an outlier.

Additionally, a simplified algorithm based on the idea of Local Outlier Factor (sLOF) was implemented. This outlierness measure is defined as an average distance from given point to few of its nearest neighbors. Similarly to $L1$ and $L2$ norms, big average distance is a result of point being far from any cluster and thus it is an outlier.

It is important to notice that none of these outlierness measures define absolutely at which threshold value one shall decide whether, according to specific measure, a point is outlier or not. This is what we attempt to overcome with procedure defined in Sect. 2.

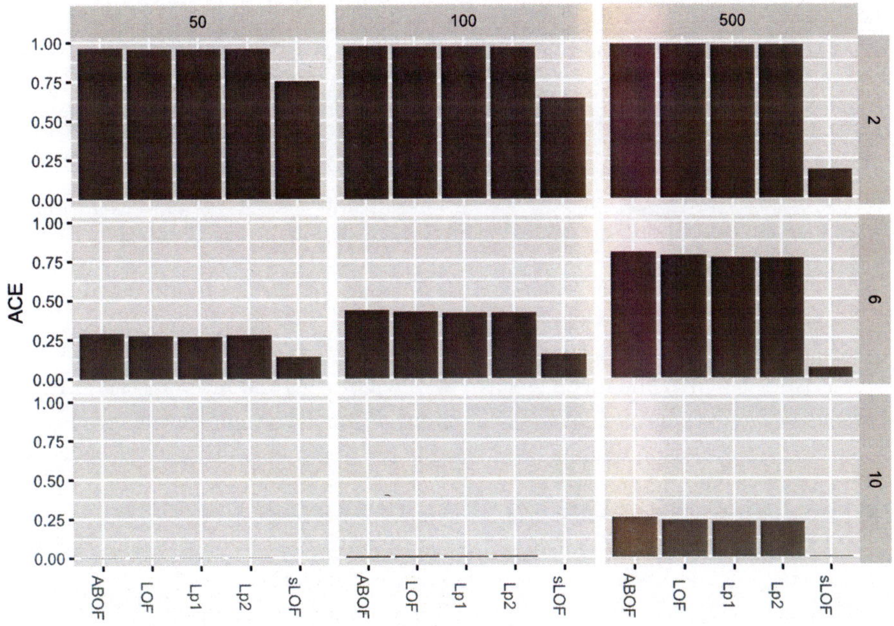

Fig. 1. The ACE cluster separability measure as a function of data dimensionality $p = 50, 100, 500$ (columns) and intercluster shift $\mu = 2, 6, 10$ (rows), for different outlierness factors, and for uncorrelated features. Small values of ACE indicate good separability of clusters.

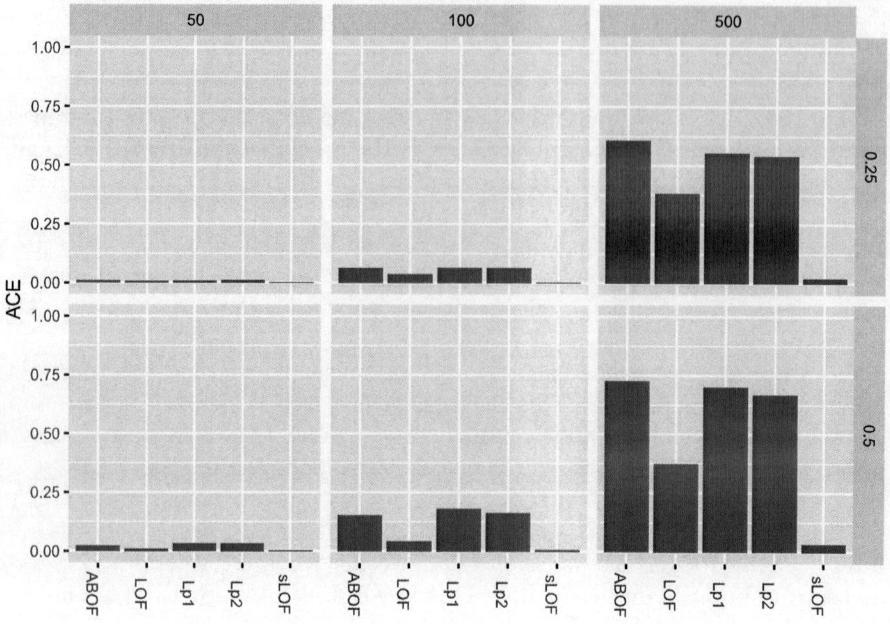

Fig. 2. The effect of correlation of features on the ACE cluster separability measure as a function of data dimensionality $p = 50, 100, 500$ (columns), for different outlierness factors. 25% of features are correlated with the value of correlation coefficient shown in rows. Results for intercluster shift $\mu = 10$. Small values of ACE indicate good separability of clusters.

3.2 Organization of the Experiment

The organization of the numerical study is as follows:

- The reference cluster is generated as N points from the multivariate normal distribution MVN with the mean 0 and the $p \times p$ identity matrix used as the covariance matrix. The testing cluster L is generated as N points from MVN distribution with the mean $[\frac{\mu}{\sqrt{p}}, ..., \frac{\mu}{\sqrt{p}}]$, where μ is the intercluster shift. The covariance matrix of L consists of the fraction m of correlated components, with the correlation equal ρ, the remaining components are uncorrelated, with variance of all components as in K.
- In the experiment we vary p, μ, m, and ρ.
- We quantify outlierness of points from L with respect to K using the ACE measure (defined below) using different functions OF.
- We also report type I and type II error (defined below) in the task of open-set classification as a function of the r parameter (see step 2 of the open-set classification procedure).

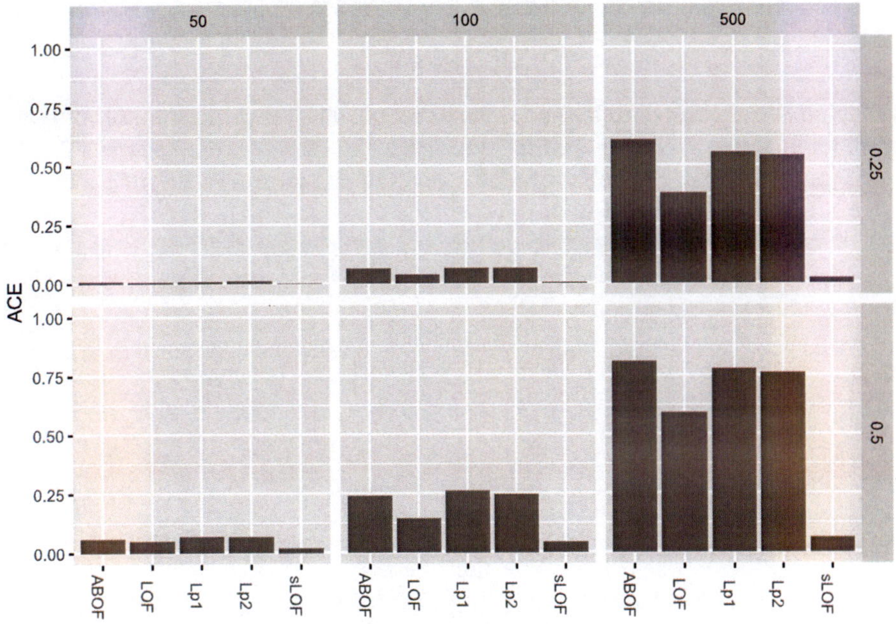

Fig. 3. The effect of correlation of features on the ACE cluster separability measure as a function of data dimensionality $p = 50, 100, 500$ (columns), for different outlierness factors. The fraction of correlated features is shown in rows, the value of correlation coefficient is 0.25. Results for intercluster shift $\mu = 10$. Small values of ACE indicate good separability of clusters.

We repeat the experiment for several different values of the random number generator (used to generate K and L) and the averaged results are reported.

The ACE (average cross-entropy) measure is defined as follows. We calculate the histogram (distribution) f_K based on the set of values $(OF(x, K) : x \in K)$, and histogram (distribution) f_L based on the set of values $(OF(x, K) : x \in L)$. The ACE is used to quantify/observe how f_K and f_L overlap for different OF methods. For each observation $x_i \in K \cup L$, we define $q_i = 1$ for $x_i \in K$ and $q_i = 0$ for $x_i \in L$. For each $x_i \in K \cup L$ we estimate whether x_i is more likely as a member of K or L, respectively, as

$$p_{K,i} = \frac{f_K(OF(x_i, K))}{f_K(OF(x_i, K)) + f_L(OF(x_i, K))} \tag{1}$$

and

$$p_{L,i} = \frac{f_L(OF(x_i, K))}{f_K(OF(x_i, K)) + f_L(OF(x_i, K))} \tag{2}$$

Knowing q_i, we can quantify if the larger value of $p_{K,i}$ and $p_{L,i}$ correctly identifies the actual class of x_i using the term $ce_i = q_i \cdot log(p_{K,i}) + (1 - q_i) \cdot log(p_{L,i})$. Then the overall separability of the clusters L and K can be measured using

$$ACE = \frac{\sum ce_i}{n} \tag{3}$$

where n is the number of observations in $K \cup L$. The interpretation of ACE is as follows: if for a given pair of different clusters K and L two measures OF1 and OF2 produce different values of ACE, then the one with smaller ACE allows for better separability between clusters.

Performance of different OF methods is also compared in terms of the type I and type II errors, defined in the task of open-set classification, and shown as a function of the r parameter (step 2 of the algorithm).

We define as false positive (type I error) the event when a point from K is considered an outlier. The false negative (type II error) occurs when a point from L is recognized as belonging to K. In the Figs. 4 and 5 we denote as *sensitivity* the true positive rate (TPR) equal (1 - type II error), and as (1-*specificity*) or false positive rate (FPR) the type I error.

3.3 Results and Discussion

In Table 1 we present results for $L = K$. In this experiment we show specificity of outlier detection using different measures OF. Since points in L are generated as non-outliers with respect to K, we expect the ACE values close to 1. We observe that the sLOF method reports L as different than K, with this effect increasing with dimensionality p. This indicates low specificity of this method and poor specificity for high-dimensional data.

In Fig. 1 we report performance of different OF measures for uncorrelated data. We observe the (obvious) behaviour that with growing intercluster shift the separability between clusters improves (smaller values of ACE for growing μ). We also observe that growing dimensionality leads to worse separability. We also observe the remarkable sensitivity of the sLOF method (however, sLOF is least successful in terms of specificity, as shown in Table 1).

Table 1. The ACE cluster separability measure for intercluster shift $\mu = 0$ for different outlierness factors. Dimensionality of data equals $p = 50, 100, 500$, sample size equal 500. Small values of ACE indicate that clusters are separated, large values denote overlapping clusters.

Dimensionality p	50	100	500
ABOF	0.988	0.987	0.988
LOF	0.985	0.988	0.994
Lp1	0.986	0.990	0.985
Lp2	0.988	0.987	0.986
sLOF	0.853	0.724	0.206

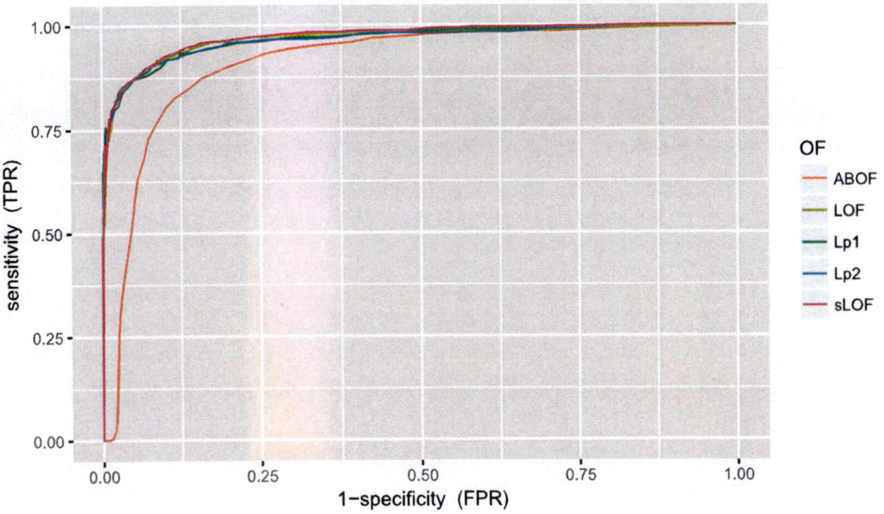

Fig. 4. The ROC curves for uncorrelated features for different outlierness factors. Results for dimensionality $p = 50$ and intercluster shift $\mu = 6$.

In Figs. 2 and 3 we report performance of different OF measures for with different correlation patterns in L. These results are obtained for the shift $\mu = 10$ (which corresponds to the last row in Fig. 1). Generally we observe that growing

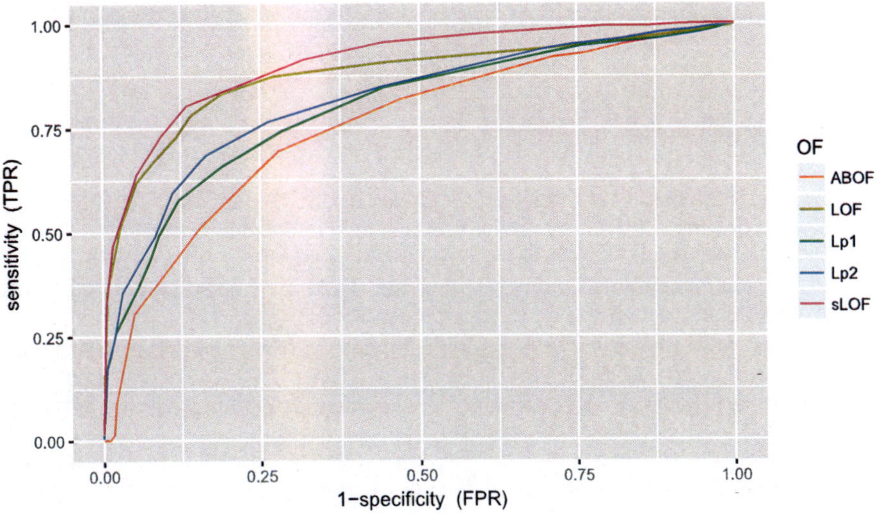

Fig. 5. The ROC curves for correlated features for different outlierness factors. Results for dimensionality $p = 50$ and intercluster shift $\mu = 6$, fraction of correlated features equal 0.5 and the correlation coefficient equal 0.5.

correlation among features leads to the worse separability between clusters, i.e. sensitivity in outlier detection decreases. This effect is least significant for the LOF methods.

In Figs. 4 and 5 we report the type I and type II errors as the ROC curves generated by varying the r parameter in step 2 of open-set classification. Contrary to results reported by the authors on ABOF, this method realizes lower specificity/sensitivity than LOF, and is also most affected by growing correlation among features.

4 Conclusion

In this work we showed a procedure of open-set classification which allows us to avoid spurious assignments of recognized observations to unrelated classes if the observation comes from a class unknown to the closed-set model. This is done by quantifying the observations as an outlier with respect to the training data representing the class pointed to by the closed-set model. In this work we evaluated performance of several measures (methods) of outlier detection proposed in literature as a function of dimensionality of data, distance of the outliers from the typical data and the correlation structure in data. The analyses allows us to suggest the preferable method of outlier detection. Contrary to the results reported by the authors of the angle-based (ABOF) methods, our study shows that the local density-based LOF method outperforms the ABOF method, and also seems to be less affected by the growing correlation among components of feature vectors. In the future work we plan to research further this phenomenon using more similarity measures, such as the Mahalanobis distance, as well as comparing the results with applications on feature vectors from the real-world data, such as in subject classification of text documents.

References

1. Amodei, D., Olah, C., Steinhardt, J., Christiano, P., Schulman, J., Mané, D.: Concrete problems in AI safety. arXiv preprint arXiv:1606.06565 (2016)
2. Bendale, A., Boult, T.: Towards open world recognition. In: Proceedings of the IEEE Conference on Computer Vision and Pattern Recognition, pp. 1893–1902 (2015)
3. Breunig, M.M., Kriegel, H.P., Ng, R.T., Sander, J.: LOF: identifying density-based local outliers. In: ACM SIGMOD Record, vol. 29, pp. 93–104. ACM (2000)
4. Chandola, V., Banerjee, A., Kumar, V.: Anomaly detection: a survey. ACM Comput. Surv. (CSUR) **41**(3), 15 (2009)
5. Ding, X., Li, Y., Belatreche, A., Maguire, L.P.: An experimental evaluation of novelty detection methods. Neurocomputing **135**, 313–327 (2014)
6. Doan, T., Kalita, J.: Overcoming the challenge for text classification in the open world. In: 2017 IEEE 7th Annual Computing and Communication Workshop and Conference (CCWC), pp. 1–7. IEEE (2017)

7. Fei, G., Liu, B.: Breaking the closed world assumption in text classification. In: Proceedings of the 2016 Conference of the North American Chapter of the Association for Computational Linguistics: Human Language Technologies, pp. 506–514 (2016)
8. Jain, L.P., Scheirer, W.J., Boult, T.E.: Multi-class open set recognition using probability of inclusion. In: Fleet, D., Pajdla, T., Schiele, B., Tuytelaars, T. (eds.) ECCV 2014. LNCS, vol. 8691, pp. 393–409. Springer, Cham (2014). https://doi.org/10.1007/978-3-319-10578-9_26
9. Kriegel, H.P., Schubert, M., Zimek, A.: Angle-based outlier detection in high-dimensional data. In: Proceedings of the 14th ACM SIGKDD International Conference on Knowledge Discovery and Data Mining, pp. 444–452. ACM (2008)
10. Pimentel, M.A., Clifton, D.A., Clifton, L., Tarassenko, L.: A review of novelty detection. Signal Process **99**, 215–249 (2014)
11. Prakhya, S., Venkataram, V., Kalita, J.: Open set text classification using convolutional neural networks. In: Proceedings of the 14th International Conference on Natural Language Processing (ICON-2017), pp. 466–475. NLP Association of India, Kolkata, India, December 2017. http://www.aclweb.org/anthology/W/W17/W17-7557
12. Scheirer, W.J., Jain, L.P., Boult, T.E.: Probability models for open set recognition. IEEE Trans. Pattern Anal. Mach. Intell. **36**(11), 2317–2324 (2014)
13. Walkowiak, T., Datko, S., Maciejewski, H.: Algorithm based on modified angle-based outlier factor for open-set classification of text documents. Appl. Stochastic Models Bus. Ind. 1–12 (2018). https://doi.org/10.1002/asmb.2388
14. Zimek, A., Schubert, E., Kriegel, H.P.: A survey on unsupervised outlier detection in high-dimensional numerical data. Stat. Anal. Data Mining ASA Data Sci. J. **5**(5), 363–387 (2012)

Performance of Modern Video Codecs in Real-Time Communication WebRTC Protocol

Aleksander Dawid(✉) and Paweł Buchwald

Department of Transport and Computer Science, WSB University, 1c Cieplaka St., 41-300 Dąbrowa Górnicza, Poland
{adawid,pbuchwald}@wsb.edu.pl

Abstract. Real-time multimedia communication technology is becoming the most significant part of our modern world. It can be a holy grail for stopping environmental degradation. Most people travel mainly to chat with other people. The COVID-19 pandemic showed that we could communicate using existing telecommunication technology without any traveling. A crucial part of today's video real-time peer-to-peer communication is a good video-stream compression algorithm. The publication uses WebRTC communication technology. It is a service that runs in HTML5 compliant web browsers. The service is designed for a peer-to-peer video connection on the local network. We conducted several audio-visual sessions with different bandwidth limits. For each network throughput, parameters responsible for the video transmission quality were measured, such as delay, packet loss, or the send and receive frame rate. In the experiment, four video compression codecs for real-time communication were tested. The obtained results indicate potential applications of AV1 compression codec in the construction of e-learning and game streaming websites.

Keywords: WebRTC · e-learning · LAN · Game streaming · Video transmission · Delay · Frame rate

1 Introduction

The past two years of lock-downs made information and communications technology (ICT) the most desirable system for remote work, learning and entertainment. These systems require multimedia data transmission services. Such services include video chat, audio-video streaming, and peer-to-peer information sharing. The video chat service is particularly demanding as it requires real-time data exchange. Many such ICT solutions are offering real-time communication. There are many native applications in a given operating system for videoconferencing or e-learning [1], remotely controlling the UAVs [2], and medical robots [3]. The only environment that provides universal access to multimedia Internet content is the Internet browser environment. Currently, both stationary and mobile devices can use this environment to communicate with web services. The only technology that has not been implemented so far in the web browser was direct client to client communication. The situation has changed with the introduction of the WebRTC

W. Zamojski et al. (Eds.): DepCoS-RELCOMEX 2022, LNNS 484, pp. 32–41, 2022.
https://doi.org/10.1007/978-3-031-06746-4_4

protocol developed by the World Wide Consortium (W3C) and the Internet Engineering Task Force (IETF). This protocol allows for real-time audiovisual communication of the browser-browser type. The WebRTC protocol has been used in communication software since 2011. One of the solutions that use multimedia communication is the identification and authorization of users [4], without the need of installing extra components, but only having access to a web browser. Additional features presented in the system are the functionalities allowing for single-cast and multicast transmission. The authors of this study implemented a web application for group communication using audio, video, and text messages. Although the RTC protocol does not provide session maintenance mechanisms and does not guarantee the quality of services, it can be the basis for high availability telecommunication systems. The possibility of using additional methods of session control and monitoring the quality of data acquisition enables the use of the WebRTC protocol also in such areas. A full description of these issues is in the study of using this protocol for telemedicine [5]. Teleconference is another area of telecommunication that uses WebRTC technology [6]. Multimedia data acquisition systems are a proper choice to extend the offer of educational units. This area of WebRTC applications is presented in the study [7]. The authors analyze the possibilities of integrating the created application with the Moodle e-learning platform. The proposed solution ensures the functioning of the so-called Virtual rooms, audiovisual broadcasts, and chat functionality. The WebRTC protocol allowed us to obtain a solution for the most popular web browsers. The created system is also capable of running on smartphones [8]. The need for real-time communication is especially visible in game streaming implementation. Nowadays, existing game streaming services operate on native apps (GeForce Now, PlayStation Now, Shadow) and web browser (Stadia or Amazon Luna). In this case, real-time action is more important than the quality of received video transmission. To achieve both quality and real-time control, we need a high-speed data connection and optimal video codec. If we look at the current offer of Internet providers, we can expect Internet bandwidth between 2 and 600 Mb/s. This offer depends on the used technology. If we plan to use this Internet connection for teleconferencing, we need a high-speed connection in both directions. In the case of low data transfer bandwidth, the real-time call quality depends on the video compression algorithms. Sending raw HD video (1920×1080) of 24-bit color depth will take about 50 Mb for one frame. So, if we want to watch a video with 25 frames per second, we need a transmission bandwidth at least at the level of 1.25 Gbps. The solution to this problem is a data compression algorithm. Picture and video compression algorithms belong to lossy compression methods. It means that the original stream is not equal to the compressed stream in a digital sense. We talk here about the acceptable visual quality of the multimedia stream. In practice, human visual perception cannot distinguish between colors at 24-bit color depth. The algorithms for video compression allows us to use slower Internet connections for video conferencing.

In this work, we test the quality of video conferencing using four different video compression codecs available in the WebRTC distributed service library. We want to choose the best solution for remote web-based communication system. The parameters characterizing this system should be a low cost of implementation, the possibility of using it without installing additional software on the client's side, quality of data acquisition allowing for ongoing communication to present didactic content, and the possibility of archiving the collected audiovisual content.

2 Video Compression and Transmission Methods Available in WebRTC Service

The basic video compression supported in WebRTC is H.264 method, also known as Advanced Video Coding (AVC) [9] and patented as ISO/IEC 14496-10: 2004. The last patent for h.264 will expire in 2027. According to the general rules of the h.264 license, you can decode any video stream for free, but if you want to encode a video stream, you must pay for it. It means that most of the license cost covers video editing software developers. In 2010 Google corporation proposed the VP8 codec as a royalty-free video compression format. The compression ratio of VP8 and H.264 is comparable, but VP8 is free of any licence. The description of the VP8 method is in RFC 6386 document. The WebM format uses the VP8 codec along with Opus audio. In 2010 VP8 replaced H.264 method in YouTube service. The WebRTC protocol supported the VP8 codec in 2011. In the same year, Google begins the development of a new VP9 codec [10]. It was direct answer to H.265 compression method. According to Google, the VP9 codec achieves 50% higher data compression than the VP8 codec. The VP9 video format is now widely used in YouTube streaming services. It works much better than VP8 codec in high latency networks [11]. In 2015 the Alliance for Open Media (AOMedia) developed a direct successor of the VP9 format named AOMedia Video 1 (AV1). On the fourth of January 2022, Google has implemented the standard of AV1 codec in the Chrome browser. Before that date, only the experimental version of AV1 named AV1X was available in the Chrome browser. The advantage of AV1 over other compressors used in WebRTC is its high data compression. In 2018, Facebook conducted a real-world condition test that shows 34% higher data compression of AV1 than VP9. The video compression method support in WebRTC depends on the web browsers. All modern browsers support H.264, VP8 and VP9 format. The only browser that supports AV1 format is the Chrome browser. It is still possible that in near future the AV1 format will be supported in all popular browsers. WebRTC enables transmission using a P2P connection. In typical e-learning applications, users rarely have a fixed public IP address. In this case, we need the STUN server. It allows to define public IP addresses and possible restrictions of communicating stations. It can help to establish a P2P connection. Thanks to this solution, workstations can send IP addresses to each other via an intermediary server (signaling server). Establishing a session requires the exchange of multimedia settings information. The exchange relies on sending an offer and receiving a response in the Session Description Protocol (SDP) format. The set of parameters and properties determined by this protocol is called a session profile.

SDP is an extensible and open protocol that allows extensions to support new media and transmission formats. Due to its properties, it is now the standard protocol for session settings. Session parameters are in groups of text format fields. For more information on this protocol, see RFC4556.

For this research, we have developed a WebRTC web application. The application has been written in JavaScript language (front-end) and in PHP language (back-end). The

JavaScript code is responsible for video codec negotiation between peers. Manipulating the SDP offer, we can set the desired receiving bandwidth. If we want to make a video connection using the given video codec, we have to reorder the codec list. We use the PHP language to store SDP offers and responses as files. The PHP module serves as an exchange server for SDP. The video call application is available at https://adawid.com.pl/vchat.

3 Experimental Setup

The real-time response of video calls is the main target of this research. First, we have to define what is real-time video. It is the measured received packet delay (RPD). This parameter characterizes real-time video. The human reaction time to screen event is about 190 ms on average [12]. If the RPD is less than 190 ms, we can say that video call is in real-time. To conduct an experiment of WebRTC video calls, we need a network infrastructure for that. In Fig. 1, you can see our experimental setup architecture. We have used the only local network to connect two clients in WebRTC protocol. Both clients are in the same IP subnet governed by router 1 (TP-LINK TL-WR1043ND). Router 1 also provides an Internet connection. The maximum bandwidth available here is 100 Mb/s. The latency of data packets in the case of workstation1 and workstation2 connection is less than 1 ms. The latency is a very significant parameter for WebRTC connections. Google Congestion Control (GCC) method changes the video quality by this parameter [13]. The GCC method will not change the video quality because of low and stable latency in the experimental network. The WWW and ICE servers are connected to the network by the second router (TP-LINK TL-WR743ND). As an HTTP server, we choose Apache software with PHP enabled module. In the experiment, we use HTTPS encrypted web protocol. The WebRTC protocol required ICE or TURN server to operate.

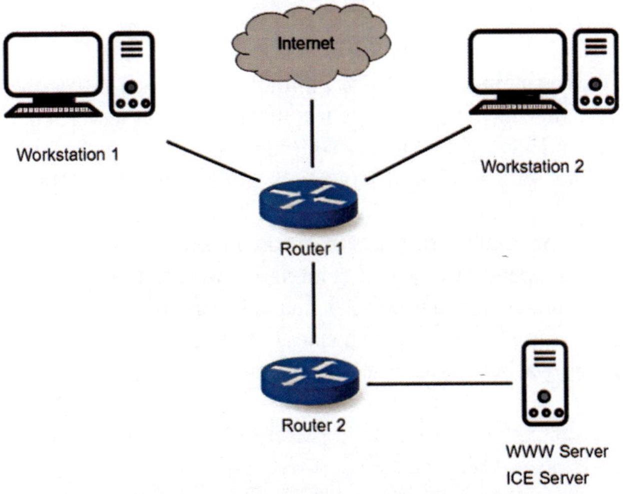

Fig. 1. The local network architecture.

We decided to run the ICE/TURN server in our network to accelerate a connection negotiation. Coturn TURN server was used in this research. Both Apache and Coturn run on Linux based machines. The WWW server is also the signalling server. Configuration of the workstations and server used in the video call experiment is shown in Table 1. The server configuration ensures low energy consumption in continuous operation.

Table 1. Workstations and server specification details.

Name	Description
Workstation1	HP OMEN: CPU i5-6300HQ, Memory 8 GB, disk 1TB, graphics GTX960M, Display 15,6″ MattFHD slim IPS 250nit, HP Wide Vision HD Camera with two digital microphone array. Network interface: Realtec RTL640 × 64
Workstation2	NVIDIA Jetson Nano: CPU Quad-core ARM® A57, GPU 128-core NVIDIA Maxwell™ architecture-based GPU, Memory: 4 GB 64-bit LPDDR4, Video 4K @ 30 fps (H.264/H.265)/4K @ 60 fps (H.264/H.265) encode and decode, Camera Logitech C270 video HD 720p/30 fps
Server	Orange Pi Zero: 512MB DDR3 SDRAM, CPU H2 Quad-core Cortex-A7, Video codec support H.265/HEVC 1080P, 64 GB SD memory card

The UBUNTU 18.04 operating system is used on both workstations with the Chrome web browser in version M95. The presented network infrastructure architecture made it possible to evaluate the transmission time efficiency for various video codecs and bandwidth configurations. The experimental setup can be used as a platform for the direct conversation between student and teacher, for example, in the case of an oral exam.

4 Results

The presentation of the results starts from the description of the video call connection procedure. In the beginning, workstation2 initializes the video call connection in our test setup (Fig. 1). It connects to our web application on the www server. Then, we have to set up a video codec and bandwidth limit. In addition to that, we can set up a period where the measurement of network and video parameters will be conducted. Next, we generate an SDP offer for another connected client. The client on workstation2 can cancel the offer if no connection is pending. If the offer exists on the server, the other client can respond to this offer. The responding client is running on workstation1. Then, the clients can talk using video streams. From now on, we can measure the packet delay and incoming frame-rate speed to evaluate the effectiveness of chosen video codec. The duration of each measurement was 5 min, and the acquisition took place every second, which gives a total of 300 measurement points. The obtained measurement results are in the form of diagrams showing the dependence of Cumulative Distribution Function (CDF) on parameters determining the quality of video transmission using VP8, VP9, H.264 and AV1X codecs. We were using bandwidth limits from 100 kbps to 1000 kbps. We have been using a 720p frame resolution set as the constraint in WebRTC protocol. The default frame rate for the experimental set without bandwidth limitation is 30 frames/s. To probe a pessimistic scenario in tests of codec performance, we use the fast-changing content of the transmitted video. We have achieved the fast-changing video content by permanently hand waving in front of the camera. The default fallback codec in WebRTC is the VP8 codec. This codec is available in all web browsers supporting WebRTC.

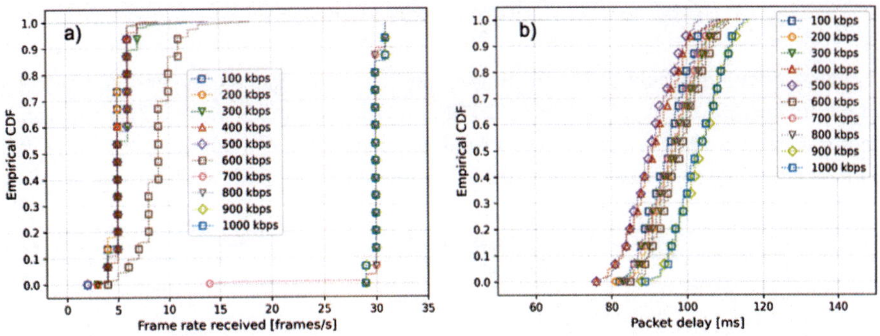

Fig. 2. The CDFs of a) incoming frame rate and b) delay of video packets when applying VP8 codec.

The CDF of incoming frame rate for VP8 codec shows that for the 50% connection time, the video recipient experiencing five frames/s for the bandwidth limit 100–500 kbps and 30 frames/s for the measured bandwidth limit 700–1000 kbps (Fig. 2a). The region between 500–700 kbps bandwidth limit can be called a transition region. In this narrow band, the frame rate intensively accelerate. The CDF at 600 kbps shows nine frames/s for 50% of connection time. As you can see, the CDFs of the incoming frame rate are perfectly well defined, which means that there is a negligible deviation from the mean value. In the CDFs of video packet delay, we see that 50% of video packets have a delay of less than 100 ms (Fig. 2b). We can observe the slight increase of the delay at 900 and 1000 kbps throughput. As we can see from this practice test, the smooth video playback when applying the VP8 codec requires at least 700 kbps network throughput. The direct successor of the VP8 codec is the VP9 codec which in its assumptions was to improve the video transmission quality. For the bandwidth 600 kbps, we see that the frame rate of VP9 is improved to VP8 on average by 66.5% (Fig. 3a).

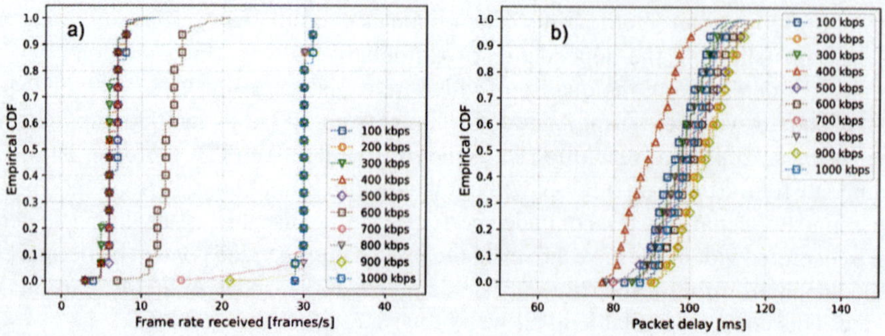

Fig. 3. The CDFs of a) incoming frame rate and b) delay of video packets when applying VP9 codec.

The observed packet delay when applying the VP9 codec is, in general, about 10% higher than for the VP8 codec (Fig. 3b). Its maximum value at 120 ms is acceptable for real-time video transmission. Using visual inspection, we can tell that the video quality in VP9 encoding is better than in VP8 encoding at the same network throughput.

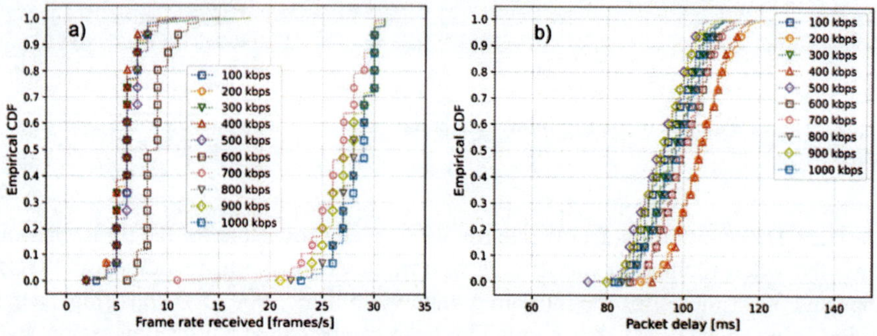

Fig. 4. The CDFs of a) incoming frame rate and b) delay of video packets when applying H.264 codec.

H.264 is the popular codec used in many video streams. The WebRTC protocol implemented in modern web browsers supports this codec. This codec has its hardware implementation. All modern graphics cards support H.264 video encoding and decoding internally. Both graphics cards used in this experiment support H.264 encoding and decoding. The CDF shape of the incoming frame rate for H.264 differs from those for VP8 and VP9 (Fig. 4a). There is still a wide gap between 500 kbps and 700 kbps, but the frame rate for throughput higher than 700 kbps is not so well localized. For example, in 1000 kbps throughput, 50% video connection time was at the frame rate equal to 29 frames/s. The minimum and maximum values registered for this connection bandwidth limit were 23 and 31 frames/s, respectively. The results show that the codec is more sensitive to video content than VP8 and VP9. In Fig. 4b, we see the CDFs dependency on packet delay when applying H.264 codec to video transmission. The delay between 75 and 122 ms ensures a real-time video connection.

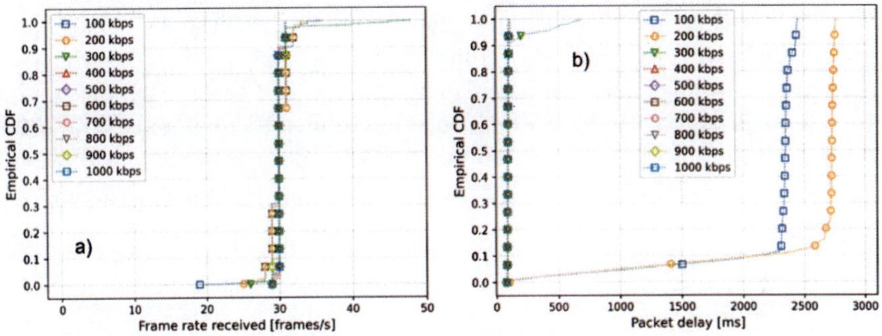

Fig. 5. The CDFs of a) incoming frame rate and b) delay of video packets when applying AV1X codec.

The next codec to test is the AV1 codec. The only available version of the AV1 codec was the AV1X experimental codec during measurement. So far, this codec for WebRTC protocol supports only Google Chrome and Chromium browsers. We have tested this codec in ranges from 100 to 1000 kbps bandwidth. Interestingly the frame rate does not change so much during these tests (Fig. 5a), but the delay at low throughputs reaches 2300 ms for 100 kbps and even 2800 ms for 200 kbps (Fig. 5b).

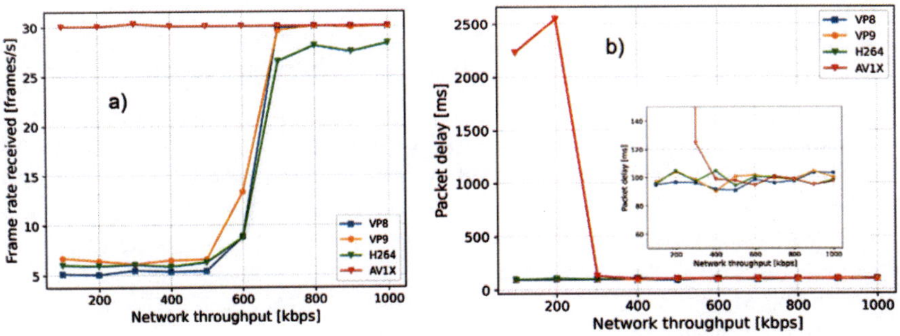

Fig. 6. Four WebRTC video codecs a) frame rate, b) packet delay dependency on bandwidth.

A delay higher than 200 ms is not acceptable for a real-time video connection. Despite this, the AV1 codec handles 30 frames/s at a delay equal in average to 100 ms using a network throughput of 300 kbps. In normal conditions of videoconferencing 300 kbps throughput using AV1 codec is enough to receive smooth real-time video. This result is about two times better than VP8, VP9 and H.264. If we do not care about a real-time response in video transmission, this codec offers a smooth video transmission at 30 kbps throughput with a 3–4 s delay. To summarize our results, we have compared the performance of each codec, calculating the average value of frame rate for given throughput and drawing the plot (Fig. 6a). In this plot, we see that the frame rate of VP8, VP9 and H.264 is almost the same. At lower values throughput (100–500 kbps), they can display up to 5–6 frames/s, but at higher tested throughputs (700–1000 kbps), they can perform with a maximum of 30 frames/s. The AV1 codec behaviour is slightly different.

The frame rate is the most significant parameter in this case. The AV1 codec produces a smooth video with a visible delay at the low throughputs. In Fig. 6b, we can see the dependence of average delay on bandwidth limit. We can observe the transition from high to low values of delay for AV1 codec at the bandwidth limit range 200–300 kbps. From the inset of this plot, we can tell that the delay does not change with bandwidth for VP8, VP9, and H.264 codecs. The value of the delay is osculating around 100 ms for these codecs.

5 Conclusions

In conclusion, we have found that VP8, VP9, and H.264 codecs are suitable for smooth video calls in the network of throughput higher than 700 kbps. If there be some lags in this network, the frame rate will drop to 5 frames/s, which is not acceptable for streaming games. The H.264 codec, despite hardware support, does not reach the maximum frame rate at the throughput of 1000 kbps. Nevertheless, these codecs will prove themselves in electronic education where in general, we do not present fast-changing video content. A new AV1 standard of video encoding and decoding proves to be the best in our test. This standard allows smooth video streaming at a network of 30 kbps bandwidth. AV1 needs a bandwidth higher than 300 kbps for real-time video calls. The application of the AV1 codec should be even more common if the hardware will be supporting this codec.

References

1. Radha, R., Mahalakshmiv, M.: E-learning during lockdown of covid-19 pandemic: a global perspective. Int. J. Control Autom. **13**, 1088–1099 (2020)
2. Mhatre, V., Chavan, S., Samuel, A., Patil, A., Chittimilla, A., Kumar, N.: Embedded video processing and data acquisition for unmanned aerial vehicle. In: 2015 International Conference on Computers, Communications, and Systems (ICCCS), pp. 141–145 (2015). https://doi.org/10.1109/CCOMS.2015.7562889
3. Rostański, M., Buchwald, P., Mączka, K., Kostka, P., Nawrat, Z.: The development of internetwork channel emulation platform for surgical robot telemanipulation control system (INSeRT). In: 2015 Federated Conference on Computer Science and Information Systems (FedCSIS), pp. 1081–1085 (2015). https://doi.org/10.15439/2015F239
4. Nayyef, Z., Amer, S., Hussain, Z.: Peer to Peer multimedia real-time communication system based on WebRTC technology. Int. J. Hist. Eng. Technol. **2**(9), 125–130 (2019)
5. Antunes, M., Silva, C., Barranca, J.: A telemedicine application using WebRTC. Procedia Comput. Sci. **100**, 414–420 (2016). https://doi.org/10.1016/j.procs.2016.09.177
6. Pasha, M., Shahzad, F., Ahmad, A.: An analysis of challenges face by WEBRTC videoconferencing and a remedial architecture. Int. J. Comput. Sci. Inf. Secur. **14**, 698–705 (2016)
7. Ouya, S., Sylla, K., Faye, P.M.D., Sow, M.Y., Lishou, C.: Impact of integrating WebRTC in universities' e-learning platforms. In: 2015 5th World Congress on Information and Communication Technologies (WICT), pp. 13–17 (2015). https://doi.org/10.1109/WICT.2015.7489664
8. Buchwald, P., Dawid, A.: Open concurrent network communication methods in building distributed web applications. TASK Q. **25**, 397–406 (2021). https://doi.org/10.34808/tq2021/25.4/c

9. Kwon, S., Tamhankar, A., Rao, K.R.: Overview of H.264/MPEG-4 part 10. J. Vis. Commun. Image Represent. **17**, 186–216 (2006). https://doi.org/10.1016/j.jvcir.2005.05.010
10. Mukherjee, D., et al.: The latest open-source video codec VP9 - An overview and preliminary results. In: 2013 Picture Coding Symposium (PCS), pp. 390–393 (2013). https://doi.org/10.1109/PCS.2013.6737765
11. Fouladi, S., Emmons, J., Orbay, E., Wu, C., Wahby, R.S., Winstein, K.: Salsify: {Low-Latency} Network Video through Tighter Integration between a Video Codec and a Transport Protocol. Presented at the 15th USENIX Symposium on Networked Systems Design and Implementation (NSDI 2018) (2018)
12. Jain, A., Bansal, R., Kumar, A., Singh, K.: A comparative study of visual and auditory reaction times on the basis of gender and physical activity levels of medical first year students. Int. J. Appl. Basic Med. Res. **5**, 124–127 (2015). https://doi.org/10.4103/2229-516X.157168
13. Vucic, D., Skorin-Kapov, L.: The impact of packet loss and google congestion control on QoE for WebRTC-based mobile multiparty Audiovisual Telemeetings. In: Kompatsiaris, I., Huet, B., Mezaris, V., Gurrin, C., Cheng, W.-H., Vrochidis, S. (eds.) MMM 2019. LNCS, vol. 11295, pp. 459–470. Springer, Cham (2019). https://doi.org/10.1007/978-3-030-05710-7_38

Improved Software Reliability Through Failure Diagnosis Based on Clues from Test and Production Logs

Wojciech Dobrowolski[1,2](✉), Maciej Nikodem[1], Marek Zawistowski[2], and Olgierd Unold[1]

[1] Politechnika Wroclawska, Wroclaw, Poland
{wojciech.dobrowolski,maciej.nikodem,olgierd.unold}@pwr.edu.pl
[2] NOKIA, Wroclaw, Poland
{wojciech.dobrowolski,marek.zawistowski}@NOKIA.com

Abstract. Growing demand for software reliability requires developers to analyze many production logs under time pressure. Unfortunately, some failures cannot be detected during the testing phase in large complex systems because they are specific to deployment, configuration parameters, non-deterministic system behaviour, and real-life user input. This article presents a novel and light approach to failure diagnosis based on natural language processing techniques. The aim is to extract as much information as it is possible from data available in standard logs attached to problem description. The approach uses unit test logs (test suites) to gather knowledge about the system. This knowledge is then used to analyze the production log and determine the test suites and the corresponding code block that most likely describes the runtime scenario. The experiments on Apache Hadoop HDFS and NOKIA systems show that the hints given by the framework are helpful to locate the fault.

Keywords: Failure diagnosis · Log analysis · Replay system · Natural language processing

1 Introduction

Software system requirements for uptime and reliability are growing every year. It poses a significant challenge to meet expectations for software vendors. Among all, the most demanding are failures in production, and available data on such failure is limited. Most often, these are logs. Reproducing it in the local environment might be hard or even impossible. There are a few reasons for this. Configurations of the field environment are often very complex. System behaviours depend on many conditions and are not deterministic. Even reproducing the same failure on the client setup might be impossible. Input to the system might be hard to capture. We focus our attention on extracting as much information as possible from existing resources. The motivating example is issue HDFS-10453 [2] visible in the Hadoop Distributed File System(HDFS) [1] in version 3.1.1.

W. Zamojski et al. (Eds.): DepCoS-RELCOMEX 2022, LNNS 484, pp. 42–49, 2022.
https://doi.org/10.1007/978-3-031-06746-4_5

A user frequently observed that after a data node is decommissioned, HDFS fails to replicate any data blocks that are under-replicated. HDFS outputs an exception log statement that complains that there is not enough capacity on any data node, and the thread responsible for block replication, Replication Monitor, freezes. This issue took months to resolve. Many debug sessions with instrumented software were required. Every reproduction was time-consuming. It took two years to deliver all patches. As in the above example, logs and source code are always available sources of knowledge. Runtime/production failure diagnosis is based on production log analysis, reproduction log analysis, and source code analysis. Reproduction and source code analysis are time-consuming and require the developer to know the system and its operation. However, the same developer knowledge is reflected in unit tests which define the correct and incorrect behaviour of system functions, classes, and components. While the knowledge represented in the unit test is not explicit, it can be extracted and used to replay the failure scenario. Contributions of this article include:

- the method of finding similarities between unit test logs and production logs,
- the approach to use similarities to replay the scenario in the production log.

This article is organized as follows. Section 2 reviews the most important works on reliability and failure diagnosis software systems. Section 3 describes the architecture of our framework. Section 4 presents the experimental results, and Sect. 5 presents the conclusions.

2 Related Works

There are two major approaches to failure diagnosis. SherLog [14], CSight [3], and ShiViz [4] are deriving clues from logs that can help the developer reason about the scenario from production logs. SherLog tries to reconstruct the exact execution path returning must and may paths. It is costly and might contain thousands of lines, limiting its usefulness. CSight and ShiViz reconstruct the system model. On the other hand, Kairus [17] objective is to help find the exact root cause of the failure by replaying the failure behaviour and comparing it to the correct one. Anomaly detection also uses production logs, but the aim is to detect which file contains an anomalous sequence of logs. Recently, this field has switched its attention to NLP models. Such models are used to extract features from logs. Log sequence, the relation between adjacent log events, is the source of information about the anomaly. Some analyze the parameters of log events [7], some concentrate on the event sequence itself [13]. NLP techniques are used to deal with whole files [9,16] or periodically collected sequences [6]. Wang et al. [13] consider the log event sequence as a sequence of words. Words are obtained by a sliding window of fixed size over the production log. A vector representation for every event is built using Bary, TfIdf and Log Sequence Matrix. Similar log event sequences will have similar vector representations. Similar log representation of log event In LogRobust [16], the whole file is treated as one sequence. It parses logs with Drain, then, using TfIdf and FastText [5], create a vector representing

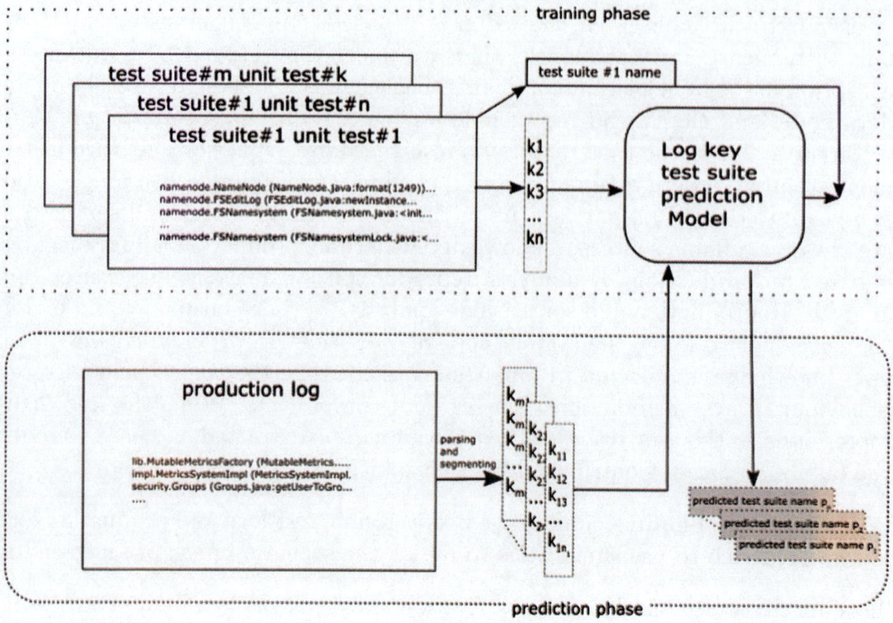

Fig. 1. Architecture overview. Test suite logs are used to train model. Production log is segmented and predictions are done on segments.

the log events. Vectorization of a sequence of log events is done by averaging the log event vectors. Here, information about the similarity between log event sequences is also kept. However, here it represents the whole file. Inspired by the observation from [7] and [13], where log event sequence is the base to extract features, we develop a framework that extracts knowledge from unit test log event sequences and transfers it to the segmented production log event sequence. This approach is novel, light, and simple. The novel, because logs from unit tests were not used to give clues about the running system. Light, because it does not require substantial computational and memory resources, and is simple, consists of two steps: learning on unit test logs and predicting the production log.

3 Proposed Method

The Architecture. of our framework is shown in Fig. 1. The main component is responsible for inferring the information about the execution. The framework first needs to be trained on logs from the unit test stage. The trained model is then used to analyze chunks of production logs and return clues in a sequence of test suites' names. Clues are in the form of a sequence of test suite names. This sequence corresponds to the production sequence of log chunks. As mentioned, production logs are presented to the model in chunks that are selected by a sliding window of fixed size (Fig. 2).

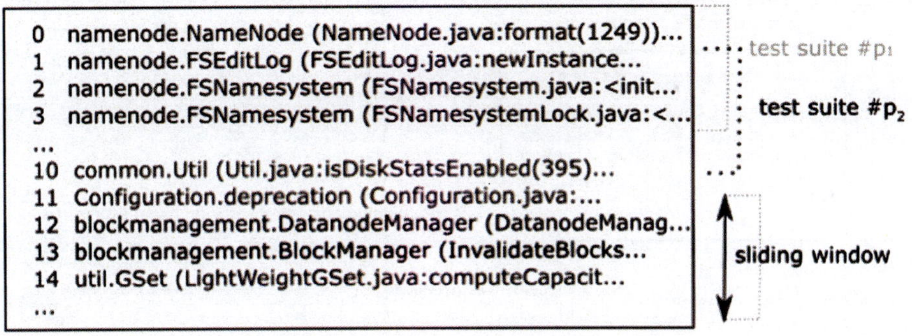

Fig. 2. Prediction phase. Window of fixed size slides across the log lines. For each window algorithm predicts test suite.

The Training Phase. requires logs from the unit tests. These logs are parsed, labelled with the test suite name, and used for the machine learning algorithm. The test logs used for training need to be collected from tests conducted on the software version corresponding to the failure production log. Every log line is parsed by Drain and transformed to a log event ID. Log event describes the constant part of the log and the variable part of the log. By matching the constant part, we can match the log lines produced by the same line of code. For example for log *Receiving from IP:127.0.0.1* code which produced it is *log.info("Receiving from IP:%s", IP)* and log event is *ID:1 Receiving from IP:<*>, [127.0.0.1]*. The framework keeps only the log event ID (Fig. 3).

The sequence of log events' IDs is a sentence in a language [13]. The test suite consists of many test cases. Every log from the test case can be transformed into a sentence. To deal with the variable length of sequences, the log event sequence is vectorized by TfIdf algorithm [12]. The training dataset consists of these vectors labelled with test suite names. We used the data balancing tool from Python library imblearn to balance the data set. We oversampled minority classes to have 0.1 the number of examples as the majority class and undersample every majority class with ratio 0.5. This dataset is used to train: Naive Bayes [15], Random Forest Classifier [8], and Gradient Boosting Classifier [10] with test suite names as classes.

In Prediction. the production log is segmented into chunks that are used to classify the test suite name (Fig. 2). The prediction likelihood needs to be over 0.5 and is stored along with the test suite name. The framework returns a sequence of test suite names and the likelihood corresponding to the log chunks.

The size of the chunk depends on the data. The production log for every software has different characteristics. That depends on the company logging policy. This is observable in the structure of logs. Segments of logs corresponding to the underlying part of the code vary in length, and this is a crucial characteristic

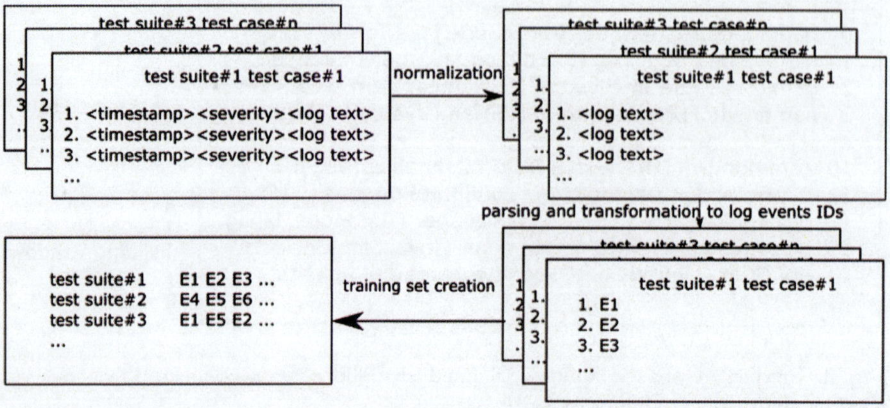

Fig. 3. Preprocessing logs

to select the most appropriate chunk size. For evaluation on HDFS, we found that the best chunk size equals 100 (Sect. 4), while on NOKIA data, the best chunk size was 10. A predicted sequence of test suites' names helps programmers understand what happened in the production. Production logs are parsed with the same instance as the unit test logs. The whole file is transformed into a log event ID sequence (Fig. 2).

Machine Learning Models. We used three machine learning (ML) models in our experiments: Naive Bayes Classifier(NBC), Random Forest Classifier (RFC), and Gradient Boosting Classifier (GBC). Application of TfIDF (Term Frequency Inverse Document Frequency) with NBC, RFC, and GBC is standard in the NLP domain. Our multiclass problem is where test suite names are target classes. For RFC and GBC, cross-validation grid search is used to select the best parameters of the model. We use sklearn GridSearchCV for this purpose. For RFC, the best parameters are the number of estimators of 100, the number of features to consider when looking for the best split is the logarithm of 2, the maximum depth not limited, and the function to measure the quality of a split is gini. For GBC, the best parameters are a depth of ten, one hundred estimators and a learning rate of 0.1. These models were used to assess learning test suite names' accuracy, precision, and recall from log event ID sequences.

4 Experimental Results

The proposed approach was implemented with the sklearn python library [11]. We evaluated our framework on data from HDFS and NOKIA production runs. It is worth mentioning that collecting logs from HDFS was much more complex than from NOKIA. Maven project do not have automatic way of collecting separate test logs for test cases. NOKIA software, a C++ system, is easily configurable to collect separate logs for a separate test case. In both cases, the developer is responsible for finding the root cause. The proposed approach was evaluated through verification if the test suites outputted correspond to the log chunk where the root cause manifests itself. Implementation and sample data are available in repository[1]. Our evaluation of NOKIA internal data shows that the best model(Random Forest Classifier) achieves high accuracy, precision, and recall during learning test suites 96%, 97% and 97% respectively. On HDFS unit test logs, RFC achieves 93%, 95%, and 93%, respectively. Clues given to the production logs are helpful.

4.1 HDFS Experiment

We used HDFS version 3.1.1 as it is where HDFS-10453 was spotted. This issue was analyzed in [17]. We want to verify if our approach would be useful in analyzing this case. We collected 5101 unit test logs corresponding to 658 different test suites. The maximum number of samples for the test suite was 78, and the minimum was 1. The longest unit test consists of 3 million log events, and the smallest is one event long. The best ML model proved to be Random Forest Classifier. It achieves accuracy of 93%, precision of 95%, and recall of 93% (Table 1). For HDFS-10453 production log, it predicted test suite org.apache.hadoop.hdfs.server.blockmanagement.TestPendingReconstruction for lines connected to the fault root cause. Fix to HDFS-10453 error was introduced in the blockmanagement module, which means that the hit outputted by the proposed approach is useful.

Table 1. Three ML models trained on HDFS data. Result show how well model can recognize test suite by looking and log sequence.

ML models results on HDFS dataset			
Model	Accuracy	Precision	Recall
RFC	93%	95%	93%
GBC	92%	94%	92%
NBC	66%	69%	66%

4.2 NOKIA Experiment

We used a version of NOKIA radio software corresponding to the PR607615 issue. There were 944 test cases and 180 test suites. Test suites have from 1 to 27 test cases. The balancing dataset was also used with oversampling to 0.1 and undersampling to 0.5. The longest test case has 588 902 lines, and the shortest was three lines long. The best model was Random Forest Classifier with the

[1] https://github.com/dobrowol/defects4all.

accuracy of 96% precision of 97%, and recall of 97% (Table 2). Moreover, log lines connected to the root cause were correctly classified with the test suite name corresponding to the root cause.

Table 2. ML models trained on NOKIA data. Result show how well model can recognize test suite by looking and log sequence.

ML models results on NOKIA dataset			
Model	Accuracy	Precision	Recall
RFC	96%	97%	97%
GBC	96%	96%	96%
NBC	91%	91%	91%

5 Conclusions

The reliability of large software systems is an iterative process. Some failures are unavoidable; thus, fast reaction time and delivery of fix improve the reliability. Our framework proves to help replay the scenario from the production log, which helps the developer understand what happened, thus increasing software reliability. It gives valuable hints to segments of log events based on expert knowledge extracted from unit tests. It is also easily aligned to the required software version, as learned from unit tests of the software version where failure was found. The process of parsing and learning can be done automatically and added to the CI job. There is also no need to store large amounts of historical data. We also can conclude that test suites should be more granular, with meaningful names and many test cases. We believe that connecting clues from the testing phase will help programmers rethink writing unit tests. It is not about code coverage only but also about understanding behaviour, separating functionality with more care. The model is biased towards the parts of code with the most intensive testing, and this is natural, as some parts of the code contribute to failures more often than others.

Acknowledgements. The authors appreciate the valuable comments provided by the anonymous reviewers. This work was supported by NOKIA company and financed by the Polish Ministry of Education and Science. Funds were allocated from "Implementation Doctorate" program.

References

1. Apache Hadoop HDFS architecture. https://hadoop.apache.org/docs/r1.2.1/hdfs_design.html
2. Apache Hadoop HDFS hdfs-10453. https://issues.apache.org/jira/browse/HDFS-10453

3. Beschastnikh, I., Brun, Y., Ernst, M.D., Krishnamurthy, A.: Inferring models of concurrent systems from logs of their behavior with CSight. In: Proceedings of the 36th International Conference on Software Engineering, pp. 468–479 (2014),
4. Beschastnikh, I., Liu, P., Xing, A., Wang, P., Brun, Y., Ernst, M.D.: Visualizing distributed system executions. ACM Trans. Softw. Eng. Methodol. (TOSEM) **29**(2), 1–38 (2020)
5. Bojanowski, P., Grave, E., Joulin, A., Mikolov, T.: Enriching word vectors with subword information. Trans. Assoc. Comput. Linguist. **5**, 135–146 (2017)
6. Chen, A.R.: An empirical study on leveraging logs for debugging production failures. In: 2019 IEEE/ACM 41st International Conference on Software Engineering: Companion Proceedings (ICSE-Companion), pp. 126–128. IEEE (2019)
7. Du, M., Li, F., Zheng, G., Srikumar, V.: DeepLog: anomaly detection and diagnosis from system logs through deep learning. In: Proceedings of the 2017 ACM SIGSAC Conference on Computer and Communications Security, CCS 2017, pp. 1285–1298. Association for Computing Machinery, New York (2017). https://doi.org/10.1145/3133956.3134015
8. Kam, H.T., et al.: Random decision forest. In: Proceedings of the 3rd International Conference on Document Analysis and Recognition. vol. 1416, pp. 278–282, Montreal, Canada, August 1995
9. Lin, Q., Zhang, H., Lou, J.G., Zhang, Y., Chen, X.: Log clustering based problem identification for online service systems. In: 2016 IEEE/ACM 38th International Conference on Software Engineering Companion (ICSE-C), pp. 102–111. IEEE (2016)
10. Mason, L., Baxter, J., Bartlett, P., Frean, M.: Boosting algorithms as gradient descent. In: Advances in Neural Information Processing Systems, vol. 12 (1999)
11. Pedregosa, F., et al.: scikit-learn: machine learning in Python. J. Mach. Learn. Res. **12**, 2825–2830 (2011)
12. Salton, G., Buckley, C.: Term-weighting approaches in automatic text retrieval. Inf. Process. Manage. **24**(5), 513–523 (1988)
13. Wang, J., et al.: LogEvent2vec: LogEvent-to-vector based anomaly detection for large-scale logs in internet of things. Sensors **20**(9), 2451 (2020)
14. Yuan, D., Mai, H., Xiong, W., Tan, L., Zhou, Y., Pasupathy, S.: SherLog: error diagnosis by connecting clues from run-time logs. In: Proceedings of the fifteenth International Conference on Architectural Support for Programming Languages and Operating Systems, pp. 143–154 (2010)
15. Zhang, H.: The optimality of Naive Bayes. Aa **1**(2), 3 (2004)
16. Zhang, X., et al.: Robust log-based anomaly detection on unstable log data. In: Proceedings of the 2019 27th ACM Joint Meeting on European Software Engineering Conference and Symposium on the Foundations of Software Engineering, pp. 807–817 (2019)
17. Zhang, Y., Rodrigues, K., Luo, Y., Stumm, M., Yuan, D.: The inflection point hypothesis: a principled debugging approach for locating the root cause of a failure. In: Proceedings of the 27th ACM Symposium on Operating Systems Principles. pp. 131–146 (2019)

Multiprocessor Tasks Scheduling. Fuzzy Logic Approach

Dariusz Dorota(✉)

Cracow University of Technology, Cracow, Poland
ddorota@pk.edu.pl

Abstract. The article concerns a problem of optimizing the execution of so-called multiprocessor tasks under the conditions of problem data uncertainty. From the implementation point of view, this is the case, where tasks require simultaneous access to many independent processors in order to realize an embedded system with increased dependability through redundancy. The uncertainty of data may concern parameters such as, for example, task execution times, task submission deadlines, number of tasks, etc., which have been perceived in this paper as a fuzzy data. Appropriate fuzzy approach has been described and analysed. There was also provided a short overview of approaches dedicated for modeling and scheduling problem solving with particular emphasis on uncertain data, especially a fuzzy approach. Starting from the deterministic off-line algorithm for multiprocessor tasks, a variant of the algorithm using fuzzy logic approach was formulated. The approach presented in this article is innovative compared to previous work focusing on uncertainty in the scheduling of multi-processor tasks. Then, the results of the experiments were presented in tables and in Gantt charts.

Keywords: Multiprocessor tasks · Uncertain scheduling · Fuzzy logic scheduling

1 Introduction

The article concerns the problems of optimizing the implementation of multiprocessor tasks under the conditions of data uncertainty. The motivation is the application of the considered problems in the management of embedded systems. The approach presented in this article is innovative compared to previous work [14] focusing on uncertainty in the scheduling of multi-processor tasks. The paper proposes to take into account the uncertainty in the scheduling process, but in this paper there is a significant change in modeling uncertainty using fuzzy logic to describe uncertainty. Computational tasks that require simultaneous access to many independent processors in order to realize an embedded system with increased dependability through redundancy are considered. Applications relate to, inter alia, for aviation and space systems, the movement of humanoid robots. The uncertainty concerns parameters such as, for example, task execution times,

W. Zamojski et al. (Eds.): DepCoS-RELCOMEX 2022, LNNS 484, pp. 50–62, 2022.
https://doi.org/10.1007/978-3-031-06746-4_6

task submission deadlines, number of tasks. Provides an overview of approaches to modeling and scheduling problem solving with particular emphasis on uncertain data problems. It was proposed to extend the three-field classification of scheduling problems with a parameter that takes into account the multiprocessing factor. Starting from the deterministic off-line algorithm for multiprocessor tasks, a variant of the algorithm with uncertain data was formulated using fuzzy logic. The system specification in the form of a task graph provides for the use of divisible tasks. Then, the results of the conducted experiments were presented in tables and in the form of Gantt charts. The article consists of five sections. The first is an introduction. In the second - we present the issue of multiprocessor tasks along with the motivation for their use. The third section provides the novel algorithm based on the MC concept (Muntz-Coffman). The fourth section discusses the concept of fuzzy logic with particular emphasis on its use in the process of scheduling tasks in the area of embedded systems. The successive section presents the computational results in the form of a table and Gantt charts. The last section is a summary and future directions of research.

2 Scheduling Multi-processor Tasks

The dynamically developing field of embedded systems stimulates the emergence of new technological hardware and software solutions in order to create the most reliable and efficient systems of this type. One of the significant components of embedded systems are effective methods of managing computational tasks, with additional restrictions, which usually are reduced to very specific scheduling issues. In general, scheduling can be viewed as a certain decision-making process of allocating resources to accomplish tasks. Each of the tasks can be characterized by specific features and limitations for a specific application. The task is to perform the service (usually activities) operation, grinding of the operation. We further assume that the elementary activity is kept in presence (in case the tasks are single-operational). Let us denote by $T = \{T_1, T_2, T_3, \ldots, T_n\}$ the set of tasks to be performed on machines (processors) defined as $M = \{M_1, M_2, M_3, ..., M_m\}$ in the deterministic case. Similar notations are also used in the case of uncertainty in the form of fuzzy logic, T^F is used to describe tasks using this approach. Basically, task T_i is assumed to contain one atomic operation used simultaneously a predefined number of processors $a_i \geq 1$ in the time $t_i \geq 0$. In the event of allowing interruption, it is assumed that the task $T_i \in T$ consists of a sequence of v_i elementary operations $O_i = (o_{i,1}, o_{i,2}, o_{i,3}, ..., o_{i,v_i})$, performed in this order, each of which requires the use of a_i processors. Of course, $\Sigma_{k=1}^{v_i} |o_{i,k}| = t_i$, where $|o_{i,k}|$ is the duration of operation $o_{i,k}$. In this case, the division of a task into operations is not fixed and it is a subject to selection. Multiprocessor tasks/operations can be classified accordingly: (1) $a_i = 1$ means single-processor tasks, (2) $a_i = k \leq m$ means that all tasks requires the same number k processors, (3) $a_i \leq k \leq m$ tasks requires at most k processors. In practice, we have $k = 2 \ldots 3$, depending on the required level of reliability or true parallelism. The idea of a multiprocessor task comes from, among others, the scheduling of tasks in multiprocessor computer systems, testing of VLSI (Very Large Scale Integration) or other devices.

Actually, testing processors by each other requires task action on at least two processors simultaneously [15]. Dependability for the entire system was calculated using the formula from the literature [13,14]

$$D_x = \frac{\sum_{i=x}^{n}(p_i^{CR})}{\sum_{i=x}^{n}(t_i^{CR})}, \quad D_x^* = \frac{\sum_{i=x}^{n}(p_i)}{\sum_{i=x}^{n}(t_i)}, \tag{1}$$

where D_x denotes level of dependability, p_i is the task priority. According to the proposed formula, reliability levels were calculated for each of the examples for which the scheduling was performed. The first version of the reliability level D_x is described by the formula (1), for the second version the following equation D_x^* in (1) was proposed. The first method considers all task designated as critical tasks and their priorities (according algorithm is a task with highest priority in each portion of time), while in the second case each tasks times and their priorities are taken into account. By definition, such tasks should be allocated first, and therefore it has been proposed that the tasks and priorities allocated in the processor with the minimum index should be considered. A similar concept of reliability can also be considered for k-processor tasks (Table 1).

Table 1. List of symbols used for task scheduling parameters

Deterministic	Fuzzy	Meaning
TG	TG^F	Graph of tasks $TG = (T, E)$, $E \subset T \times T$
T	T^F	Collection of tasks, $T = \{T_1, \ldots, T_n\}$
T_i	T_i^F	Task $T_i \in T$
a_i	a_i^F	Number of processors required by task T_i
t_i	t_i^F	Processing time of task T_i, where $t_i = \sum_{j=1}^{v_i} \lvert o_{ij} \rvert$, and $v_i = \lvert O_{ij} \rvert$
$C_{\max}$	$C_{\max}^F$	Schedule length
H_i	H_i^F	The level of task T_i in the graph TG (number of nodes from the source)
A	A^F	Task path (sequence of nodes) in the graph TG;
t_i^{CR}	$t_i^{F,CR}$	Time of execution of task T_i, when T_i has p_i^{CR}
o_{ij}	o_{ij}^F	Operation j, where i - is the number of the task T_i, j - is the number of the operation
$\lvert o_{ij} \rvert$	$\lvert o_{ij}^F \rvert$	Processing time of operation o_{ij}
S_i	S_i^F	Earliest possible start time for the task T_i
d_i	d_i^F	Latest possible completion time for the task T_i
p_i	p_i^F	Priority of task T_i
p_i^{CR}	$p_i^{F,CR}$	Task priority of T_i for highest priority in portion of time
p	p^F	Set of priorities for tasks from the set T

3 Application of Fuzzy Logic

Taking into account the nature of the real instance data, one can use the classification distinguishing the following problems [1]: (A) deterministic, (B) stochastic, (C) fuzzy. Considering the nature of the data, classes of the problem and its complexity, it is possible to make the right decision regarding the use of tools for targeting and analyzing the problem, respectively [1]: (A) deterministic scheduling theory, (B) queuing theory, stochastic processes, Markov chains, (C) fuzzy set theory. The literature review shows that there is a large group of practical problems characterized by uncertain and imprecise data. It is usually associated with a high cost of determining the optimal solution and its interpretation. Non-deterministic problems often refer to the results of analogous deterministic case. This approach allows the application of the deterministic scheduling theory to the previously presented problems, correlated with the methods of data representation, regardless of the class of these problems [1] (Fig. 1).

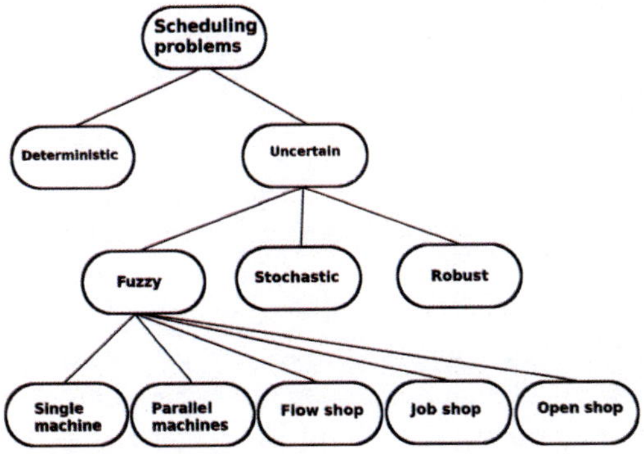

Fig. 1. Classification of uncertain scheduling problems

Deterministic scheduling problems are characterized by the symbol $\alpha|\beta|\gamma$, where α denotes the type of the problem (more precisely, the task routes in the system), β specifies additional constraints, γ determines the form of the objective function, [1–3]. Examples of criteria are: length of the schedule, maximum/average delay, weighted average delay, number of delayed tasks, weighted number of tasks delayed, sum of completion times, etc., as described in [3,4]. For the purposes of the issues discussed in the article, the range of meanings of the β symbol was extended, namely: fuzzy (p) and an additional parameter $mprc$ was proposed, which specifies the number of processors necessary to perform a single task and can assume values in the range $1 \ldots k \leq m$. This type of uncertainty is modeled by assuming that the problem data, schedule and its evaluation index are fuzzy quantities. The review [10] presents a summary of problems and scheduling methods taking into account some fuzzy parameters. It lists, inter alia, : (A) due dates; (B) processing time; (C) precedence relation;

(D) precedence delay; (E) disruption; (F) machine breakdown; (G) resource constraint. The review [10] also found that the most frequently considered objective functions in the fuzzy model are: (a) makespan; (b) total completion time; (c) flow time.

In this work, the fuzzy number theory was used along with the arithmetic consequences of this choice for this solution. Since the concept of applying fuzzy is treated as multi-valued logic, in which the considered element may belong to both sets simultaneously, the consequence is the use of unconventional arithmetic. If the multiple value of one variable is taken into account, then there must be functions determining the degree of belonging to the set. This concept, along with the most important membership functions and the most elementary arithmetic operations, has been widely discussed in the literature, see among others [5–8,12]. Taking into account the above-presented features and properties of fuzzy logic, as well as fuzzy sets, a conclusion can be drawn about the structure of the fuzzy system. In such systems, the desired values are first blurred, then classified using defined rules, to finally obtain the desired values after subjecting them to the sharpening process [8]. The construction of such a system is presented as the assembly of individual elements, as it is described in formula (2). Systems of this type are characterized by the following: (A) Fuzzification block; (B) Inference block; (C) Block of defuzzification.

Taking into account the review of the state of knowledge and the concepts mentioned above, the specification of task times as an uncertain parameter is proposed here. For the purposes of examining fuzzy logic, new markings dedicated to fuzzy values have been proposed: $A_i^F = \mu(T_i^F)$ - the result of the membership function for the i-th problem, where $A_i^F = \{(i, \mu_A(i)) : i \in X\}$, and $\mu_A : X \to [0, 1]$. In the considered case it is assumed that tasks are specified using the triangular membership function, where a triangle is used to specify the assumptions that the minimum value of the membership function is 20% lower than the mean value of this function $\mu(T_i^F) = 1$ and the membership maximum is a value 20% greater than the mean value. This is described by Formula (2), and will be denoted in the sequel as Model A (left) and Model B (right).

$$T_i^F = \begin{cases} a = b - 20\% \\ b = \mu(T_i^F) = 1 \\ c = b + 20\% \end{cases} \quad T_i^{F*} = \begin{cases} a = b - 10\% \\ b = \mu(T_i^F) = 1 \\ c = b + 30\% \end{cases} \tag{2}$$

In Tables 2 and 3 symbols mean accordingly to formula (2): $MIN = a$, $MED = b$, $MAX = c$.

4 Description of the Approach

We start our considerations from the Muntz-Coffman algorithm for the stated problem in the deterministic case, see e.g. [13,14]. The next step in the analysis is to determine which parameters in the problem can be uncertain and how the uncertainty will be modelled. From the practice point of view, one expects that processing time of the task can be uncertain, but the number of required processors for the task should be deterministic, since it follows from a reliability/safety procedure. Regarding the modeling style, the problem can be perceived as stochastic (see our previous papers [13,14]) or fuzzy (considered here first time). There are exist at least two various fuzzy approaches derived from the Muntz-Coffman algorithm. As we pinpointed previously, both approaches assume the fuzzy processing times of tasks but deterministic processor requirement. We describe the former approach in detail there, while the latter we discuss in the sequel of the paper. Let us assume that the data of the problem is given, including the description of the task graph based on the TGFF [10,11] method with the multiprocessor task specification. Note that TGFF is given in advance. The key of the approach is the method to represent times in the task graph using fuzzy logic. After loading the system specifications, the paths in the task graph are determined. For this purpose, the algorithm with the relatively lowest computational complexity A^* has been selected. Then levels are set for each of the tasks. Levels (priorities) for MC algorithm are calculated as defuzzified values of former fuzzy levels (because of fuzzy processing times) using the maximum measure method. Having the target values in the range of the deterministic approach, it is possible to adapt the prioritization from this case. After determining the levels for individual tasks, the algorithm allocates tasks with the highest priorities. Finally the fuzzy schedule is obtained with the goal fuzzy function value the makespan, which is deffuzified next. Ultimately, the scheduling can be presented using the Gantt chart. Since fuzzy logic in the form of a triangular membership function is considered here, it is necessary to present the scheduling in three variants. The scheduling representation includes the extreme and mean values of the membership function.

4.1 Algorithm

The modification in this case, in relation to the traditional algorithm, consists in the use of defuzzification and fuzzification blocks. The input data is a task graph containing the specification of task times in the form of a selected membership function. To further illustrate the scheduling procedure, an additional description of the approach is provided below. Specified data is fused, where task durations are blurred. The triangular membership function was used to describe the task times, with the specification defined by the Formula (2). Then, using the maximum measure method, the times sharpening is performed. The specified data is then passed as input to the scheduling algorithm. Tasks are prioritized taking into account the so-called multiprocessing factor. On the basis of so determined priorities, tasks are subject to scheduling in accordance with the mechanism described in Algorithm 1. The prioritization mechanism itself is presented in Algorithm 2, which, of course, takes into account the mutual relations between tasks specified in the task graph. A simplification of the model consisting in omitting the transmission time is taken into account here. The performed scheduling operations are then subjected to the fuzzification process so that it is possible to determine the extremes of the membership function. The fuzzification of the obtained results is carried out, which allows for the determination of the ranking for the values: a) minimum, b) average, c) maximum. These values relate to the values of the membership function, in this specific case of the membership function, For the following approach, two versions of the triangular membership function were considered during the experiments (see Formula (2)). This solution has been compared with the classic fuzzy logic approach, where the data (in this case, task duration) are described with the use of linguistic variables. The approach based on the use of specifications in the form of linguistic variables was also based on prioritization according to the rules of the same algorithm (see MC algorithm). A comparative experiment was carried out for dependent and independent tasks, as well as for each of the cases in two variants (divisibility and indivisibility of tasks). This comparison showed that both in the classical approach using linguistic variables (for the description of times and priorities) it gives an identical solution as to the length of the sequence, without affecting the sequence of tasks in the target schedule. Therefore, it is assumed that if these approaches give identical results in different cases, they can be considered equivalent solutions, hence the use of the approach described by the A and B algorithms. According to the adopted principle, a task schedule is proposed for various values, correlated with the values of the membership function. For the proposed algorithm, tests were carried out to the same extent as in the case of the algorithm with the specification from the scope of the deterministic approach.

The pseudocode of Algorithms 1 and 2 is shown detailed description of the implementation and operation method is presented below. We use denotations: $[x]$ for integer of x, $A = \{\ldots\}$ for the set and $|A|$ for the cardinality of set A.

Algorithm 1. *Scheduling multiprocessor tasks using fuzzy logic.*

1. $t = 0$; $TG^F_{struct} = read(TG)$;
2. $TG = defuzz(TG^F_{struct})$; /* use maximum measure method */
3. $TG_{path} = find\_paths(TG_{struct})$; /* TG_{path} is a set of paths */
4. `for` $T_i \in T$ `do` $TG_{level} = calculate\_levels(TG_{path})$; /* *calculate priorities of tasks* */
5. $MP = TG_{level}.p$; /* MP is a set of task's priorities */
6. $MMP = \max\{p_i : T_i \in MP\}$; /* find maximal priority in MP */
7. $HP = \{T_j \in T : p_j = MMP\}$ /* find all tasks with priority MMP */
8. `for` $T_i \in HP$ `do` `if` $(p_i > 1)$ /* analyse tasks with non-zero priorities */
9. `if` $(|HP| > 1)$ /* HP is the same for a few tasks */
10. $LHP = \max(TG_{level}.high)$; /* LHP is the set of tasks from HP with the highest level */
11. `if` $(|LHP| > 1)$ $p_i = \max\{p_j : T_j \in LHP\}$ /* select in LHP the one with the highest index */
12. `else` $p_i = \max\{p_j : T_j \in HP\}$
13. `else` $p_i = MMP$;
14. `end for;`
15. `for` $T_i \in T$ *do*

    if $(t_i - [t_i] == 0)$
      $x = 1$; *remove* p_i *from* MP;
    else $x = t_i - [t_i]$;
end for;
$TG^F_{struct} = fusification(TG_{struct})$;
for $T_i \in TG^F_{struct}$ **do**
    order T_i *for updated* t_i *according to Mc'Naughton algorithm;*
    if $|P(t)| > 0$ **go to** *step 5* /* Function $P(t)$ checked is exits an unused processor in time unit t */
    else $\{t_i = t_i - x,\ t = t + x;\ \}$
end for;

Algorithm 2. *Task leveling is performed by the function* $calculate\_levels(TG_{path})$:

for $A_j \in TG_{path}$ **do** /* A_j is the path with the end in T_j */
    $p_j = \sum_{k\in A_j} t_k \cdot a_k;\ time_j = \sum_{k\in A_j} t_k;\ high_j = \sum_{k\in A_j} E_k;$
    $TG_{level}.add(p_j, t_j)$;
    mark A_j *as calculated;*
end for;
for $p_i \in TG_{level}$ **do**
    if $\exists(\max(TG_{level}.time) \leq T_{iCRIT})$
      $p_i = \max(TG_{level}(p_j))$
else
      The ranking does not exist
end for;

4.2 Scheduling Procedure in Details

We will provide some details of algorithms from the previous subsection. The specification of the system is a graph of tasks, the times of which are described using fuzzy logic, then they are subjected to the process of defuzzification. For the purpose of defuzzification, the method of the maximum measure was used. Thus, the maximum value of the membership function is taken from the proposed fuzzy task durations. Next, with the use of the A* algorithm, paths are searched in the graph, which define the relationships between tasks. It should be taken into account here that each of the tasks may have a defined time limit (T^{CR}), which may translate into the determination of the critical path. The reference to the *calculate_levels* function takes the handling to the algorithm (Algorithm 2) which describes the procedure for determining the priorities. According to the proposed approach, when setting priorities, a large role is played mainly by the multiprocessing attribute of a task, as well as the possible dependence of tasks. If it is missing, then the priority determination is based on the aforementioned *mprc* argument, otherwise the total value of all priorities of the tasks on the path is taken into account, starting with the task currently being considered and ending with the last task in the path under consideration. From the designated priorities, the tasks with the highest values are selected taking into account the position of each task on the path (s). Tasks with the above-mentioned priorities are selected and then these tasks are subject to fuzzification. After the fuzzification process, tasks are allocated to selected processors for the selected values of the fuzzy function (MIN, MED, MAX), so that (if possible) there is no unused (idle) processor(s) in a given time unit. If there are no idle processors in the time unit, or it is not possible to assign any task to unused processors / processor, the time will be shifted by another unit, and the algorithm will re-run the all steps until all tasks have been scheduled.

Table 2. Schedule length. Model A. $m = 5$

Nr.	Size	Number of k-processor tasks			Schedule length		
		$k = 1$	$k = 2$	$k = 3$	MIN	MED	MAX
1	9	4	2	3	44	55	66
2	13	5	4	4	76	95	114
3	15	5	5	5	84	105	126
4	19	5	7	7	104	130	156
5	22	6	6	10	164	205	246
6	26	8	7	11	192	240	288
7	32	11	8	13	248	305	397
8	41	13	10	18	320	400	480
9	46	12	10	24	624	780	936
10	51	16	11	24	504	630	756

The schedules presented in Fig. 2 (for the sake of readability) do not show the division between tasks in a single time unit, if they are the same tasks as in the preceding unit. This increases the readability of individual schedules, especially the naming of individual tasks.

5 Discussion of the Results

The solution described in paper has been compared with the classic fuzzy logic approach, where the data (in this case, task duration) are described with the use of linguistic variables. The approach based on the use of specifications in the form of linguistic variables was also based on prioritization according to the rules of the same algorithm (MC algorithm). The comparative experiment was carried out for dependent and independent tasks as well as for each of the cases in two variants (divisibility and indivisibility of tasks). This comparison showed that both in the classical approach using linguistic variables (for the description of times and priorities) it gives an identical solution as to the length of the sequence, without affecting the sequence of tasks in the target schedule. Therefore, it is assumed that if these approaches give identical results in different cases, they can be considered equivalent solutions, hence the use of the approach described by the Algorithm 1 and Algorithm 2.

In this case, among the considered graphs, the number of tasks ranges from 9 to 51, being different combinations. The transparency of the experiments was ensured by the generalization of the conducted experiments to a sample of 10 representative task graphs. Task scheduling experiments were also carried out for the target 3-, 4- and 5-processor architecture organized using the Network On Chip idea. Data and results in Tables 2, 3: A) instance identifier, B) the number of 1-, 2- and 3-processor tasks, C) schedule length, for membership function (2) left, D) schedule length, for membership function (2) right.

Both of the approaches specified in this work use the concept of a triangular membership function. Thanks to the extension of Graham's notation with the parameters described in Sect. 2, it is possible to precisely define and systematize this aspect of scheduling. Determining the class of problems for the proposed approaches allows for the selection of algorithms adequate to the complexity, which in the future will facilitate the practical verification of the assumptions made. One of the possible directions of extension may be to consider the uncertainty of task times by examining other membership functions in the area of classical fuzzy logic. Another important aspect may be a more accurate description of the uncertainty of the data using directed fuzzy numbers and the rough set theory (Table 4).

Table 3. Schedule length. Model B. $m = 5$

Nr.	Size	Number of k processor tasks			Schedule length		
		$k = 1$	$k = 2$	$k = 3$	MIN	MED	MAX
1	9	4	2	3	50	55	72
2	13	5	4	4	86	95	124
3	15	5	5	5	95	105	137
4	19	5	7	7	117	130	169
5	22	6	6	10	185	205	267
6	26	8	7	11	216	240	312
7	32	11	8	13	244	305	366
8	41	13	10	18	360	400	520
9	46	12	10	24	702	780	1014
10	51	16	11	24	567	630	819

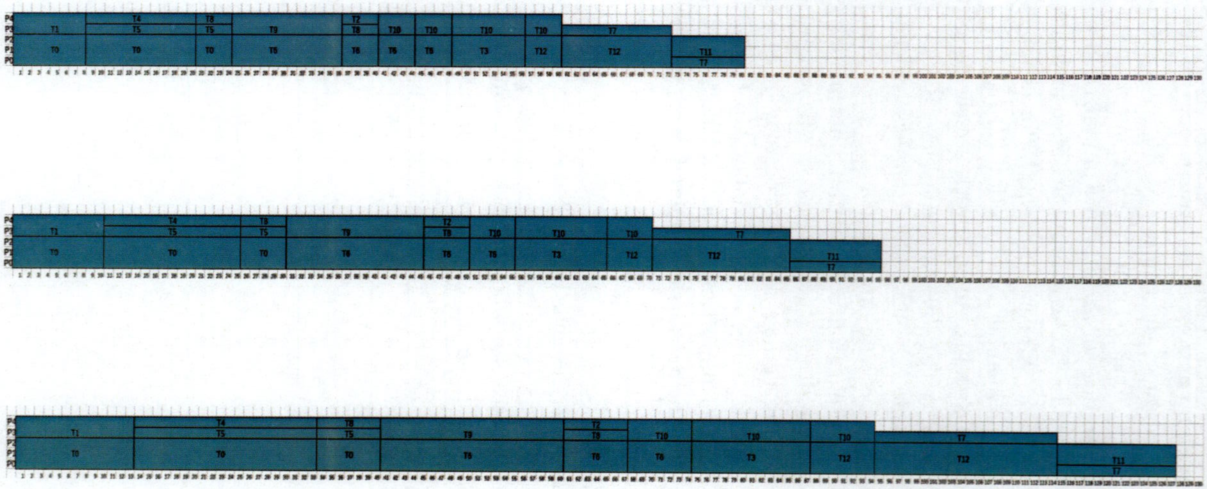

Fig. 2. Scheduling for $m = 5$, model A: $a = b - 20\%$ (top), $b = 1$ (middle), $c = b + 20\%$ (bottom).

Table 4. Reliability levels. Fuzzy approach. $m = 3, 4, 5$

Graph		Number of k processor tasks			Reliability level					
		k			$m = 3$		$m = 4$		$m = 5$	
Nr.	Size	1	2	3	D_x	D_x^*	D_x	D_x^*	D_x	D_x^*
1	9	4	2	3	2,29	2,59	2,44	2,75	2,64	2,67
2	13	5	4	4	2,32	2,59	3,04	3,38	2,95	4,63
3	15	5	5	5	2,38	2,62	2,52	3,88	2,29	4,62
4	19	5	7	7	2,33	2,58	2,59	3,35	2,27	4,77
5	22	6	6	10	2,62	2,84	2,90	3,39	3,02	4,22
6	26	8	7	11	2,61	2,81	2,82	3,30	2,85	3,92
7	32	11	8	13	2,65	2,97	2,79	3,58	2,62	4,28
8	41	13	10	18	2,59	2,85	2,80	3,50	2,84	3,47
9	46	12	10	24	2,61	2,87	2,89	3,34	3,05	3,34
10	51	16	11	24	2,59	2,99	2,68	3,51	3,21	4,26

References

1. Pinedo, M.: Scheduling. Springer, New York (2015). https://doi.org/10.1007/978-1-4614-2361-4
2. Błażewicz, J., et al.: Handbook on Scheduling: From Theory to Applications. Springer, Heidelberg (2007). https://doi.org/10.1007/978-3-540-32220-7
3. Drozdowski, M.: Scheduling for Parallel Processing. Springer, London (2009). https://doi.org/10.1007/978-1-84882-310-5
4. Drozdowski, M.: Selected Problems of Scheduling Tasks in Multiprocessor Computer Systems. Politechnika Poznańska, Poznań (1997)
5. Gottwald, S.: Fuzzy Sets and Fuzzy Logic: The Foundations of Application-from A Mathematical Point of View. Springer, Heidelberg (2013). https://doi.org/10.1007/978-3-322-86812-1
6. Klir, G., Yuan, B.: Fuzzy Sets and Fuzzy Logic. Prentice Hall, New Jersey (1995)
7. Poleshchuk, O., Komarov, E.: Expert Fuzzy Information Processing, vol. 268. Springer, Heidelberg (2011). https://doi.org/10.1007/978-3-642-20125-7
8. Aburas, A.A., Miho, V.: Fuzzy logic based algorithm for uniprocessor scheduling. In: 2008 International Conference on Computer and Communication Engineering, pp. 499–504. IEEE, May 2008
9. Bhattacharya, S., et al.: Fast recognition and control of walking mode for humanoid robot based on pressure sensors and nearest neighbor search. In: International Symposium on Intelligent Signal Processing and Communication Systems (ISPACS), pp. 331–334 (2018)
10. Vallerio, K.: Task graphs for free (TGFF V3. 0). Official version released in April, 15 (2008)
11. Vanbekbergen, P., Lin, B., Goossens, G., De Man, H.: A generalized state assignment theory for transformations on signal transition graphs. J. VLSI Signal Process. Syst. Signal Image Video Technol. **7**, 101–115 (1994)
12. Zadeh, L.A.: Fuzzy Sets, Fuzzy Logic, and Fuzzy Systems: Selected Papers by Lotfi a Zadeh 796-804 (World Scientific) (1996)

13. Dorota, D.: Scheduling tasks in a system with a higher level of dependability. In: Zamojski, W., Mazurkiewicz, J., Sugier, J., Walkowiak, T., Kacprzyk, J. (eds.) DepCoS-RELCOMEX 2019. AISC, vol. 987, pp. 143–153. Springer, Cham (2020). https://doi.org/10.1007/978-3-030-19501-4_14
14. Dorota, D.: Scheduling tasks with uncertain times of duration. In: Zamojski, W., Mazurkiewicz, J., Sugier, J., Walkowiak, T., Kacprzyk, J. (eds.) DepCoS-RELCOMEX 2020. AISC, vol. 1173, pp. 197–209. Springer, Cham (2020). https://doi.org/10.1007/978-3-030-48256-5_20
15. Błażewicz, J., Drozdowski, M., Formanowicz, P., Kubiak, W., Schmidt, G.: Scheduling preemptable tasks on parallel processors with limited availability. Parallel Comput. **26**, 1195–1211 (2000)

Anomaly Detection Techniques for Different DDoS Attack Types

Mateusz Gniewkowski(✉), Henryk Maciejewski, and Tomasz Surmacz

Faculty of Information and Communication Technology,
Wroclaw University of Science and Technology, Wrocław, Poland
{mateusz.gniewkowski,henryk.maciejewski,tomasz.surmacz}@pwr.edu.pl

Abstract. Malicious activities in computer network systems often generate patterns in network data that do not conform to normal behaviour. Since the nature of such anomalies may be different for different types of attacks, detection of these is not trivial and may require specific anomaly detection techniques. In this work, we focus on anomaly or outlier detection techniques for DDoS attacks on computer networks. Our main goal is to find such techniques that prove most appropriate for different types of attacks. We restrict our research to fully unsupervised methods, because, in real world scenarios, it is difficult to obtain examples of all possible anomalies, especially that the set of those is constantly growing. To the best of our knowledge, our work is the first that utilizes time-related features in a purely unsupervised manner and that provides a fair comparison between widely known outlier detection methods. We evaluate clustering, autoencoder and LSTM-based techniques on commonly used datasets, i.e. DARPA1998, ISCXIDS2012, CICDDOS2019. Moreover, we propose IQRPACF method that combines IQR with partial autocorrelation function. The proposed method not only does not require to be trained, but also, in most cases, outperforms the other solutions.

Keywords: Computer network intrusion detection · Machine learning · Unsupervised anomaly/outlier detection · DDoS · Cybersecurity

1 Introduction

Users of computer networks need to deal with an emerging number of security threats. Since the beginnings of the Internet we are concerned with hackers gaining remote access to computers using modem lines or network connections, theft of sensitive information, spreading of computer viruses and worms, and more recently – cryptolockers and other malware. These attacks are very often accompanied by network-level security vulnerabilities that enable these attacks to succeed. On this level, we have security holes in networking libraries that are being discovered long after the initial implementation (such as SYN-flooding attacks, Ping of Death, or Heartbleed in SSL) or general low-level security problems such

W. Zamojski et al. (Eds.): DepCoS-RELCOMEX 2022, LNNS 484, pp. 63–78, 2022.
https://doi.org/10.1007/978-3-031-06746-4_7

as Denial of Service (DoS) or Distributed DoS (DDoS) problems applied at network or protocol level, misuse of broadcasts (broadcast storms), IP spoofing and other attacks. Intrusion Detection Systems (IDS) are the de-facto standard for protection against all these attacks, but in order to function properly, they need to constantly adapt to new types of attacks. We can achieve this, by implementing machine-learning techniques for recognising both known and yet unknown methods of attacks.

In this work, we consider the detection of (D)DoS attacks as an outlier detection problem, which means that the classification is based on the difference from the majority of the data. All parameters of the utilized methods are established in a unsupervised manner. We will discuss datasets chosen for our experiments, the attack types they contain, and the features we use to tune the detection algorithms. We will describe and test several outlier detection methods (clustering-based, IQRPACF, LSTM, autoencoder) and compare the results using typical evaluation methods for outlier detection (F1-score, FPR, TPR, RoC).

The main contribution of this work is to show that the outlier detection approach may be successfully used in identification of (D)DoS attacks, where every attack is simply treated as an anomaly. We also provide a fair comparison between widely known outlier detection methods and propose our own called IQRPACF. It combines interquartile range with a partial autocorrelation function calculated on every new window that occurs in the data. This allows us to detect not only volumetric attacks, but also those that cause small but frequent changes in a given time series. The proposed method not only does not require to be trained, but also, in most cases, outperforms the other solutions.

2 Taxonomy of Defense Mechanisms

Figure 1 shows the taxonomy of DDoS Defense Mechanisms [19]. In the context of our study, the most important group is the "Classification by attack detection strategy", which describes methods of detecting DDoS attacks.

"Pattern Detection" depends on comparing network traffic with certain patterns of known attacks. This allows for their quick detection with a minimum number of false positives. It is probably the best possible method, but unfortunately requires the detection pattern to be previously defined. These can only be known after a specific attack had been well identified and yet more sophisticated ones can be difficult to describe with a set of rules. Next, "Anomaly detection" is about marking connections that deviate in some way from normal traffic. The most obvious approach – "Standard" – is to define a static set of rules, that every normal traffic should follow. If these conditions are not met, the traffic is marked as an anomaly. Finally, the "Trained" group of methods is the one we investigate in this study. These methods are based on historical data that have been already classified as normal or malicious traffic and can produce certain threshold values that allow one to determine whether something is an attack. This group could be further divided into several subcategories:

1. Traditional Statistical Approaches (e.g. thresholding certain parameters obtained from a distribution)
2. Machine Learning approaches
   (a) Supervised Learning – learning a function that maps features into labels, which were used in the training process.
   (b) Unsupervised Learning – learning a function without previously defined labels to detect unknown patterns or groups.
   (c) Semi-supervised Learning – a hybrid method that most often uses only a few labeled samples in the training process

These methods can be used to detect specific attacks (especially supervised learning), but also in outlier detection (also known as "anomaly detection"), which is a process of identifying samples (called "anomalies") that differ from all the others. This will be the main topic of this work. Besides, we are also interested mostly in unsupervised learning as it does not require labels in the process of training. Those methods can also potentially detect new types of attacks better.

3 Datasets

Based on existing research [8], we decided to use the following datasets for both training and later for evaluation and comparison of our methods with those that are already known.

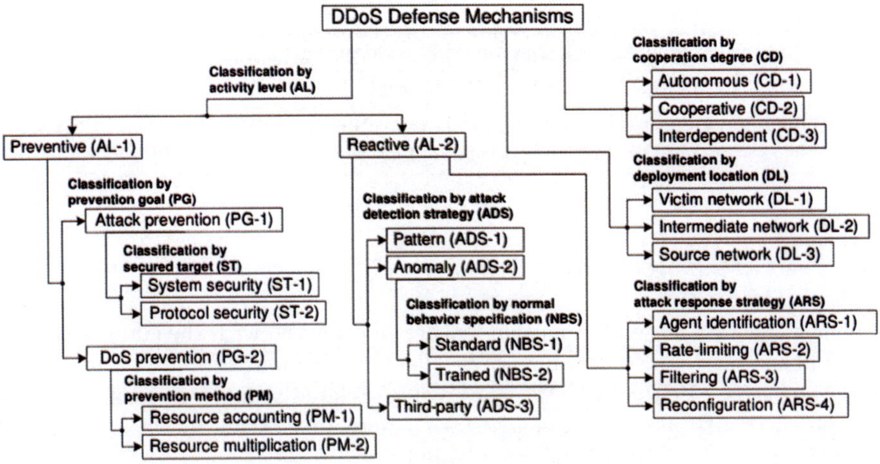

Fig. 1. Taxonomy of DDoS Defense Mechanisms [19]

1. DARPA1998 (Friday Data) [1,15] – DARPA1998 is one of the most famous and commonly used datasets in IDS area. Unfortunately, the characteristics of modern attacks and network traffic have changed significantly since then, so this dataset should no longer be used nowadays. Moreover, there is also some criticism about how the dataset was made. In our work, we decided to use Friday of week five to be able to compare our methods with the ones described in [22]. This day mainly includes a Neptune attack (SYN Flood) and port scans (Portsweep).
2. ISCXIDS2012 [3,21] – ISCXIDS2012 dataset consists of 7 days of real-life traffic. Two of these days (used in our experiments) include the following DDoS attacks: Slowloris and HTTP GET Flood performed by an IRC botnet. The dataset was used for example in [23].
3. CICDDOS2019 [2,20] – one of the newest and biggest datasets that contains exclusively DDoS attacks. Most of them are so-called reflective attacks: the packets are sent to a server with source IP address set to the victim's IP. This way the attacker not only hides his identity, but also significantly amplifies the volume of the attack (the server response is much larger than the attacker's request). The other group of attacks is described as exploitation-based – these attacks aim to exploit the rules of how a given protocol works. This group includes Syn, UDP Flood, UDP-Lag. The authors of this dataset also evaluated it against various supervised approaches [20].

Table 1. Attack list for the used datasets

Datasets	Attacks
DARPA1998	Neptune, Portsweep (not an actual attack)
ISCXIDS2012	Slowloris, HTTP GET Flood
CICDDOS2019	PortMap, NetBIOS, LDAP, MSSQL, NTP, DNS, LDAP, SNMP, SSDP, UDP Flood, UDP-Lag, WebDDoS, SYN, TFTP

4 Feature Selection

In our work, we decided to focus on features not related with the content of packets, but on temporal dependencies in clients' behaviour. For this purpose, the Internet traffic (TCP and UDP packets only) related to a given client (identified by the client's IP address) was split into 5-sintervals. Then, for each window we computed several features shown in Table 2.

The entropy features are calculated as follows:

$$H(X) = -\sum_i p(x_i) \log p(x_i) \tag{1}$$

$$H(Y|X) = -\sum_i p(x_i) \sum_j p(y_j|x_i) \log(p(y_j|x_i)) \tag{2}$$

Table 2. Features used in this work

Feature	Description
n_packets	Number of packets in a given time interval
total_length	Total length of packets in a given time interval
$H(dst\_ips)$	Entropy of destination IPs
$H(src\_ports)$	Entropy of source ports
$H(dst\_ports)$	Entropy of destination ports
$H(dst\_ports\|dst\_ips)$	Conditional entropy of destination ports given destination IPs

A similar feature set was proposed, among others, in [4,11,17]. In [16] one can read more about conditional entropy in the context of detecting DDoS attacks.

To detect (D)DoS attacks, researchers occasionally use aggregated features for all the traffic rather than for individual clients (hence a conditional entropy of source IPs given destination was used in previously mentioned papers). This approach might help to identify DDoS attacks, but used solely (without any additional steps) does not allow to identify attackers. In our work, we decided to use only the data obtained from individual clients, but all the methods described later could also be used for any time-related features.

4.1 Behaviour of Attacks in the Context of Features

Most of the attacks listed in Table 1 are volumetric. It means that they might be detectable by observing large and sustainable spikes in the features such as *n_packets* or *total_length*. On the other hand, the attacks such as Syn, UDP Flood and most of the port scans should be noticed by looking at $H(dst\_ports)$ or $H(dst\_ports|dst\_ips)$. Finally, some of them might only be caught by detecting short-term patterns (certain repeatability) in the data. After all, the attacks are performed automatically, so the randomness caused by human behaviour is none and yet the frequency of those patterns is probably much higher than generated by any legitimate service. The most certain representatives of this group are: Slowloris (this attack is presumably undetectable without the time context), partially Syn attack and UDP Lag.

Figure 2 shows the behaviour of two features in the context of Syn attack in CICDDOS2019 (the end of the first day). The attack could be hard to detect using only the momentary values of features, because the change of their values is not significant. On the other hand, the attack may be indicated by the peak frequency of $H(dst\_ports|dst\_ips)$, which needs to be taken into account by the selected outlier detection methods. Note that in timeslices, when no packet has been observed, the traffic is marked as normal, which can lead to an increased number of false positives.

5 Methods

The anomaly detection problem is present within various research areas. Therefore, a considerable number of methods have emerged over time to address this

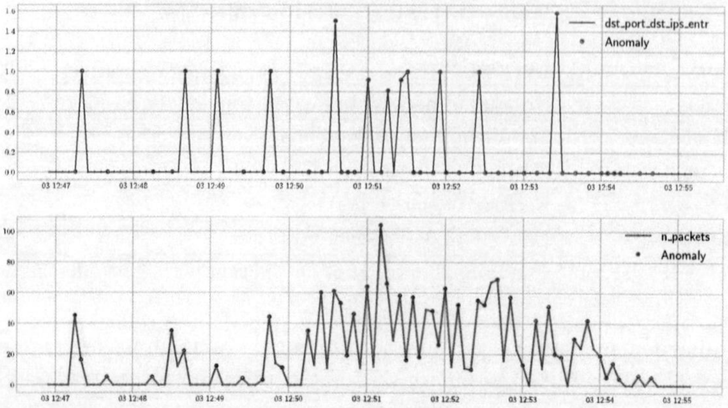

Fig. 2. Example of features' behaviour for SYN attack in CICDDOS2019

problem. In [5] the authors describe in detail many of the available methods and areas of their application (including IDS systems). Work [10] is devoted mainly to unsupervised approaches, which were also evaluated using various datasets, including KDD99 (transformed DARPA1998). In [9,13,24] one can read more about anomaly detection techniques used only in the context of networks.

The following sections briefly describe the anomaly detection methods that we use in our research.

5.1 IQRPACF

The algorithm (see Fig. 3) consists of two elements:

1. detecting large and persistent spikes in features values,
2. detecting repeating patterns in short time intervals.

For the first part of the algorithm, we transform a time series (a vector of values of one of the previously introduced features) using two sliding windows of a given length (*sw_length*) that are rolled side-by-side. For every position of these windows we calculate a difference of their median values. Next, if the difference is too large, we mark that moment (the last moment from the second sliding window) as an anomaly. We consider the difference to be too large if it falls outside of the range from $Q_1 - IQR \cdot c$ to $Q_3 + IQR \cdot c$ ($Q1$ – first quartile, Q_3 – third quartile, IQR – interquartile range, c – parameter). All of the statistics (Q_1, Q_3, IQR) are obtained from at most 300 timestamps (samples from all clients) preceding the first window. This method should be able to detect all attacks that cause spikes in values and yet it is not sensitive to instantaneous and short changes, which is likely to have a positive effect on the number of false positives.

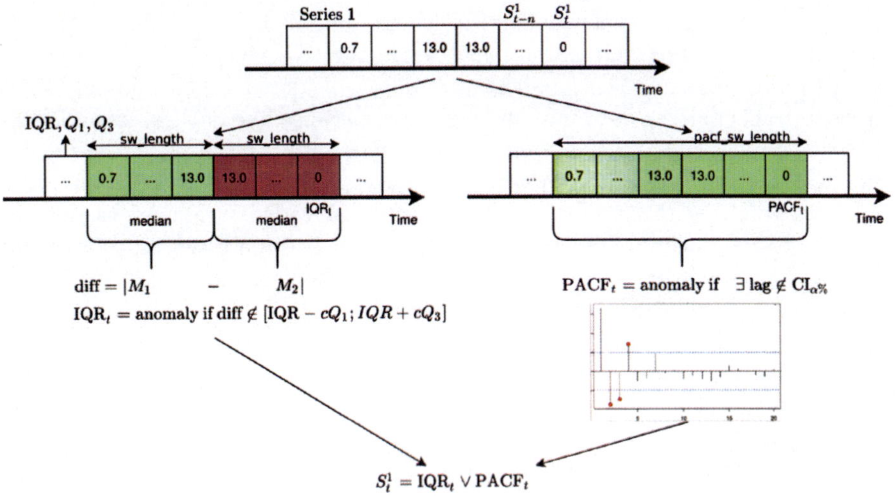

Fig. 3. The IQRPACF method. A moment t is considered to be anomalous if the algorithm has established an anomaly in any of the series. For some features, it might be better to only use values returned by IQR part.

Our intuition tells us that some attacks should cause frequent repetitive patterns in a time series. On the other hand, most of the normal traffic is rather random, because it depends on people's behaviour. In order to detect such patterns we decided to calculate the Partial Auto-Correlation Function (PACF) for every window (sliding window of a given length *pacf_sw_length*). If any of the lagged values (previous values of the time series) correlates with the current one, we consider it as an anomaly (otherwise or if it is impossible to compute, we mark the sample as normal). We say that a correlation occurs if its value is outside the confidence interval depending on the *alpha* parameter. We expect that this method should be able to detect attacks such as SYN flooding.

The last step of the algorithm is to combine the predictions of every feature and every sub-method using the OR operator. In other words, if at any given moment one of the algorithms has determined an anomaly, we consider that moment to be anomalous. Later, to further reduce the number of false positives, we marked those moments where no packet appeared as normal (it is connected with how we prepared the data to our experiments).

5.2 Clustering-Based Approach (k-means and AC)

In this approach, we generate points (samples to be clustered) using a sliding window algorithm. In the scenario, where the window length is one, we simply use feature vectors. As the length of the window increases, more and more of these vectors are concatenated to create a final sample. The information about the anomaly is taken from the last timepoint in a window.

After generating a set of samples, we use one of two predefined clustering algorithms (k-means and hierarchical agglomerative clustering) to divide the samples into groups. The smallest group is considered to be anomalous. We decided to cluster the data into three groups, because one of them should detect normal samples, the other – points close to zero, and the last one (hopefully the smallest) should contain anomalies. Later, we verified this assumption (trying different values of the parameter) and it actually leads to the best results.

Due to the nature of clustering algorithms, it is often essential to transform or scale features (these algorithms are based on distance measures – in our case, Euclidean). We decided to use Min-Max scaling:

$$\hat{x}_i = \frac{x_i - X_{min}}{X_{max} - X_{min}} \tag{3}$$

5.3 k-Nearest Neighbors

In this algorithm, a certain subset of the dataset (normal traffic only) is used as neighbours. Then in the prediction process, we calculate a maximum distance between a given sample and five closest neighbours. If it is greater than a given threshold (parameter), then the sample is marked as anomaly. Later, we use that threshold to calculate ROC curve.

5.4 Autoencoder (AE)

Autoencoder is a relatively simple neural network that learns to recreate its input using a smaller, intermediate representation. Beyond the dimensionality reduction, it can also be used to recreate damaged signals, or, just like in our case, to detect outliers. If the coded and then decoded sample varies greatly from its original form, then it probably represents an anomaly. In our work, we used a ready-made autoencoder implementation taken from [14]. The architecture is similar to the one described in [6] as *Conventional Autoencoder*, but with more layers. The input/output layer is the size of the sequence length[1] multiplied by the number of features. The error prediction (outlier score) is calculated using normalized *Mean Absolute Error* (MAE).

5.5 LSTMED

The LSTMED network proposed in [18] (ready-made implementation from [14]) is also an autoencoder, but both the coding and decoding parts use the LSTM layer. This way not only the hidden state is used in the prediction process, but also the values from previous steps. As with the previous outlier detection method, we use 5-min long time slices. The outlier score is computed exactly in the same way.

---

[1] Arbitrarily set to 5 min, so the network has a chance to detect anomalies correlated with time.

6 Experiments

In order to evaluate the AE, LSTMED and kNN models, we used several "normal" clients (without any suspicious activity) from a given dataset and then test them using the rest of the dataset. The size of the training set is about 20% of the total (this value varies depending on the dataset as we always used all the traffic for a given client). The IQRPACF method has always been tested on whole datasets. Due to memory constraints, for clustering algorithms we had to use only part of the "normal" data.

We obtained the best results using only two of previously introduced features ($n\_packets$ and $H(dst\_ports|dst\_ip)$) on which we have based all our experiments presented in the next section. Moreover, in IQRPACF method only the second feature is analyzed using PACF.

6.1 Evaluation

Methods such as AE, kNN and LSTMED generate a so-called outlier score (see Sect. 5), which means that only a single threshold is required to conclude whether a sample is an anomaly. This allows us to generate a receiver operating characteristic (ROC) curve, which is an excellent tool for evaluating binary classifiers (outlier detection problem is a kind of binary classification). The ROC curve can easily be used to control the sensitivity of the algorithm. One way to do it is to select $N\%$ of attacks we would like to detect (on the *TPR* axis). Then, the corresponding point on the *FPR* axis shows how many false alarms we should expect. This measure is called FPRN [12] and should be kept as small as possible.

In our experiments we also use quality measures like: *AUC*, *TPR*, *FPR*, *TPR-FPR*, *Precision*, *Recall*, *F1-score*, *Accuracy* and *weighted F1-score* (wF1). All of them were calculated with the assumption that anomaly is a "positive" sample (e.g. *TPR* refers to the rate of correctly identified anomalies).

Before we get into the results, let us consider which of the presented measures are best suited to evaluate the quality of the algorithm in an outlier detection problem (a problem with strongly unbalanced labels). The *AUC* score is comparatively reliable because it takes into account a certain threshold and the percentage of FPs and TPs, but unfortunately, for some ROC curves, the score might be overestimated (see kNN).

The measure of *Precision* decreases significantly as the absolute number of FPs increases, which might be slightly misleading if there are only a few anomalies in the dataset. *Recall* simply shows the ratio of properly classified anomalies. *F1-score* is the harmonic mean of the other two, so it also has the disadvantages of *Precision*. Even though all of them can be successfully used in the comparison of outlier detection algorithms, it is the shape of ROC curve, *TPR* and *FPR* that are best suited for this. Before we discuss why, it is worth to notice that *Accuracy* score and *weighted F1-score* measures carry no significant information with them, due to a large number of normal labels and yet they are often used for similar problems.

The ROC curve makes it possible to select such a threshold that would make it possible to adjust the sensitivity of the given method. One of the options for selecting such a threshold is maximizing *TPR-FPR* (all the tables with results were generated on the basis of this parameter). Another one could be *FPRN* metric, which show us the probability that the normal sample raises an alarm when $\%N$ of anomalies are detected, which simply represents a point on ROC curve. For example, the point ($FPR = 0.953$, $TPR = 0.073$) obtained using IQRPACF method (Table 5, CICDDOS2019) can be understood as $FPR95 = 0.073$ which is an excellent result.

7 Results

Figure 4 shows ROC curves for all the datasets. Each of the curves grows rapidly to a certain common point, and then their behavior starts to differ. The kNN goes straight to the final point, which actually means that the algorithm failed to detect any more anomalies, specifically those that actually depended on time and not instantaneous values (mind that kNN method threats time series as independent points). This feature causes the overestimation of the AUC value. As a result, it would seem that kNN works better than AE, which is not actually true. Bearing in mind the behavior of the kNN method, from Tables 3 and 4, it can be concluded that *Syn*, *Portmap*, *Portsweep* and *Slowloris* are especially time-dependent. Considering only the ROC curves LSTMED method turns out to be better suited for CICDDOS2019 dataset than AE and kNN. It is mainly because its ability to detect mentioned time-related *Syn* attack. For the other datasets, the simple autoencoder outperformed the other two solutions (this can be easily seen from the AUC).

Table 3. Prediction table for DARPA1998 and ISCXIDS2012

Method	Dataset	DARPA1998			ISCXIDS2012		
	Pred	True					
		Normal	Neptune	Portsweep	Normal	Slowloris	HTTP GET flood
IQRPCAF	Anomaly	254	581	0	103493	495	4464
	Normal	319315	63	112	921634	498	178
kmeans	Anomaly	6307	**630**	0	11520	641	**4562**
	Normal	55344	0	112	32643	316	78
AC	Anomaly	**5**	**630**	0	**164**	4	45
	Normal	61646	0	112	43999	953	4595
LSTMED	Anomaly	4312	**644**	1	55646	321	4356
	Normal	259195	0	111	624011	636	285
knn	Anomaly	5983	**644**	1	7709	46	3423
	Normal	257524	0	111	671948	911	1218
AE	Anomaly	8251	**644**	**112**	113193	**839**	4561
	Normal	255256	0	0	566464	118	80

To compare the operation of methods for which ROC curves have not been generated, it is probably best to use the measure *TPR-FPR* and *F1-score*. As it can be seen in Table 5, the proposed IQRPACF method works best with CICDDOS2019 and exceeds baseline results achieved by the authors of this dataset [20]. In the case of the AE algorithm, the ROC curve shows that at some point

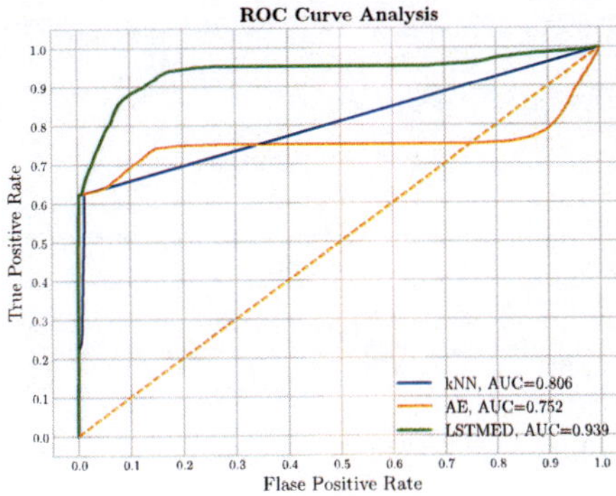

(a) CICDDOS2019

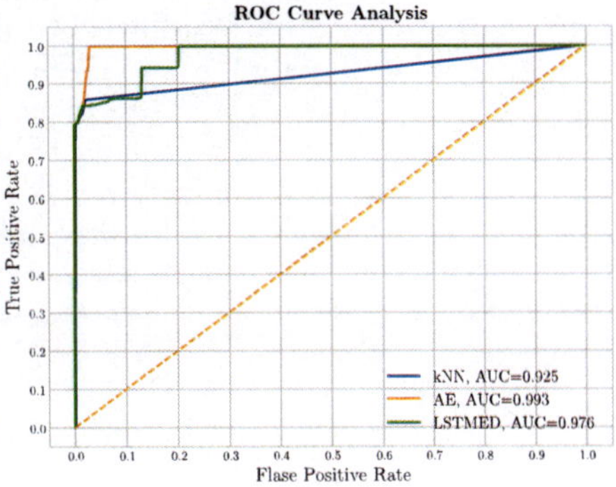

(b) DARPA1998

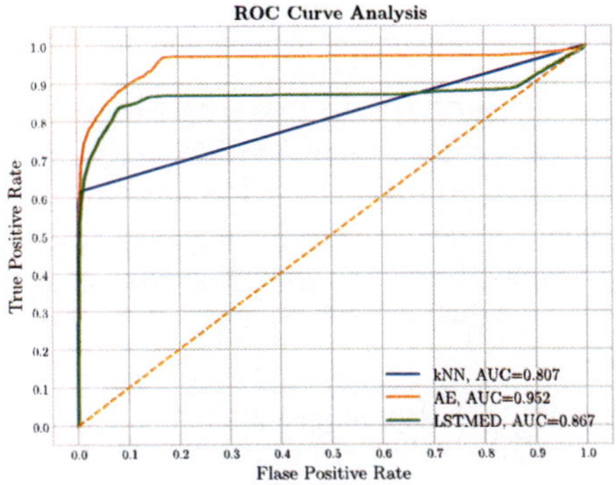

(c) ISCXIDS2012

Fig. 4. ROC curves for all datasets

Table 4. Prediction table for CICDDOS2019

Method	Pred	True													
		Normal	DNS	LDAP	MS-SQL	NTP	Net BIOS	Port-map	SNMP	SSDP	Syn	TFTP	UDP	UDP Lag	Web DDoS
IQRPACF	Anomaly	6403	**417**	**238**	**407**	**932**	270	**245**	276	150	**2113**	**2137**	**540**	**376**	**55**
	Normal	102629	3	15	14	57	33	15	25	10	104	29	13	1	0
k-means	Anomaly	14812	348	197	348	841	264	88	259	151	1140	1753	523	364	48
	Normal	32589	72	56	73	135	39	166	42	9	1077	413	30	13	7
AC	Anomaly	1131	345	200	352	795	266	94	262	151	1148	1660	525	352	47
	Normal	46270	75	53	69	181	37	160	39	9	1069	506	28	25	8
LSTMED	Anomaly	9224	390	208	388	866	**293**	206	**286**	**159**	1910	1915	506	324	31
	Normal	81912	30	45	33	123	10	54	15	1	307	251	47	53	24
kNN	Anomaly	1312	345	193	349	589	271	125	265	151	995	1231	495	287	9
	Normal	89824	75	60	72	400	32	135	36	9	1222	935	58	90	46
AE	Anomaly	**195**	344	193	349	587	271	125	265	151	988	1225	495	287	9
	Normal	90941	76	60	72	402	32	135	36	9	1229	941	58	90	46

Table 5. Results summary

Dataset	Model	AUC	TPR	FPR	TPR-FPR	Prec	Recall	F1	Acc	wF1
CICDDOS2019	IQRPACF	–	**0,962**	0,059	**0,904**	0,560	**0,962**	0,708	0,943	0,950
	k-means	–	0,748	0,312	0,435	0,748	0,697	0,655	0,697	0,655
	AC	–	0,733	0,024	0,709	0,733	0,846	**0,785**	0,939	0,941
	LSTMED	**0,939**	0,883	0,101	0,782	0,448	0,883	0,594	0,897	0,912
	kNN	0,752	0,626	0,014	0,612	0,802	0,626	0,703	0,955	0,952
	AE	0,806	0,624	**0,002**	0,622	**0,964**	0,624	0,758	**0,966**	**0,963**
DARPA1998	IQRPACF	–	0,732	0,001	0,731	0,701	0,732	0,716	**0,999**	**0,999**
	k-means	–	0,835	0,102	0,732	0,835	0,095	0,170	0,897	0,859
	AC	–	0,795	**0,000**	0,795	**0,992**	0,795	**0,883**	0,997	0,997
	LSTMED	0,976	0,843	0,016	0,827	0,137	0,843	0,236	0,983	0,989
	kNN	0,925	0,858	0,023	0,835	0,104	0,858	0,186	0,977	0,986
	AE	**0,993**	**0,998**	0,031	**0,966**	0,089	**0,998**	0,164	0,969	0,982
ISXIDS2012	IQRPACF	–	0,880	0,101	0,779	0,046	0,880	0,087	0,899	0,942
	k-means	–	0,930	0,261	0,669	**0,930**	0,311	**0,466**	0,761	0,718
	AC	–	0,009	**0,004**	0,005	0,009	0,230	0,017	0,885	0,935
	LSTMED	0,867	0,835	0,082	0,754	0,078	0,835	0,142	0,917	0,950
	kNN	0,807	0,620	0,011	0,608	0,310	0,620	0,414	**0,986**	**0,988**
	AE	**0,952**	**0,965**	0,167	**0,798**	0,046	**0,965**	0,087	0.835	0,902

the method is unable to distinguish normal traffic from abnormal, or even better reproduces (encodes and decodes) attacks than regular samples. The architecture described in [7] allowed to achieve results close to the perfect classifier, but it should be noted that proposed solution was trained in supervised manner (at least the last layer). The authors also used a subset of features provided in the dataset and trained their model using 70% of the data. Our solution, on the other hand, is fully unsupervised and uses only two features. This significant difference in results also leaves room for further attempts to improve the method (most likely through better understanding and analysis of the used features).

In case of the DARPA1998 dataset, the AC algorithm turned out to be the best one, due to the very low *FPR*. The second best method is AE, because (surprisingly) it is the only one that can detect a *Portsweep* (mind that Portsweep is actually not a DoS attack). Otherwise the second best solution would be IQRPACF. The algorithm proposed in [22] is based on a similar idea of time series analysis. The authors use ARIMA model to predict the number of packets and the number of different source IPs for the next minute. Next, based on the difference between the actual values and the predicted ones, the sample is classified as an anomaly or not. The experiments however did not test the algorithm for all of the Friday data, and the authors did not consider *Portsweep* to be an anomaly. At the end of their work, you can find a comparison with several others approaches. In our experiments we achieved comparable results.

For ISCXIDS2012 the best method is AE, although, because of the low number of false positives, it may seem that kNN performed equally good, but in reality the ROC curve shows, that AE could be easily tuned to obtain similar scores and it can also detect Slowloris. The k-means algorithm also performed surprisingly well, which means that the method is worth further investigation (e.g. by using the modified version of the algorithm proposed in [11]).

Just as we expected, IQRPACF was not able to properly detect *Slowloris*, because it looks for repeating patterns only in $H(dst\_ports|dst\_ips)$ feature. The authors of [23] evaluated their supervised solution on the same subset of ISCXIDS2012 dataset and managed to achieve better results. This was achieved mainly due to a different formulation of the problem.

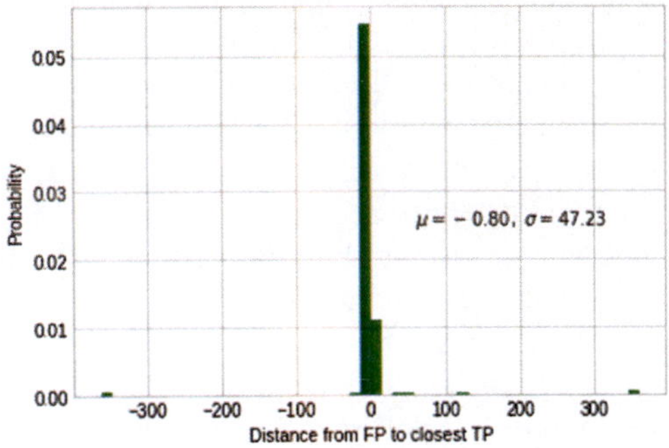

Fig. 5. Distributions of minimum distances between FPs and TPs

One of the factors that can significantly improve the effectiveness of attack detection is their tendency to appear in groups (side by side). Figure 5 shows minimal distances between false positives (FPs) and true positives (TPs) while using IQRPACF on CICDDOS2019 dataset. Most often they are very close to each other (the sample should be considered positive) and sometimes relatively far away (the sample should be rather considered negative). This approach could be implemented on the top of any of the proposed outlier detection methods.

8 Conclusions

In this paper we have shown and compared the performance of several outlier detection methods on three datasets using time-related features. The only dataset that contained only DDoS attack was CICDDOS2019 as it is a relatively new dataset. Both the attacks and background traffic are appropriate for our purpose. Thanks to its hybrid nature, our approach allowed to properly detect most of the anomalies of various types. Our IQRPACF method not only looks for large values that stand out from the rest (IQR), but also utilizes partial autocorrelation function (PACF) to find frequent and repetitive changes in the data. This combination allowed us to construct a simple, yet very effective tool for detecting anomalies that does not require to be trained.

The oldest dataset, DARPA1998, was only used for the purpose of comparing our solutions with other methods. As the dataset is over 20 years old it might be considered outdated. Moreover, throughout its existence, there have been some objections to the method of background traffic generation.

The part of CICXIDS2012 dataset contained two types DDoS attacks: HTTP Get Flood and Slowloris. We were able to detect both of these, but the number of FPs turned out to be relatively large. We believe it could be reduced by relating it to how often the method returns anomaly, choosing different hyperparameters or by changing the sampling rate (we used 5-s windows).

As the results suggest, simple methods based on statistical analysis of the data (IQRPACF) may produce better results than those that require learning. On the other hand, it is easier to adjust methods such as AE or LSTMED using RoC curve. The two methods behaved differently depending on the dataset, and since both solutions are based on a similar principle, we believe that it is possible to develop such a solution that would work well in each case. Additionally, using autoencoders with more hand-crafted features than we used, potentially requires less effort in adjusting the methods.

As the approaches we propose are based on outlier detection methods (which also means that labeled data is not needed), they are easier to implement and maintain. Moreover, they are probably better suited for detecting unknown attacks. In the real world scenario, we suggest to implement an algorithm that works on a similar principle as IQRPACF and to take into account burst anomalies (an alarm should be raised if a client acts suspiciously too often). IQRPACF provides a decent detectability of anomalies with a fair number of FPs and works out of the box. If one decides to collect the network specific data, it is also possible to train AE to further improve the results. We suggest to use it hierarchically after IQRPACF to confirm that an anomaly actually occurred, although this may result in a growth of FNs.

We believe that our results could be further improved by extending the list of used features and validating the methods trained on normal traffic from one dataset using the other. This would probably need further development of IQRPACF, but all the other methods (especially AE, LSTMED and kNN) have potential for immediate improvement.

References

1. The 1998 DARPA intrusion detection evaluation dataset. https://www.ll.mit.edu/r-d/datasets/1998-darpa-intrusion-detection-evaluation-dataset. Accessed 10 May 2022
2. DDoS evaluation dataset (CICDDoS2019). https://www.unb.ca/cic/datasets/ddos-2019.html. Accessed 10 May 2022
3. Intrusion detection evaluation dataset (ISCXIDS2012). https://www.unb.ca/cic/datasets/ids.html. Accessed 10 May 2022
4. Behal, S., Kumar, K.: Detection of DDoS attacks and flash events using information theory metrics-an empirical investigation. Comput. Commun. **103**, 18–28 (2017). https://doi.org/10.1016/j.comcom.2017.02.003. http://www.sciencedirect.com/science/article/pii/S0140366417301718
5. Chandola, V., Banerjee, A., Kumar, V.: Anomaly detection: a survey. ACM Comput. Surv. (CSUR) **41**(3), 1–58 (2009)

6. Chen, Z., Yeo, C.K., Lee, B.S., Lau, C.T.: Autoencoder-based network anomaly detection. In: 2018 Wireless Telecommunications Symposium (WTS), pp. 1–5 (2018). https://doi.org/10.1109/WTS.2018.8363930
7. Elsayed, M.S., Le-Khac, N.A., Dev, S., Jurcut, A.D.: DDoSNet: a deep-learning model for detecting network attacks. In: 2020 IEEE 21st International Symposium on "A World of Wireless, Mobile and Multimedia Networks" (WoWMoM), pp. 391–396. IEEE (2020)
8. Gniewkowski, M.: An overview of DoS and DDoS attack detection techniques. In: Zamojski, W., Mazurkiewicz, J., Sugier, J., Walkowiak, T., Kacprzyk, J. (eds.) DepCoS-RELCOMEX 2020. AISC, vol. 1173, pp. 233–241. Springer, Cham (2020). https://doi.org/10.1007/978-3-030-48256-5_23
9. Gogoi, P., Bhattacharyya, D., Borah, B., Kalita, J.K.: A survey of outlier detection methods in network anomaly identification. Comput. J. **54**(4), 570–588 (2011)
10. Goldstein, M., Uchida, S.: A comparative evaluation of unsupervised anomaly detection algorithms for multivariate data. PLoS One **11**(4), e0152173 (2016)
11. Gu, Y., Li, K., Guo, Z., Wang, Y.: Semi-supervised K-means DDoS detection method using hybrid feature selection algorithm. IEEE Access **7**, 64351–64365 (2019)
12. Hendrycks, D., Mazeika, M., Dietterich, T.: Deep anomaly detection with outlier exposure. arXiv preprint arXiv:1812.04606 (2018)
13. Kwon, D., Kim, H., Kim, J., Suh, S.C., Kim, I., Kim, K.J.: A survey of deep learning-based network anomaly detection. Clust. Comput. **22**(1), 949–961 (2017). https://doi.org/10.1007/s10586-017-1117-8
14. Li, Y., Zha, D., Zou, N., Hu, X.: PyODDS: An end-to-end outlier detection system (2019)
15. Lippmann, R.P., et al.: Evaluating intrusion detection systems: the 1998 DARPA off-line intrusion detection evaluation. In: Proceedings DARPA Information Survivability Conference and Exposition. DISCEX 2000, vol. 2, pp. 12–26. IEEE (2000)
16. Liu, Y., Yin, J., Cheng, J., Zhang, B.: Detecting DDoS attacks using conditional entropy. In: 2010 International Conference on Computer Application and System Modeling (ICCASM 2010), vol. 13, p. V13-278. IEEE (2010)
17. Ma, X., Chen, Y.: Ddos detection method based on chaos analysis of network traffic entropy. IEEE Commun. Lett. **18**(1), 114–117 (2014). https://doi.org/10.1109/LCOMM.2013.112613.132275
18. Malhotra, P., Ramakrishnan, A., Anand, G., Vig, L., Agarwal, P., Shroff, G.: LSTM-based encoder-decoder for multi-sensor anomaly detection. arXiv preprint arXiv:1607.00148 (2016)
19. Mirkovic, J., Reiher, P.: A taxonomy of DDoS attack and DDoS defense mechanisms. ACM SIGCOMM Comput. Commun. Rev. **34**(2), 39–53 (2004)
20. Sharafaldin, I., Lashkari, A.H., Hakak, S., Ghorbani, A.A.: Developing realistic distributed denial of service (DDoS) attack dataset and taxonomy. In: 2019 International Carnahan Conference on Security Technology (ICCST), pp. 1–8 (2019). https://doi.org/10.1109/CCST.2019.8888419
21. Shiravi, A., Shiravi, H., Tavallaee, M., Ghorbani, A.A.: Toward developing a systematic approach to generate benchmark datasets for intrusion detection. Comput. Secur. **31**(3), 357–374 (2012). https://doi.org/10.1016/j.cose.2011.12.012
22. Tabatabaie Nezhad, S.M., Nazari, M., Gharavol, E.A.: A novel DoS and DDoS attacks detection algorithm using ARIMA time series model and chaotic system in computer networks. IEEE Commun. Lett. **20**(4), 700–703 (2016). https://doi.org/10.1109/LCOMM.2016.2517622

23. Yuan, X., Li, C., Li, X.: DeepDefense: identifying DDoS attack via deep learning. In: 2017 IEEE International Conference on Smart Computing (SMARTCOMP), pp. 1–8. IEEE (2017)
24. Zhang, J., Zulkernine, M.: Anomaly based network intrusion detection with unsupervised outlier detection. In: 2006 IEEE International Conference on Communications, vol. 5, pp. 2388–2393. IEEE (2006)

Nonparametric Tracking for Time-Varying Nonlinearities Using the Kernel Method

Purva Joshi and Grzegorz Mzyk(✉)

Wroclaw University of Science and Technology, Wroclaw, Poland
{purva.joshi,grzegorz.mzyk}@pwr.edu.pl

Abstract. In the real world, there is a number of systems that are nonlinear and time-varying. Modelling and nonparametric tracking of time-varying nonlinear characteristics are famous and playing important role in system identification. The weighted least-squares method has been used for the identification of such systems. The main objective of this paper is to track the time-varying nonlinearities using the kernel estimation method and balance between variance and bias. The optimal solution of the bias-variance trade-off has been evaluated by the zero-one forgetting weighted kernel method. The simulation results indicate that by tuning the bandwith and time horizont, the mean square error (MSE) can be reduced.

Keywords: Time-varying system · Kernel estimation · System identification · Bias variance trade-off

1 Introduction

Identification of time-varying processes tackles the most important topic in adaptive systems: tracking time-varying systems [11]. The combination of linear time-variant and nonparametric method gives new insight into the statistical problems to solve uniquely. In this paper, kernel estimation theory is applied to identify non-linearities for a time-varying static system. The tracking of time-varying nonlinearities can be easily specified using previous data knowledge and kernel estimate. This paper proposes a general idea for the non-parametric kernel estimation method and tuning of time-varying variables in the nonlinear system.

This article looks at the identification problem for a class of nonlinear time-varying systems with the following scenario: the system's behavior is determined by a set of adjustable parameter attributes. Several different experiments, corresponding to different parameter qualities are carried out and different results are collected to study the effect of modifying characteristics.

The main goal of this paper is to track the time-varying non-linear characteristics using the nonparametric kernel estimation method [12]. Traditional representation of nonlinear systems using Volterra an Wiener series (see [13]) is

W. Zamojski et al. (Eds.): DepCoS-RELCOMEX 2022, LNNS 484, pp. 79–87, 2022.
https://doi.org/10.1007/978-3-031-06746-4_8

often successfully replaced with conception of block oriented models. The subject of block-oriented nonlinear dynamic system identification in the presence of random disturbances was addressed in [1,4,15]. The Hammerstein system, Wiener system, Wiener-Hammerstein ("sandwich") system, and additive NARMAX systems with feedback are examples of systems with varied interconnections of linear dynamic blocks and static nonlinear parts. First attempts to nonparametric identification of time-varying systems have been made in [8] and [9], where the orthogonal series expansion approach has been proposed.

Many difficult applications in various fields benefit from time-varying identification approaches [2,10,14]. Kernel smoothing is a quick and easy approach to detect structure in data sets without having to use a parametric model. The simple regression problem, where paired observations for each of two variables are given, is one of the most fundamental contexts where kernel smoothing principles can be applied [3].

The book [4] shows how to find nonlinear subsystems by combining block-oriented system identification with nonparametric regression estimation. Although random signals unexpectedly evolve in time, their average statistical features demonstrate considerable regularity, which is the core characteristic of random signals [5]. Based on (second-order moment) residual analysis and covariance estimation, the typical machinery for system identification of linear time-invariant (LTI) models produces a nominal model and a confidence (uncertainty) region around it. These should reflect the distance to the genuine nonlinear, time-varying system, as well as provide a solid foundation for LTI control design [6]. By permitting the linear approximation of the speech signal to be time variable, the formulation presented in [7] aims to accommodate nonstationarity more automatically.

In this paper, Sect. 2 describes the problem statement with some assumptions and figures of the time-varying system. In Sect. 3, the kernel method is implemented for tuning of time-varying parameters, while Sect. 4 is illustrated with optimal solution of bias and variance trade-off. Section 5 and 6 elaborated as simulation study and conclusion respectively.

2 Statement of the Problem

Here, let us consider a static nonlinear system (Fig. 1), where input u_k is known. Now, according to the time varying regression model,

$$y_k = \mu_k(u_k) + z_k, \tag{1}$$

where,

$$\mu_k(u) = \mu(u) + f(k). \tag{2}$$

Here, there are a few assumptions, which should be taken into consideration based on the nonlinear system, where $\mu_k(.)$ is the function of u_k, and u_k is considered as finite numbers of observations.

In this article, we assume that:

(A1) The input signal $u_k \sim \mathcal{U}[-1, 1]$ denotes an i.i.d. bounded random process with a uniformly distributed probability density function for $k = 1, 2, 3.....$
(A2) Let us consider $\mu(u)$ is a Lipschitz function, which means that $|\mu_k(u_k) - \mu_k(u)| \leq L|u_k - u|$ and L is the Lipschitz constant.
(A3) We also assume that z_k is a zero-mean ergodic random disturbance independent of the input u_k. So, $\mathrm{E}z_k = 0$ and var $z_k < \infty$.

The main goal of this paper is for a given u we want to estimate current $\mu_N(u)$ using $\left\{(u_k, y_k)\right\}_{k=1}^{N}$. Here, we assume that $f(k)$ is linearly dependent with respect to time k and i.e. $f(k) = \beta k$. (The $\mu(u)$ is stationary part of $\mu_k(u)$.)

3 The Algorithm

For the model in Eq. (2), the kernel estimation method has been used. In our case, we have considered the following,

$$\hat{\mu}_N(u) = \frac{\sum_{k=1}^{N} y_k \alpha_k}{\sum_{k=1}^{N} \alpha_k} \tag{3}$$

where, α_k are some weights that describe the relative importance of the measures $y_k^{\prime}s$. The weights k's are affected by the distance between the kth input observation u_k and the specified estimation point u in this case. By above Eq. (3), the two ideas can be approached.

- If we consider the exponential forgetting factor then it is,

$$\alpha_k = \lambda^{N-k} K\left(\frac{u_k - u}{h}\right) \tag{4}$$

  where $0 < \lambda \leq 1$, and

$$K(v) = \begin{cases} 1 & \text{as } |v| \leq 1 \\ 0 & \text{otherwise} \end{cases} \tag{5}$$

  is Parzen kernel function.
- The static nonlinear system's zero-one weights have been considered. So that it is clear that by the weighted least squares minimum,

$$\alpha_k = \begin{cases} K(.) & \text{for } k = N, N-1, ..., N-(H-1) \\ 0 & \text{for } k \leq N - H \end{cases} \tag{6}$$

For α_k in Eq. (6), the estimate (3) can be written in the form as below

$$\hat{\mu}_N(u) = \frac{\sum_{k=N-(H-1)}^{N} y_k K\left(\frac{u_k - u}{h}\right)}{\sum_{k=N-(H-1)}^{N} K\left(\frac{u_k - u}{h}\right)}. \tag{7}$$

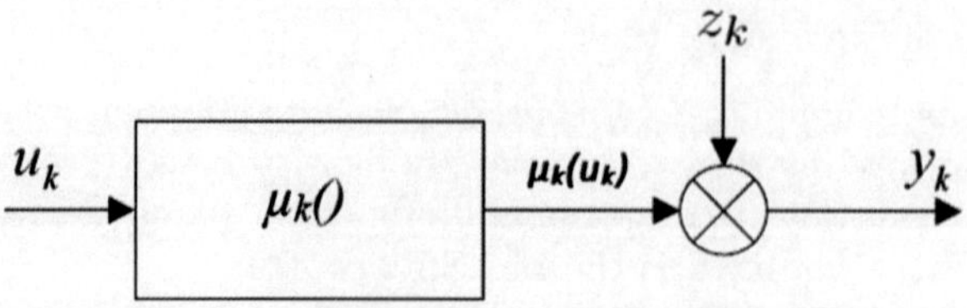

Fig. 1. Nonlinear static time-varying system

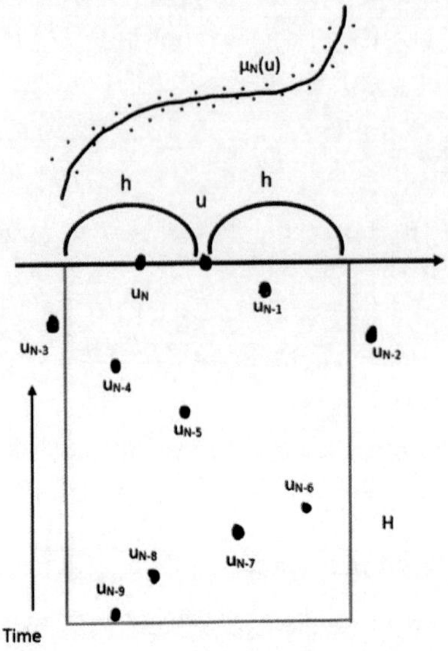

Fig. 2. Illustration of weights assignment (rectangle 2 h by H represents the set of selected observations)

4 Optimal Bias-Variance Trade-off

Let's analyze the tracking procedure's quality index. We get the bias associated with parameter changes if the time horizont H is too lengthy, i.e. the weights decline too slowly. When the horizont H is short, however, the estimate becomes sensitive to noise and the variance error emerges.

If we define quality index in terms of bias and variation as MSE (Mean Square Error), it must be

$$MSE = var\hat{\mu}_N(u) + bias^2\hat{\mu}_N(u)$$

The static nonlinear characteristic is illustrated in Fig. 2, and it is evident that for any given point u, it will be approximated using current data, $\{(u_k, y_k)\}_{k=N-(H-1)}^{N}$. Figure 2 shows the time changing function $\mu_N(u)$, as well as the horizont H (green color) and bandwidth h.

To find the optimal solution, we need to tune the two factors. 1) Horizont (H) and 2) Bandwidth parameter (h). Regarding the bias we have that

$$\begin{aligned}|E(y_k - \mu_N(u))| &= |E(\mu_k(u_k) + z_k - \mu_N(u))| \\ &= |E(\mu(u_k) + \beta k + z_k - \mu(u) - \beta N)| \\ &= |E(\mu(u_k) - \mu(u) + \beta(k - N))| \\ &\leq Lh + \beta|N - k| \end{aligned} \tag{8}$$

Here, it is evident that the bias has equally depends on L and β such a way that,

$$|bias\hat{\mu}(u)| \leq Lh + \beta\left(\frac{H-1}{2}\right) \tag{9}$$

Here, where, $k = N, N-1, ..., N-(H-1)$. The goal is to find a good balance between bias and variance, i.e., a good trade-off between tracking ability and volatility. Here H denotes horizont of the tracking.

Our goal is to find the optimal value of H and h, which depends on β and L. However, to find the optimal value of H is complex task nevertheless if we find the balance between bandwidth parameter h and variance bias trade-off then it can be possible. Since we have assumed that probability density function of u_k is

$$\vartheta(u) = \begin{cases} 1/2 & \text{as } u \in [-1, 1] \\ 0 & \text{elsewhere} \end{cases}$$

the probability of selection is $2h\vartheta(u) = 2h\frac{1}{2} = h$ and expected number of selections in terms of probability density function is Nh. Therefore, it is clear that, $varz_k \triangleq v =$ constant can be bounded as follows

$$var\hat{\mu}(u) = \frac{v}{Hh} and bias \leq Lh + \beta\left(\frac{H-1}{2}\right)$$

It is also possible to use a limited (hard) collection of the fixed number N of the most recent measurements. Assume that, the genuine regression function $\hat{\mu}_N(u)$ meets the Lipschitz condition and let us define the mean squared error, which can be upper-bounded as follows:

$$\begin{aligned} MSE\hat{\mu}_N(u) &= bias^2\hat{\mu}_N(u) + var\hat{\mu}_N(u) \\ &\leq \left(Lh + \beta\left(\frac{H-1}{2}\right)\right)^2 + \frac{v}{Hh} = Q(H, h) \end{aligned} \tag{10}$$

Equation (9) is defining the quality index Q with respect to horizont H and bandwidth h. For example we can set $H = N$ when $\beta = 0$ since, $H_{opt} = N$ and $bias\hat{\mu}(u) \leq Lh$, i.e.,

$$\hat{\mu}_N(u) = \frac{\sum_{k=1}^{N} y_k K\left(\frac{u_k - u}{h}\right)}{\sum_{k=1}^{N} K\left(\frac{u_k - u}{h}\right)}. \tag{11}$$

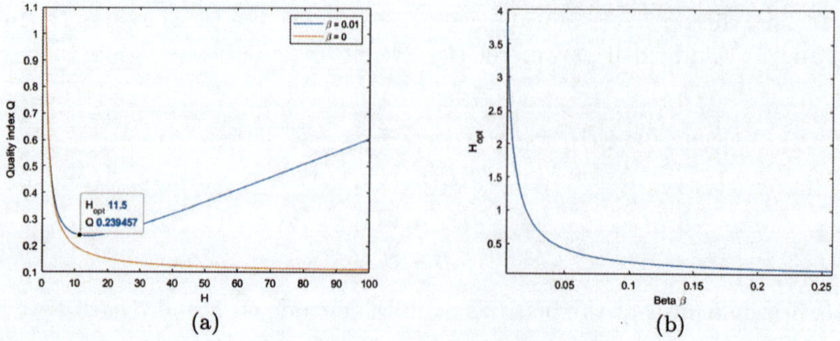

Fig. 3. (a) Quality index value vs horizont H ($v = 0.1, L = 1, \beta = 0.01$ and 0) and (b) H_{opt} with respect to β

5 Simulation Results

The results of a basic experiment in which a time-varying static characteristic was used are presented in the figures below. In Fig. 3, it is clear that the optimal value of H depends on the value of β, Lipschitz constant L and v. For example, if $\beta = 0$ then, the H_{opt} can be defined and this case is stationary because MSE is only depends on L and v.

Let's assume we have given β, Lipschitz constant L and v and we can see the behaviour of time-varying system in Fig. 3(a). If horizont H will be higher, the old number of measurements tends to be more bias and low variance because bias is the summation of weights with respect to horizont (H) and $H \propto \frac{1}{\sigma^2}$ where, $v = \sigma^2$ is variance. This reduces the influence of old measurements on the final estimate and increases the influence of fresh observations.

Figure 3(b) indicates that, the various optimal values of H and β depend on each other inversely, which means that for smaller values of β, higher the H_{opt} and the exponential behaviour is showcased in Fig. 3(b).

In order to illustrate the operation of the proposed method, an exemplary nonlinear system was simulated with the stationary part of nonlinear characteristic $\mu(u) = |u|$ and linear time changes βk with $\beta = 0.0005$. For $N = 10^4$ the $\mu_N(u) = |u| + 5$ was estimated using historical data pairs $\{(u_k, y_k)\}_{k=1}^{N}$. Traditional Parzen window kernel (refer to Eq. (5)) was applied with the bandwidth $h = 0.1$. Both input and noise were mutually independent and uniformly distributed, *i.e.*, $u_k \sim \mathcal{U}[-1, 1]$ and $z_k \sim \mathcal{U}[-2, 2]$. Figure 4 shows the results for various values of horizont H. Deactivation of forgetting, by putting $H = N$ (see Fig. 4) reduces the variance the best, but it comes at the expense of increasing the bias. On the other hand, if H is too small (see Fig. 4), the bias is reduced, but the estimator loses the ability to reduce the impact of disturbances. Reasonable compromise is shown in Fig. 4. Numerical version of the mean integrated square error

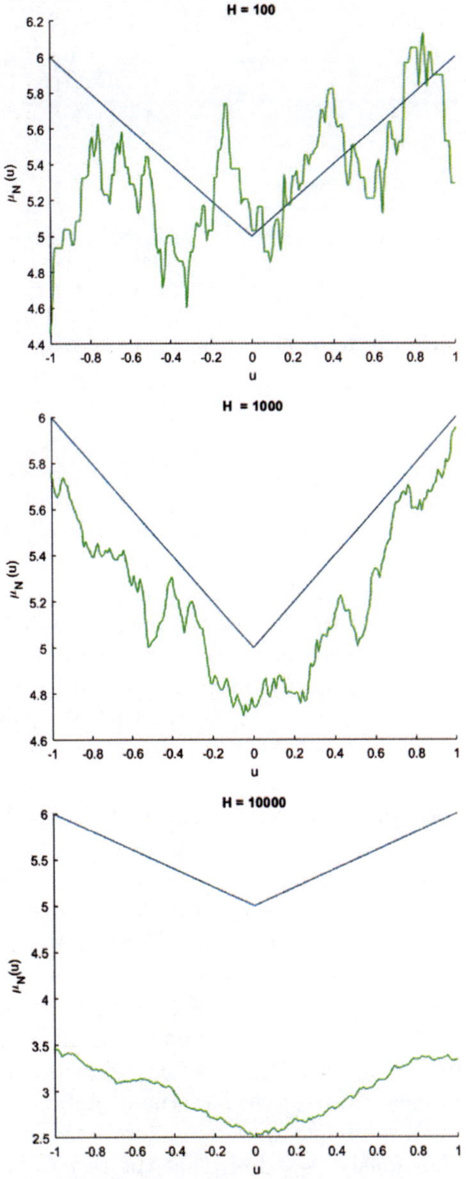

Fig. 4. True characteristic (blue) and its estimate (green) for $H = 100, H = 1000$ and $H = 10000$

$$MISE = \int_{-1}^{1} E\left(\widehat{\mu}_N\left(u\right) - \mu_N\left(u\right)\right)^2 du$$

has been computed and shown in Fig. 5.

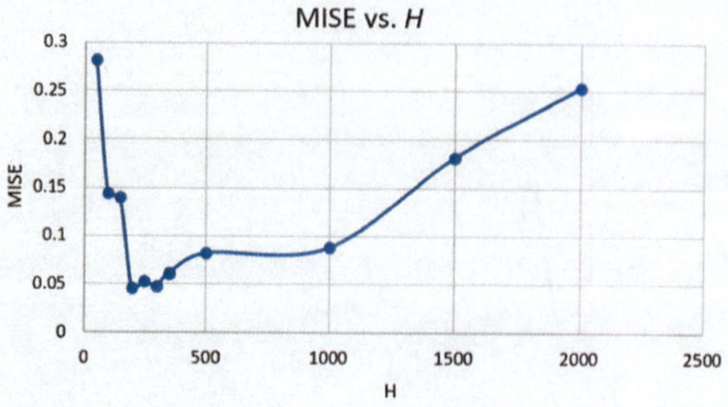

Fig. 5. Mean integrated square error vs horizont

6 Summary

There is an endless number of nonlinear and time-varying systems in the reality. Modelling and nonparametric tracking of time-varying nonlinear systems are well-known and vital in system identification. For the identification of such systems, weighted local least square approaches have been applied. The major goal of this study is to use the kernel estimation approach to monitor time-varying nonlinearities and to compare the variance and bias trade-offs. Zero-one weighted least square approach was used to determine the best solution for the bias-variance trade-off. According to the theoretical results, we have derived upper bound mean square error which can be reduced by fine-tuning the time-varying components.

Compared to traditional non-parametric regression estimation techniques, where the system is assumed to be stationary, we generalized the weights by incorporating forgetting factors apart from kernels. Presented algorithm is tuned in two dimensions: (i) with respect to bandwidth, h, which is responsible for neighborhood selection in the input domain, and (ii) time horizont, H, which allows for faster reactions on system changes. The tracking quality indeks $Q(H, h)$ can be optimized by proper selection of H and h. Obviously, the slower changes of characteristics, the longer optimal horizont H_{opt}. On the other side if the system is changing rapidly, H_{opt} is short. Optimal bandwidth depends on the relation between local slope of characteristic (L) and variance of the noise. Increasing H and h allows for variance reduction, however, it is associated with an increase of bias. Obviously, we admit that the results presented in the paper require full prior knowledge about β, L and $var z_k$, which can be problematic in practice and cross-validation-based methods need to be used.

References

1. Mzyk, G.: Combined Parametric-Nonparametric Identification of Block-Oriented Systems. LNCIS, vol. 454. Springer, Cham (2014). https://doi.org/10.1007/978-3-319-03596-3

2. Niedzwiecki, M.: Identification of Time-Varying Processes. Wiley, Chichester (2000)
3. Wand, M., Jones, M.C.: Kernel Smoothing. Chapman and Hall, London (1995)
4. Greblicki, W., Pawlak, M.: Nonparametric System Identification. Cambridge University Press, Cambridge (2008)
5. Manolakis, G., Ingle, K., Kogon, M.: Statistical and Adaptive Signal Processing. Artech House Inc., McGraw-Hill Book, Norwood (2005)
6. Ljung, L.: Estimating linear time-invariant models of nonlinear time-varying systems. Eur. J. Control **7**, 203–219 (2001)
7. Liporace, A.: Linear Estimation of Nonstationary Signals. Acoustical Society of America (1975)
8. Rutkowski, L.: On nonparametric identification with prediction of time-varying systems. IEEE Trans. Autom. Control **29**(1), 58–60 (1984)
9. Rutkowski, L.: On-line identification of time-varying systems by nonparametric techniques. IEEE Trans. Autom. Control **27**(1), 228–230 (1982)
10. Niedźwiecki, M.J., Ciołek, M., Gańcza, A.: A new look at the statistical identification of nonstationary systems. Automatica **118**, 109037 (2020)
11. Bruce, L., Goel, A., Bernstein, S.: Convergence and consistency of recursive least squares with variable-rate forgetting. Automatica **119**, 109052 (2020)
12. Zhang, T., Wu, W.B.: Time-varying nonlinear regression models: nonparametric estimation and model selection. Inst. Math. Stat. Ann. Stat. **43**(2), 741–768 (2015)
13. Schetzen, M.: Nonlinear system modeling based on the Wiener theory. In: Proceeding of the IEEE, vol. 69, issue number 12, December 1981
14. Niedźwiecki, M.: First-order tracking properties of weighted least squares estimators. IEEE Trans. Autom. Control **33**(1), 94–96 (1988)
15. Giri, F., Bai, E.W.: Block-Oriented Nonlinear System Identification. Springer-Verlag, Heidelberg (2010). https://doi.org/10.1007/978-1-84996-513-2

Safety Assessment of the Two-Cascade Redundant Information and Control Systems Considering Faults of Versions and Supervision Means

Vyacheslav Kharchenko[1], Yuriy Ponochovnyi[2], Eugene Ruchkov[3], and Eugene Babeshko[1](✉)

[1] National Aerospace University KhAI, Kharkiv, Ukraine
{v.kharchenko,e.babeshko}@csn.khai.edu
[2] Poltava State Agrarian Academy, Poltava, Ukraine
[3] Research and Production Company Radiy, Kropyvnytskyi, Ukraine
rev@radiy.com

Abstract. This paper studies functional safety and availability models of information and control system (ICS) with two-cascade 2oo3/1oo2 redundancy. First cascade has a two-version structure which is conventional for reactor trip systems. The structure being studied includes supervision means for each subsystem of the first redundancy cascade, as well as means for their outputs comparison. A peculiarity of developed models is that they consider failures of supervision means along with failures caused by different (single and multiple) faults of software and hardware versions of main and diverse subsystems. Markov models describe behavior of repairable system considering failures caused by physical faults of channels and version design fault supervision, detection and elimination means. During analysis of models the peculiarities of FPGA-based RadICS platform used for ICSs of nuclear power plants (NPPs) and other safety critical systems development were considered.

Keywords: Information and control system · Safety assessment · Availability function · Markov model · Two-cascade redundant structure

1 Introduction

Safety-related industrial information and control systems are being developed using CPU and FPGA-based platforms. Application of such system, in particular reactor trip systems (RTS), in nuclear power engineering is regulated by international standards and normative documents [1, 2]. IEC 61508-6 [3] provides a number of mathematical models and methods for safety assessment of NPP ICS based on electronic and programmable components and platforms [4]. The most safety critical systems have structure involving two (main and diverse) subsystems that implement three-channel 2oo3 voting. These subsystems are built using different software (SW) and hardware (HW) implementations, platform solutions and settings, and must conform to SIL3 safety integrity level

W. Zamojski et al. (Eds.): DepCoS-RELCOMEX 2022, LNNS 484, pp. 88–98, 2022.
https://doi.org/10.1007/978-3-031-06746-4_9

requirements [5]. Such ICSs designated as multi-version systems are covered in [6–10] where metric based techniques of diversity are suggested [6, 7], Bayesian [8] and Markov's [9, 10] models of NPP reactor trip system (RTS) are developed and studied.

Different scenario of ICS channels and subsystems failures, including failures caused by insider interventions and cyber attacks, necessitate particular emphasis placed on development process of adequate and complete safety and availability models. Despite extensive protective measures of NPP ICS, their isolation from environment and external information systems, cases of intervention into operation of such critical systems must not be ruled out [11]. Internet of Things technologies application in the industrial systems, advancement of Industry 4.0 concept [12, 13] are attended with rise of new threats and vulnerabilities that could be exploited for attacks [14, 15]. This fact motivates not only incorporation of through supervision and protection of systems, but their operation test as well.

In this work we present NPP ICS operation model with consideration for failures of supervision means in either (main and diverse) subsystem. These means are constituents of diagnostics module described in [16].

The object is the research of ICS operation during failure occurrence conditions, caused not only by software and hardware faults, but by supervision means failures as well. To accomplish this, extended failure model with consideration for their different combinations is being developed.

The rest of paper is organized as follows: Sect. 2 is dedicated to the description of enhanced system structure with embedded subsystems and channels supervision means, set of states with consideration for failure combinations; systems availability Markov models are proposed and studied in Sect. 3, as well as values of simulation input parameters are justified. Analysis of modelling results has been the objective of Sect. 4. Finally, in Sect. 5 we summarize findings, provide guidance on usage of developed models and discuss the future work.

2 The Structure and Faults Model of the Two-Cascade Redundant ICS

2.1 The Two-Cascade Redundant Structure

The structure of two-version NPP ICS with embedded supervision means is shown in Fig. 1. This structure is modified in reference to one described in [17] through the use of cascade 2oo3 and 1oo2 redundancy. It allows eliminating hardware failures by majority voting at the first cascade, as well as tolerating version design faults (of main and diverse subsystems) by using 1oo2 version redundancy at the second cascade. The 1oo2 voting is made possible by the fact that subsystems generate single-bit signals with the priority for reactor trip signal. Utilization of supervision means admits detection of version faults in cases when the output channels retain identical states.

Such redundant architecture incorporates reasonable reliability margin allowing to withstand failures adequately. When assessing traditional two-version system [9, 10] it is often admitted that supervision means for both subsystems have the perfect reliability.

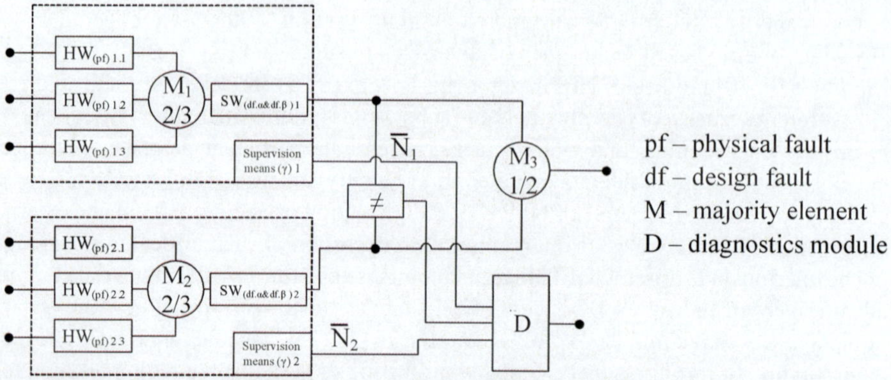

Fig. 1. RTS structure chart to consider physical faults (pf), design faults (df) and supervision system faults

Modular implementation of the system (for example, using RadICS platform [4, 5]) makes it possible to utilize individual modules of supervision system for either of diverse channels and supervise operability signals from these modules by means of diagnostics module D. Doing so along with an additional module of output signals comparison (module «≠») allows to extend space of diagnostic states. Application of additional signals requires extension of ICS state space.

2.2 The Faults Model

Figure 2 depicts space state that considers failures of software versions as well as supervision systems failures of first and second diverse channels. To reduce dimension, hardware failures state sets studied in [16] are not shown in Fig. 2.

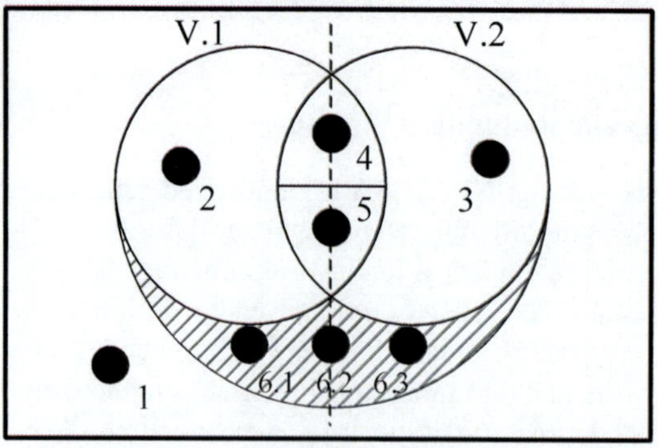

Fig. 2. Generation of RTS ICS failures intersection

Table 1 provides signals combinations that generate ICS failure space with consideration of software version failures and supervision system failures.

Table 1. Signal states at the diagnostics module input and ICS operability

№	State	Notation	Parameter	State of ≠ signal	State of N1 signal	State of N2 signal	General ICS operability	Number of operative versions
1	Operative state	S_0	–	0	0	0	1	2
2	Failure of main (first) version/system	S_1	$\alpha 1$	1	1	0	1	1
3	Failure of diverse (second) version/system	S_2	$\alpha 2$	1	0	1	1	1
4	Failure of both versions/systems with coincidence of output results	S_3	$\beta 1$	0	1	1	0	0
5	Failure of both versions/systems with different output results	S_4	$\beta 2$	1	1	1	0	0
6.1	Supervision means failure	S_K	γ	0/1	1	1	0	–
6.2	Supervision means failure	S_K	γ	0	1	0	0	–
6.3	Supervision means failure	S_K	γ	0	0	1	0	–

Two-version and two-cascade redundant structure with subsystems supervision is shown in Fig. 1. Supervision means from either subsystem generate operability signal. Generally they are designed in the similar way and have resembling hardware implementation, hence their reliability (γ parameter) could be treated as identical.

In this work we will not concern ourselves with situation related to operative state display error at the output of supervision means ($N1 = 0, N2 = 0$). When a failure occurs in either of supervision modules, system state should be changed to the safe failed state. Therefore, three states denoted as 6.1, 6.2 and 6.3 could be treated as one safe failed state so as to decrease the dimension.

3 Markov's Availability Model of Two-Cascade Redundant ICS Considering Faults of Supervision Means

3.1 The Model of Two-Cascade Redundant ICS Considering Faults of Supervision Means and Hardware Faults of Channels

Elements of base Markov model described as marked graphs are shown in Fig. 3. During model construction ICS design features were taken into consideration, namely, utilization of identical hardware in both diverse channels (hardware in every channel is characterized by reliability parameters λp and μp). In such a manner Markov model presented in Fig. 3a was obtained with operative states S1-S5 and safe failed state S6. The peculiarities of such models development are described in [16, 17]. To consider supervision systems failures this model has been modified (Fig. 3b).

The main point of modification shown in Fig. 3 lies in the fact that every operative state obtains corresponding inoperative state such that supervision system failure occurred with failure rate λγ (states S7…S11). After recovery of supervision means to the operative state with the rate μγ, system goes back to the corresponding operative state.

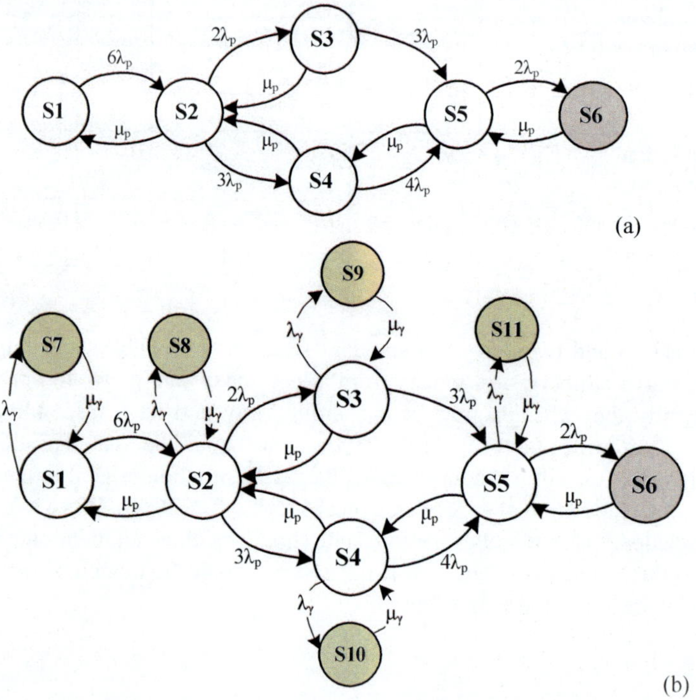

Fig. 3. Orgraph of RTS ICS base model fragment with consideration of design faults (a) and its modification for taking into account supervision means failures and recovery (b)

The availability function for orgraph presented in Fig. 3b is defined as (1):

$$A(t) = \sum_{i=1}^{5} P_i(t). \quad (1)$$

Baseline conditions are the following: $t = 0$, $P_1(0) = 1$, $P_2(0) \ldots P_{11}(0) = 0$.

3.2 The Model of Two-Cascade Redundant ICS Considering Hardware and Software Faults and Faults of Supervision Means

Occurrence of relative (individual) software faults causes hardware where such fault occurs be idled. After detection of such failure, isolating and elimination actions are being taken leading to the change of failure rate $\lambda\alpha$. In the model such events are described using multi-fragmentation modelling mathematical apparatus [9].

Considering ICS design features, the following assumption is being made: reliability parameters of both used software versions are equal ($\lambda\alpha 1 = \lambda\alpha 2 = \lambda\alpha$, $\mu\alpha 1 = \mu\alpha 2 = \mu\alpha$).

Figure 4 presents inter-fragmental transitions resulted from occurrence and elimination of one relative design fault of the first software version. For the case of fault occurrence in the second version of software, inter-fragmental states would be identical. Considering that during occurrence of one relative design fault of hardware channels with another software version remain operative, in the inter-fragmental space hardware physical faults continue to occur. This fact justifies transitions between states S7–S8; S8–S9; S10–S11; S11–S12. Moreover, for all operative states it is necessary to further model occurrence of supervision system faults (not shown in Fig. 4). Such model is shown in Fig. 5, where the following marking is applied: white color is used to present operative states, red color is used for safe failed states caused by software design faults, and blue color is used for safe failed states caused by supervision system faults.

The availability function for orgraph in Fig. 5 is defined as (2):

$$A(t) = \sum_{i=1}^{5} P_i(t) + \sum_{i=12}^{13} P_i(t) + \sum_{i=15}^{16} P_i(t) + \sum_{i=22}^{26} P_i(t) \quad (2)$$

Baseline conditions are the following: $t = 0$, $P_1(0) = 1$, $P_2(0) \ldots P_{32}(0) = 0$.

3.3 Models Assumptions and Input Parameters

The main assumptions of this Markov models are:

- the flow of events that transfers the system from one functional state to another one within the same fragment has the properties of stationarity, ordinariness and the absence of aftereffect, the model parameters within one fragment are assumed to be constant;
- during the elimination of software faults, new faults are not introduced;

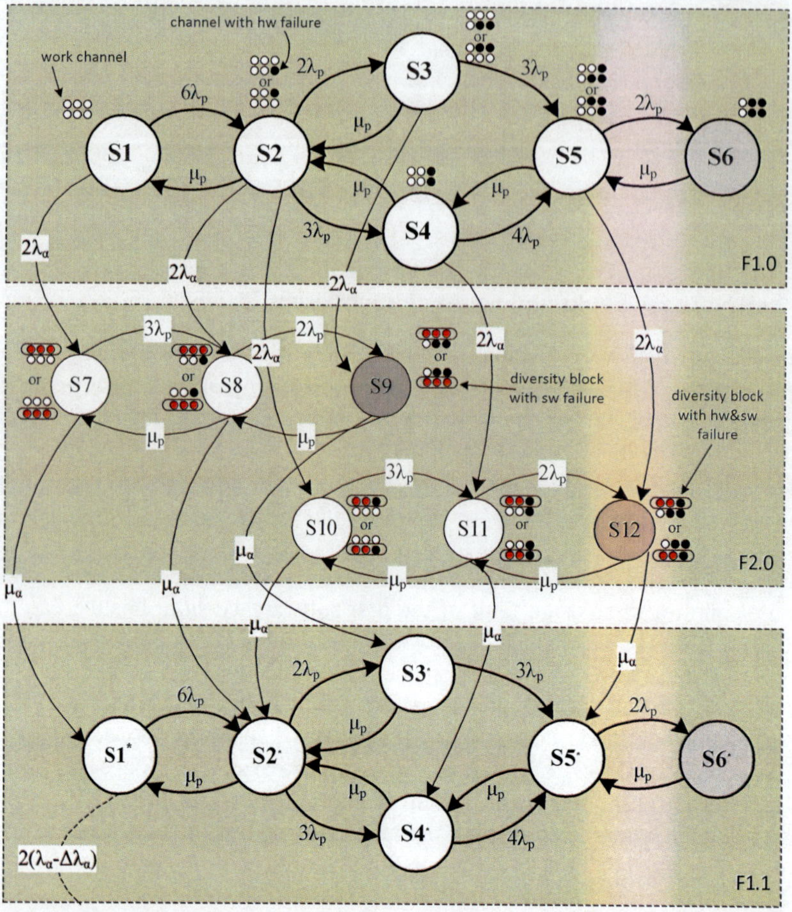

Fig. 4. Transitions between fragments resulted from relative design fault occurrence of one of the software versions

– after elimination of relative software fault in F1.1 fragment, the intensity of the occurrence of software failures is specified as $\lambda_{\alpha\, i+1}$, and is defined as:

$$\lambda_{\alpha\, i+1} = \lambda_{\alpha\, i} - \Delta\lambda_{\alpha}. \quad (3)$$

The primary input parameters of Markov models were determined on the basis of certification data [5, 16, 17] for the previous NPP ICS versions samples. Their values are presented in Table 2.

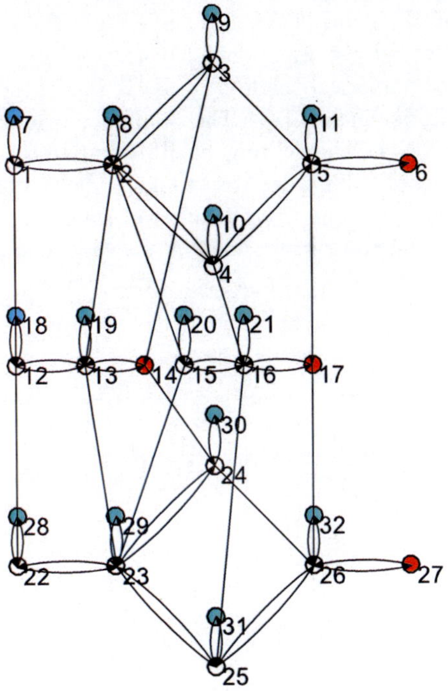

Fig. 5. Orgraph of model with fragments, resulted from occurrence of the relative design fault in the one of software versions

Table 2. Values of simulation processing input parameters

#	Sym	Parameter	Value
1	λp	HW failure rate due to physical faults (pf)	1e–4 (1/hour)
2	λα	SW failure rate due to design faults (df)	5e–4 (1/hour)
3	λγ	Supervision means failure rate	1e–6 (1/hour)
4	μp	HW recovery rate after failure	1 (1/hour)
5	μα	SW recovery rate after failure	2 (1/hour)
6	μγ	Supervision means recovery rate after failure	0.25 (1/hour)
7	Δλα	SW failure rate change after fault elimination	1.25e–4 (1/hour)
8	Nα	Assumed number of design faults	4

4 Simulation and Analysis

To build the matrix of Kolmogorov-Chapman system of differential equations (SDE) in Matlab, matrix A function was used [18]. The SDE can be solved using analytical methods (substitutions, Laplace transforms, etc.). But this approach is applicable for systems of small dimension. In this paper, we consider a single-fragmentation model of medium dimension (11 states in Fig. 3b) and a multi-fragmentation model of large dimension (95 states for Nα = 4, Fig. 4, Fig. 5), therefore, an approach that is universal for both models to the numerical solution of SDEs was chosen. The SDE solution is obtained using the ode15s function [19]. The simulation results are shown in Fig. 6.

The graphs of the NPP ICS model (Fig. 6a) illustrate the typical nature of the change in the availability function with a decrease to a stationary coefficient A = 0.999996 during the first 30 h of operation. Consideration of supervision means failures allows improving availability function assessment by ΔA = 4e–6 (shown as difference between red and blue levels at the chart). As this takes place, impact of failures resulted from relative software version faults is not traced in the chart.

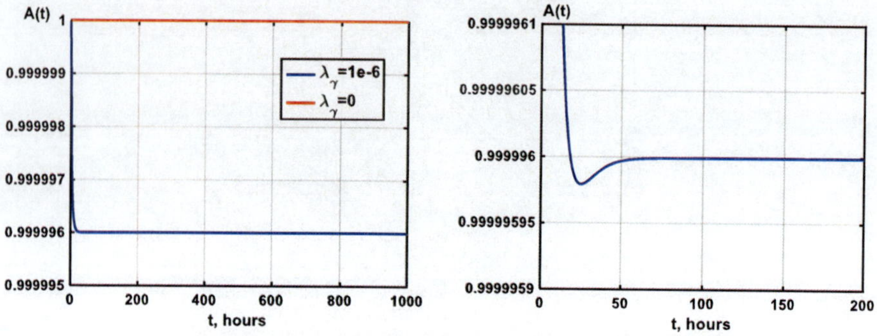

(a) (b)

Fig. 6. Results of availability simulations of NPP ICS: a) from parameters value in Table 1, b) from λα = 5e1 (1/hour)

To assess the impact of relative software failures on availability, their occurrence rate was increased by three degrees. Chart shown in Fig. 6b depicts decay of availability curve on the interval 20..70 h by ΔA = 2e–8 when λα = 5e–1 (1/hour).

5 Conclusion

Paper provides development and research of models two-version safety system with two-cascade redundancy, making it possible to consider supervision means failures. System structure with self-supervision is presented, comprising of subsystem supervision means, inter-channel comparison and analysis means. Such structure makes it possible not only identify subsystem failures states, but also provide a check of supervision means and decrease risk of unsafe and undetected states.

Construction of Markov models that describe occurrence of physical faults in the hardware channels and supervision means failures is discussed by stages. To simulate system behavior considering version (subsystems) design faults occurrence and elimination the multi-fragmentation Markov model has been suggested and researched. Main its particularity is detail analysis of fault combination described by inter-fragmental part of the model.

Considering supervision means failures allows improving accuracy of stationary availability factor by $\Delta A = 4e{-}6$ (from $A = 0.999999999673858$ to $A = 0.999995999689761$). It is important taking into account high requirements to availability of RTS (more than 0.99999). Therewith impact of relative software failures with rate $\lambda\alpha = 5e{-}1$ to the general availability of the whole system is within $\Delta A = 2e{-}8$ due to two-version structure. However, the overlap of version and control failures can be more critical.

Further research should be devoted to the development of analytical models intended for assessment of impact of the whole set of design faults (including 4 and 5, Fig. 2) to the availability and safety, as well as to possible attacks to the system [14, 20]. In case of attacks or any intrusions results of the vulnerability analysis should be taken into account so as to parametrize and redevelop Markov model or develop combined technique of assessment.

References

1. IEC 61513:2011. Nuclear power plants – Instrumentation and control important to safety – General requirements for systems (2011). https://webstore.iec.ch/publication/5532
2. IEC 60880:2006. Nuclear power plants. Instrumentation and control systems important to safety. Software aspects for computer-based systems performing category A functions (2006). https://webstore.iec.ch/publication/3795
3. IEC 61508-6:2010 Functional safety of electrical/electronic/programmable electronic safety-related systems - Part 6: Guidelines on the application of IEC 61508-2 and IEC 61508-3 (2010). https://webstore.iec.ch/publication/5520
4. Sklyar, V.: Safety-critical certification of FPGA-based platform against requirements of US nuclear regulatory commission (NRC): industrial case study. In: Proceedings of the 12th International Conference on on ICT in Education, Research and Industrial Applications (2016)
5. D7.24-FSC(P3)-FMEDA-V6R0. Exida FMEDA Report of Project: Radiy FPGA-based Safety Controller (FSC). 69 p. (2018)
6. NUREG/CR-7007. Diversity Strategies for Nuclear Power Plant Instrumentation and Control Systems (2010). https://www.nrc.gov/reading-rm/doc-collections/nuregs/contract/cr7007/index.html
7. Kharchenko, V., Siora, A., Sklyar, V., Volkoviy, A., Bezsaliy, V.: Multi-diversity versus common cause failures: FPGA-based multi-version NPP I&C systems. In: Proceedings of the 7th Conference NPIC&HMIT (2010)
8. Littlewood, B., Popov, P., Strigini, L. DISPO project: a summary of CSR work on modelling of diversity. Centre for Software Reliability. City University, London, UK (2006)
9. Kharchenko, V., Butenko, V., Odarushchenko, O., Sklyar, V.: Multifragmentation Markov Modeling of a Reactor Trip System. J. Nuclear Eng. Radiat. Sci. **1** (2015). https://doi.org/10.1115/1.4029342
10. Kharchenko, V., Ponochovnyi, Y., Boyarchuk, A., Andrashov, A., Rudenko, I.: Multi-fragmental Markov's models for safety assessment of NPP I&C system considering migration of hidden failures. In: Ermolayev, V., Mallet, F., Yakovyna, V., Mayr, H.C., Spivakovsky, A. (eds.) ICTERI 2019. CCIS, vol. 1175, pp. 302–326. Springer, Cham (2020). https://doi.org/10.1007/978-3-030-39459-2_14
11. Johnson, G.: Cyber Robust Systems: The Vulnerability of the Current Approach to Cyber Security (2020). https://www.igi-global.com/chapter/cyber-robust-systems/258678

12. Bhamare, D., Zolanvari, M., Erbad, A., Jain, R., Khan, K., Meskin, N.: Cybersecurity for industrial control systems: a survey. Comput. Secur. **89**(101677) (2020). https://doi.org/10.1016/j.cose.2019.101677
13. Thames, L., Schaefer, D. (eds.): Cybersecurity for Industry 4.0: Analysis for Design and Manufacturing. Springer, Cham (2017)
14. Kolisnyk, M.: Vulnerability analysis and method of selection of communication protocols for information transfer in internet of things systems. Radioelectron. Comput. Syst. **1**(97), 133–149 (2021). https://doi.org/10.32620/reks.2021.1.12
15. Morozova, O., Nicheporuk, A., Tetskyi, A., Tkachov, V.: Methods and technologies for ensuring cybersecurity of industrial and web-oriented systems and networks. Radioelectron. Comput. Syst. **4**, 145–156 (2021). https://doi.org/10.32620/reks.2021.4.12
16. Ponochovniy, Y., Bulba, E., Yanko, A., Hozbenko, E.: Influence of diagnostics errors on safety: indicators and requirements. In: IX International Conference on Dependable Systems, Services and Technologies, pp. 53–57 (2018). https://doi.org/10.1109/DESSERT.2018.8409098
17. Kharchenko, V., Ponochovnyi, Y., Andrashov, A., Brezhniev, E., Bulba, E.: Modelling and safety assessment of programmable platform based information and control systems considering hidden physical and design faults. In: Zamojski, W., Mazurkiewicz, J., Sugier, J., Walkowiak, T., Kacprzyk, J. (eds.) DepCoS-RELCOMEX 2019. AISC, vol. 987, pp. 264–273. Springer, Cham (2020). https://doi.org/10.1007/978-3-030-19501-4_26
18. Solve stiff differential equations and DAEs – variable order method – MATLAB ode15s. https://www.mathworks.com/help/matlab/ref/ode15s.html. Accessed 1 Jan 2022
19. Kharchenko, V., Ponochovnyi, Y., Boyarchuk, A.: Availability assessment of information and control systems with online software update and verification. In: Information and Communication Technologies in Education, Research, and Industrial Applications, pp. 300–324 (2014). https://doi.org/10.1007/978-3-319-13206-8_15
20. Shelekhov, I., Barchenko, N., Kalchenko, V., Obodyak, V.: A hierarchical fuzzy quality assessment of complex security information systems. Radioelectron. Comput. Syst. **4**(96), 106-115 (2020). https://doi.org/10.32620/reks.2020.4.10

Network Anomaly Detection Based on Sparse Representation and Incoherent Dictionary Learning

Tomasz Kierul[1](✉), Tomasz Andrysiak[2](✉), and Michał Kierul[1]

[1] Research and Development Center SOFTBLUE S.A., Jana Zamoyskiego 2B, 85-063 Bydgoszcz, Poland
{tkierul,mkierul}@softblue.pl

[2] Institute of Telecommunications and Computer Science, Faculty of Telecommunications, Computer Science and Electrical Engineering, Bydgoszcz University of Science and Technology, Al. prof. S. Kaliskiego 7, 85-796 Bydgoszcz, Poland
andrys@pbs.edu.pl

Abstract. This article presents the use of sparse representation of signal and incoherent dictionary learning for solving the problem of anomaly detection. The signals which are analysed in the article represent selected features of network traffic. To detect the proper structure of the dictionary two modified algorithms were used in the learning process: Method of Optimal Directions and K-means Singular Value Decomposition. The algorithms were modified by inserting a decorrelation step in the loop of the dictionary update. Such dictionaries constituted ground for sparse representation of the analysed signals. Anomaly detection is performed by estimating parameters of the analysed signal and their comparative analysis towards network traffic profiles. Experimental results confirmed the effectiveness of the solution proposed for searching of anomalies in the analysed network traffic.

Keywords: Network traffic analysis · Sparse signal representation · Incoherent dictionary learning · Anomaly detection

1 Introduction

One of the key social and civilization problems of today is the increasing number of threats and incidents violating the security of computer systems and networks. The scale and dynamics of the phenomenon are constantly increasing, affecting not only single users but also big corporations and state institutions [1].

The answer to this type of threats are the currently emerging systems of monitoring and protection of different types of IT infrastructures and single mobile devices. The systems mentioned above used for detection and prevention of attacks, breaking or abuse, most often are able to protect systems and IT networks from known threats defined by means of previously acknowledged patterns. However, the absence of a security breach that matches the identified signatures does not imply that there is no threat. In this context the biggest problem may be created by the so called "zero-day attacks", i.e. attacks which

W. Zamojski et al. (Eds.): DepCoS-RELCOMEX 2022, LNNS 484, pp. 99–108, 2022.
https://doi.org/10.1007/978-3-031-06746-4_10

have not occurred so far thus there are no previously recognized patterns. Therefore, one should not seek threat signature but detect abnormal behavior which is a deviation from normal standard characteristics of the analyzed signal [2].

An anomaly may be any deviation from the adopted rule or a specific profile describing the "normal" variability of the analyzed characteristics of the tested signal. Idea formulated in this way was a contribution to numerous research and thesis, in particular those connected to its implementation based on the methods of computational intelligence or expert systems [3, 4]. The currently developed techniques of anomaly detection use different methods of the signal's processing and analysis. Most often these solutions are based on adaptive decomposition of the analyzed signal, in particular with the use of redundant dictionaries [5].

The present article suggests the use of incoherent dictionary learning method and sparse signal representation for given time series of the analyzed network traffic. In the learning process two modified algorithms were used to detect proper dictionary structure, namely, Method of Optimal Directions (MOD) and K-means Singular Value Decomposition (KSVD). So defined dictionaries were basis for sparse representation of the analysed signal. Anomaly detection is performed by means of comparison between the normal behavior parameters (estimated parameters of the sparse signal representation) and those of real network traffic.

This paper is organized as follows: after the introduction, in Sect. 2, the overview of sparse and redundant representation is shown. In Sect. 3, sparse signal representation for date traffic prediction is described. Next, in Sect. 4, the dictionary learning methods based on incoherent KSVD and MOD algorithm estimation are discussed. Details of implementation and experimental results are described in Sect. 5. Conclusions are presented thereafter.

2 Overview of Sparse and Redundant Representation

The signal representations implemented as linear expansions to a certain set of base functions well located with respect to time and / or frequency are usually computationally efficient and are characterized by a relatively simple interpretation of the results. In many cases these attributes are not sufficient though. It occurs when the analyzed signals require more precise and flexible as well as sparse representations [6].

In general, depending on the choice of the base function, the extension coefficients represent different features of the examined signal. If the size of the signal's structural elements differ significantly from the base function's scaling constant then coefficients of linear expansion do not constitute optimal representation of the subject signal. If the structures of the signal are complicated, it is impossible to define their optimal parameters for the given basic functions. In such case, the most effective solution may be implementation of more diverse and numerous function sets, described as dictionaries with redundancy. These functions are chosen to closely correspond to the nature of the analyzed signal. The dictionary used in the process can be chosen in two ways: (i) by building the dictionary on a mathematical model of data (adoption of already known forms of models, such as: wavelets, wavelet packets, contourlets, curvelets or Gabor), or (ii) learning the dictionary on the ground of a training data set (the most common part of the processed signal) [7]. The Method of Optimal Directions is the most typical method of dictionary learning. It was proposed by Engan et al. [8] and became one of the first applied methods, known also as the process of sparsification. Another equally favored method, despite being different in nature, is K-means Singular Value Decomposition algorithm, proposed by Ahron et al. [9]. The most significant difference between the two methods is the mode of updating the dictionary, i.e. some update all atoms simultaneously (MOD), others update atoms one after another (KSVD).

The methods and techniques based on sparse signal representation with the use of redundant dictionaries are becoming a promising approach towards analysis and detection of different types of anomalies [5]. In the process of adaptive decomposition, they allow to extract essential structural features of the analyzed signal respectively to the nature of the used dictionary [10].

3 Sparse Representation of a Signal

Generally, representations of the analyzed signal, performed in the form of linear extensions to a definite set of base functions properly located in time and/or frequency, are often described as not sufficiently precise and optimal. Therefore, a better solution is to use redundant dictionaries, i.e. numerous and diverse sets adjusted to the nature of the signal. In result, more universal and flexible representations are acknowledged.

Sparse representation of the analyzed signal continuously seeks limited solution of sparse representation coefficients C describing the signal S concerning the over complete dictionary when the residue signal is smaller than the given threshold value δ which we may assume as [11]:

$$\min\|C\|_0 \qquad subject\ to \qquad \|S - \textstyle\sum_{m=0}^{M-1} c_m d_m\| < \delta \tag{1}$$

where $\|\bullet\|_0$ describes ℓ_0 norm counting the non-zero entries of a vector, $c_m \in C$ showing a group of decomposition coefficients, $d_m \in D$ are the atoms of the over complete dictionary D, and δ is the constant specifying the exactness of the representation.

The Eq. (1) shows sparse representation of signal S, obtained by means of the minimal number of decomposition coefficients c_m and the d_m atoms of D dictionary corresponding to them (after assuming limited level of δ precision). Optimal representation of the analyzed signal is defined as a subset of dictionary D elements the linear combination of which resembles the biggest rate of the signal S energy among all the subsets of the same count. Identifying such a representation is computationally NP-hard. A suboptimal expansion can be defined in an iterative procedure, by means of greedy algorithms, e.g. the matching pursuit algorithm [12].

4 Dictionaries for Sparse Representations

While describing the analyzed signal, redundancy appears when the dictionary is bigger than the orthogonal base. Conciseness is possible to obtain, however, by agreeing to approximate the signal but only if we utilize as few functions as possible. If the number of dictionary functions (atoms) chosen for representation of the signal is called the size of representation, then we usually pursue a situation, where [7].

$$\text{size of representation} < \text{dimension of base} \ll \text{size of dictionary}.$$

However, decomposition of the signal s_i towards the dictionary requires then continuous seeking and matching its appropriate atoms which reflects best the necessary attributes of the analyzed signal. This matching ought to be performed to maximize correspondence between the chosen dictionary's atoms and the remaining part of the analyzed signal. Thus, it is strongly recommended to define a measure of quality and an algorithm which would ensure finding the best signal's representation for that measure. However, to perform such an aimed goal it is advisable, first of all, to define dictionary elements (atoms) which show the necessary attributes of the decomposed signal in the best possible manner. These dictionaries can be created with the use of properly parametrized analytic functions (e.g. Gabor functions [13]) or formed as approximated reflections of defined analyzed signal structures (e.g. Method of Optimal Directions [8] and K-means Singular Value Decomposition [9]).

4.1 Gabor's Functions Dictionary

As far as the proposed method is analyzed, it was suggested to use a waveform (atoms) from a Gabor dictionary in the form of translation, dilation and modulation of a Gaussian window function $g(x)$. We may then define a set of dictionary atoms' indices, i.e. $\theta_n \in \{\eta_n, \vartheta_n, \omega_n, \phi_n\}$, where index $\theta_n \in Z$ and $Z = R^+ \times R^3$, parameter η is responsible for dilation, ϑ is translation, whereas ϕ is modulating frequency [12].

For so defined parameters we obtain the following form of real Gabor function (so called atoms)

$$g_{\theta_n}(x) = K_{\theta_n} g\left(\frac{x - \vartheta_n}{\eta_n}\right) sin\left(\frac{2\pi\omega_n}{N}(x - \vartheta_n) + \phi_n\right) \tag{2}$$

and

$$g(x) = \frac{1}{\sqrt{2\pi}} e^{-\frac{x^2}{2}}, \tag{3}$$

where N describes the size of the processed signal for which the dictionary is built, K_{θ_n} is the normalizing constant used to achieve $\|g_{\theta_n}\| = 1$ and $\theta_n = \{\eta_n, \vartheta_n, \omega_n, \phi_n\}$ describes parameters of the Gabor's dictionary functions [13].

In our work, we used the idea of a dictionary initially proposed in the quoted work [14]. The parameters of the atoms referred to above were picked from dyadic sequences of integers. In order to create an over-complete set of Gabor functions, dictionary D was constructed by varying subsequence atom parameters $\{\eta_n, \vartheta_n, \omega_n, \phi_n\}$.

4.2 Methods of Dictionary Learning

Numerous researchers believe sparse representation used for dictionary learning is an interesting topic [7]. The reason why the field of dictionary learning algorithms surpasses other approaches is the adjustment process, i.e. some update all atoms simultaneously (e.g. Method of Optimal Directions), other update atoms one after another (e.g. K-means Singular Value Decomposition).

For these solutions it is essential to search for the best dictionary D which will reflect the signal S as sparse representation, as a solution to the Eq. (4)

$$\min_{D,C}\left[\|S - DC\|_F^2\right] \quad \textit{subject to} \quad \forall_i \; \|c_i\|_0 \leq T, \quad i = 1, 2, \ldots, M, \tag{4}$$

where $\| \bullet \|_F^2$ is the Frobenius norm and T is a fixed and predetermined number of nonzero entries. The commonly applied strategy to solve this problem is to start with an initial dictionary and alternate between the following steps: sparse coding and dictionary update [11].

The MOD and KSVD are utilized for solving the optimization issue, presented in Eq. (6). This process is performed by iterative minimization of the objective function over one variable, while the other two remain fixed. Firstly, D is initiated. In the following step, minimization over C is done – the process of iterative optimization begins. The standard procedure of initializing D requires a predefined dictionary, e.g. Gabor's [10], otherwise the dictionary is constructed of atoms randomly chosen from the training signals. The second solution is not proper for our process because certain outliers might be wrongfully taken as atoms, which could affect the whole process (in the sequel iterations).

The MOD and KSVD algorithms have the following two general stages [8, 9]:

Sparse Coding: at this stage, the decomposition coefficients c_i are collected in the over-complete dictionary D and signal S. The aim of each phase is to find the smallest possible number of coefficients which fulfil the Eq. (5). The given D is known. The Orthogonal Matching Pursuit (OMP) algorithm [15] is used to calculate M sparse coefficients c_i for each signal S, by estimation of

$$c_i \leftarrow \textit{arg}\min_{c_i}\|\mathrm{S} - Dc_i\|_2^2 \quad \textit{subject to} \quad \|c_i\|_0 \leq T, \quad i = 1, 2, \ldots, M. \tag{5}$$

Dictionary Update: at this stage, the atoms of dictionary D are updated (see Eq. (6)). Alternative values of atoms and decomposition coefficients are calculated to decrease the possibility of an error within the range of the signal S and its sparse representation $D * C$ with outliers.

$$\mathrm{D} \leftarrow \mathrm{arg}\min_{D}\sum_{i=1}^{L}\|S - Dc_i\|_2^2 \; \textit{subject to} \; \|c_i\|_0 \leq T, \quad i = 1, 2, \ldots, M. \tag{6}$$

In the dictionary update stage, the following problem needs to be solved

$$\min_{\mathrm{D}}\|\mathrm{S} - D\mathrm{C}\|_{\mathrm{F}}^2 \tag{7}$$

In general, all dictionary atoms are updated in this manner. Iteration, resulting from the two above stages, enables creation of a dictionary which estimates the signal S in a sparse and concise way. It is then followed by the MOD and KSVD algorithms, being a dictionary D composed of frequently correlated atoms which reflect the examined signal S with reference to its sparse representation $D * C$. What differs the two methods is the manner of updating the dictionary, i.e. some update all atoms simultaneously (MOD), other update atoms one after another (KSVD). A detailed description of the presented algorithms can be found in the works [8, 9].

The problem of frequently correlated atoms can be solved by introduction of the decorrelation condition, i.e. incoherent dictionary learning step [16]. To every iteration of the dictionary learning algorithm that includes sparse approximation, preceded by dictionary actualization, it is suggested to use the optimization problem described below.

4.3 Incoherent Dictionary Learning

The general problem of the incoherent dictionary learning is to find the closest dictionary $\widehat{D}$ to a given dictionary $\overline{D}$ with a coherence lower than the given μ_0. The dictionary $\overline{D}$ is described as [16]

$$\widehat{D} = arg \min_{D \in \Gamma} \|D - \overline{D}\|_F^2 \tag{8}$$

and

$$\Gamma = \left\{ D | \mu(D) \leq \mu_0 \wedge \|d_i\|_2 = 1, i \in \{1, .., M\} \right\}. \tag{9}$$

Good strategy for this problem can be including a decorrelation step to the iterative scheme in Sect. 4.2. At each iteration of the dictionary learning algorithm which consists of sparse approximation followed by dictionary update, we add the following optimization problem

$$\widehat{D} = arg \min_{D \in \Gamma} \mu(D) \tag{10}$$

and

$$\Gamma' = \left\{ D | \|D - \overline{D}\|_F^2 \leq \theta \wedge \|d_i\|_2 = 1, i \in \{1, .., M\} \right\}, \tag{11}$$

where θ the unknown minimum value reached by the criterion (8).

In our implementation the problem is solved by inserting a decorrelation step in the KSVD and MOD loop after the dictionary update. The modified algorithms are called the INC-KSVD [17] and INC-MOD.

5 Experimental Results

Evaluation of dictionaries (i. e. Gabor, INC-MOD and INC-KSVD) used for anomaly detection was made by means of SNORT [18] which gave an interface for capturing 25 traffic features TFC1-TFC25 (see Table 1). Efficiency of proposed solutions based on anomaly detection algorithm was evaluated by simulating different real world attacks on test LAN network. Kali Linux distribution was utilized in order to simulate certain attacks, such as: Application specific DDoS, various port scanning, Syn Flooding, DoS, DDoS, pocket fragmentation, spoofing and other [19].

Table 1. The list of TFC1-TFC26 network traffic features.

Feature	Traffic feature description	Feature	Traffic feature description
TFC1	Number of TCP packets	TFC14	Out TCP packets (port 80)
TFC2	In TCP packets	TFC15	In TCP packets (port 80)
TFC3	Out TCP packets	TFC16	Out UDP datagrams (port 53)
TFC4	Number of TCP packets in LAN	TFC17	In UDP datagrams (port 53)
TFC5	Number of UDP datagrams	TFC18	Out IP traffic [kB/s]
TFC6	In UDP datagrams	TFC19	In IP traffic [kB/s]
TFC7	Out UDP datagrams	TFC20	Out TCP traffic (port 80) [kB/s]
TFC8	Number of UDP datagrams in LAN	TFC21	In TCP traffic (port 80) [kB/s]
TFC9	Number of ICMP packets	TFC22	Out UDP traffic [kB/s]
TFC10	Out ICMP packets	TFC23	In UDP traffic [kB/s]
TFC11	5n ICMP packets	TFC24	Out UDP traffic (port 53) [kB/s]
TFC12	Number of ICMP packets in LAN	TFC25	In UDP traffic (port 53) [kB/s]
TFC13	Number of TCP packets with SYN and ACK flags		

Table 2. DR[%] results for three methods of anomaly/attacks detection.

Feature	GABOR	INC-KSVD	INC- MOD	Feature	GABOR	INC-KSVD	INC- MOD
TFC1	5.87	8.44	11.26	TFC14	8.87	5.44	11.26
TFC2	10.12	12.15	13.52	TFC15	6.26	14.86	15.52
TFC3	9.12	12.15	13.52	TFC16	0.00	0.00	0.00
TFC4	7.45	10.44	10.52	TFC17	5.41	8.40	11.26
TFC5	10.63	14.11	14.52	TFC18	6.33	14.11	16.52
TFC6	0.00	0.00	0.00	TFC19	5.17	8.24	10.26
TFC7	0.00	0.00	0.00	TFC20	7.42	15.52	18.26
TFC8	27.27	35.89	36.58	TFC21	10.37	14.20	17.52
TFC9	78.84	98.48	98.91	TFC22	0.00	0.00	0.00
TFC10	87.58	96.22	97.73	TFC23	0.00	0.00	0.00
TFC11	7.24	10.12	15.26	TFC24	0.15	0.00	0.00
TFC12	78.53	86.34	89.95	TFC25	2.41	0.08	0.02
TFC13	6.11	15.23	17.52				

Table 3. FP[%] results for three methods of anomaly/attacks detection.

Feature	GABOR	INC-KSVD	INC- MOD	Feature	GABOR	INC-KSVD	INC- MOD
TFC1	5.12	4.21	4.03	TFC14	4.14	3.44	3.08
TFC2	5.32	5.01	4.99	TFC15	5.36	3.56	2.97
TFC3	4.45	4,24	3.96	TFC16	1.32	0.06	0.02
TFC4	4.24	4,06	4.01	TFC17	1.22	0.20	0.19
TFC5	5.01	4.21	2.62	TFC18	5.52	3.71	2.74
TFC6	3.72	3.05	2.14	TFC19	3.78	4.44	4.36
TFC7	6.43	3.57	3.33	TFC20	3.11	3.20	3.50
TFC8	6.78	5.09	4.28	TFC21	3.16	3.15	3.09
TFC9	7.21	6.18	5.13	TFC22	2.47	2.50	3.08
TFC10	1.45	0.42	0.28	TFC23	2.77	2.85	3.07
TFC11	5.34	4.12	3.96	TFC24	0.00	0.00	0.00
TFC12	4.86	4.24	4.01	TFC25	0.03	0.02	0.02
TFC13	4.98	4.53	3.87				

In this work, the idea of dictionary is used, initially suggested in the quoted work [6]. The parameters of the above mentioned atoms were pointed out from dyadic sequences of integers. For the sake of creating an over-complete set of Gabor functions, dictionary D was constructed by varying subsequence atom parameters. Base functions dictionary D were formed with the use of 10 different scales, 50 diverse modulations and 20 translations.

In learning processes we apply modified INC-KSVD and INC-MOD algorithm to detect incoherent dictionary on the basis of network traffic which does not contain anomalies. The classification is performed with the use of normal network traffic profiles and sparse representation parameters of the analyzed signal.

In order to classify anomalies, profiles of normal traffic behavior are created on the basis of network traffic features with an assumption that there is no attack in this traffic. For algorithms' evaluation 25 traffic features were extracted from network traffic (see Table 1). Traffic features are represented as one dimensional – 1D vector of values. In Table 2 and Table 3 there are results for DR[%] (detection rates) and FP[%] (false positive) for the proposed method. Results of DR achieved up to 98.91%, while FP up to 7.21%.

Some values in Table 2 and Table 3 have 0 because the attacks simulated in the scenario do not have impact on every traffic feature from Table 1. Such traffic features did not provide usable data for anomaly/attack detection process. Values of DR and FP are acceptable so it can be stated that INC-MOD and INC-KSVD can be useful for anomaly detection in TCP/IP LAN networks and can be complementary to IDS systems based on databases of known attacks.

6 Conclusion

The article describes the complete procedure of building sparse signal representation and suggests how to identify anomalies based on network traffic prediction. The learning process is based on implementation of the extended MOD and KSVD algorithms including incoherent dictionary learning step in order to obtain proper dictionary structure. The classification process is done by means of normal network traffic profiles and sparse representation parameters of the analyzed signal. An extended set of test traces from real network traffic enables to examine efficiency of the proposed method.

Summarizing the anomalies (abuse) and attacks created with the use of Kali Linux tools, it can be inferred that the proposed INC-MOD algorithm provides better results than INC-KSVD and Gabor dictionary used to illustrate the analyzed traffic features. For the examined set of traffic tracks, the best outcomes were obtained for TFC9 and TFC10 traffic features, where DR [%] changes from 98.91−97.73, while FP [%] changes within the range 5.13−0.28 for INC-MOD base algorithm. The results confirm that abnormal activities presented in the traffic signal can be detected by the proposed methods.

Acknowledgements. This research was supported by the National Centre for Research and Developmen under the realized fast track - Intelligent Development 2014–2020 (Project POIR.01.01.01-00-0633/21).

References

1. KandanMani, M., Kandan, G., Jaspher, W., Melvin, A.: Network attacks and prevention techniques - a study. In: Conference: 2019 IEEE International Conference on Electrical, Computer and Communication Technologies (2019)
2. Biggio, B., Fumera, G., Roli, F.: Security evaluation of pattern classifiers under attack. IEEE Trans. Knowl. Data Eng. **26**(4), 984–996 (2014)
3. Chondola, V., Banerjee, A., Kumar, V.: Anomaly detection for discrete sequences: a survey. IEEE Trans. Knowl. Data Eng. **24**(5), 823–839 (2012)
4. Scherrer, A., Larrieu, N., Owezarski, P., Borgnat, P., Abry, P.: Non-gaussian and long memory statistical characterizations for internet traffic with anomalies. IEEE Trans. Depend. Secure Comput. **4**(1), 56–70 (2007)
5. Adler, A., Elad, M., Hel-Or, Y., Rivlin, E.: Sparse coding with anomaly detection. J. Signal Process. Syst. **79**(2), 179–188 (2014). https://doi.org/10.1007/s11265-014-0913-0
6. Mallat, S.G.: A theory for multiresolution signal decomposition: the wavelet representation. IEEE Trans. Pattern Anal. Mach. Intell. **11**(7), 674–693 (1989)
7. Rubinstein, R., Bruckstein, M., Elad, M.: Dictionaries for sparse representation modeling. Proc. IEEE **98**(6), 1045–1057 (2010)
8. Engan, K., Aase, S.O., Husoy, H.J.: Method of optimal directions for frame design. In: Proceedings of IEEE International Conference on Acoustics, Speech, and Signal Processing, pp. 2443–2446 (1999)
9. Aharon, M., Elad, M., Bruckstein, A.: K-SVD: an algorithm for designing overcomplete dictionaries for sparse representation. IEEE Trans. Signal Process. **54**(11), 4311–4322 (2006)
10. Elad, M.: Sparse and Redundant Representations: From Theory to Applications in Signal and Image Processing (2010)

11. Bruckstein, A.M., Donoho, D.L., Elad, M.: From sparse solutions of systems of equations to sparse modeling of signals and images. J. SIAM Rev. **51**(1), 34–81 (2009)
12. Mallat, S., Zhang, Z.: Matching pursuits with time-frequency dictionaries. IEEE Trans. Signal Process. **41**(12), 339–415 (1993)
13. Gabor, D.: Theory of communication. J. Inst. Electr. Eng. **93**(26), 429–457 (1946)
14. Durka, P., Matching Pursuit and Unification in EEG Analysis. Artech House (2007)
15. Pati, Y.C., Rezaiifar, R., Krishnaprasad, P.S.: Orthogonal matching pursuit: recursive function approximation with applications to wavelet decomposition. In: Proceedings of 27th Asilomar Conference on Signals, Systems and Computers, pp. 40–44 (1993)
16. Barchiesi, D., Plumbley, M.D.: Learning incoherent dictionaries for sparse approximation using iterative projections and rotations. Signal Process. IEEE Trans. Signal Process. **61**(8), 2055–2065 (2013)
17. Andrysiak, T., Saganowski, Ł: Incoherent dictionary learning for sparse representation in network anomaly detection. Schedae Informaticae **24**, 63–71 (2015)
18. Snort – intrusion detection system. https://www.snort.org, Accessed 10 Jan 2022
19. Kali Linux. https://www.kali.org, Accessed 10 Jan 2022

UAV Fleet with Battery Recharging for NPP Monitoring: Queuing System and Routing Based Reliability Models

Ihor Kliushnikov[1], Vyacheslav Kharchenko[1,2], Herman Fesenko[1], Kostiantyn Leontiiev[2], and Oleg Illiashenko[1](✉)

[1] National Aerospace University "KhAI", st. Chkalov, 17, Kharkiv 61070, Ukraine
{i.kliushnikov,v.kharchenko,h.fesenko,o.illiashenko}@csn.khai.edu

[2] Research and Production Company Radiy, Kropyvnytskyi, Ukraine
research@radiy.com

Abstract. Reliability models of the unmanned aerial vehicle (UAV) fleet with battery recharging for the system of monitoring critical objects as nuclear power plants (NPPs) and developed. The general model for describing and assessing availability of a UAV monitoring station (UAVMS) by use of a queuing system model is developed and analyzed in detail. These models consider UAV failures considering a mission plan. Battery recharging is carried out either at the depot or by using autonomous battery maintenance stations (ABMSs) deployed at certain points. Paper presents and discuss a general and partial UAVMS reliability models considering parameters of a ABMS and routings for monitoring. The results of research of the partial reliability models of the UAVMS with the ABMS (UAVMS-ABMS) set considering reliability, structure and routing parameters are described.

Keywords: Unmanned aerial vehicle · Routing · Reliability · Monitoring station · Nuclear power plant · Autonomous battery maintenance stations

1 Introduction

Modern technologies of onboard equipment allow usage of small-scale UAVs e.g. quadrotors monitoring mission of NPPs [1–3]. Such mission with the use of the quadrotors can possibly involve covering the whole target monitoring stations of the NPP in order to gather data from them on meteorological or radiological parameters during post-accident period. It can be characterized by damaging the wired networks which connect the monitoring stations and the crisis center (CC). However, the small battery life time (from 8 to 40 min) is a significant barrier to utilize quadrotors for the long-term missions.

In order to close this gap, the UAV's battery has to be charged or fully changed either at the UAV's quadrotor depot (QD) or with the use of ABMSs. Normally, an ABMS serves for quickly charging or changing a depleted UAV battery and simultaneously recharging other batteries to form a battery replacement pool.

W. Zamojski et al. (Eds.): DepCoS-RELCOMEX 2022, LNNS 484, pp. 109–119, 2022.
https://doi.org/10.1007/978-3-031-06746-4_11

The analysis of open sources shown several methods of using autonomous battery change/charge stations (ABMSs) which are used in ensuring of the long-term missions via UAVs. In study [4] the UAV replacement procedure which is used to guarantee persistent operation of UAV-based aerial networks providing Internet connectivity to ground users is shown. Paper [5] describes option for battery charging. Such cases may be considered by a network operator while using simulations for demonstration of the performance impact of incorporating the options into a cellular network where UAV infrastructure provides wireless service. An algorithm allowing a UAV fleet to provide continuous uninterrupted missions related to structural inspection is given in paper [6]. For implementation of the algorithm, the special protocol MAVLink should be enhanced with a set of ad hoc messages and commands. The operability of the algorithm is proven by the simulation results. In [7] the results of the three scenarios with different types of targets evaluation are given. The strategy of patrolling was able to successfully perform the mission as well as for detection of the targets and safe return to the recharging stations even for a several times.

An approach focused on changing and charging UAV battery simultaneously with the help of an autonomous battery maintenance mechatronic system (BMMS) is presented in the paper [8]. The study [9] focuses on issues which are related to the routing of a particular UAV fleet and organizing the battery replacement process in a way which can guarantee the maintenance of the desired production tact time. From the point of view of authors [10] when a drone has a problem linked with landing on a charging station, the charging station should comprise power transmitters and a receiver for charging the drone's battery. They also provide the results of different approaches research, which justify the structure and composition of UAV-based monitoring systems in accordance with a given requirements.

In [11] authors describe an algorithm of justifying composition and utilization of monitoring systems for critical infrastructure objects which make it possible to define the monitoring system composition taking into account the requirements that are put forward to the monitoring system, namely: the size of the monitoring area (the number of ground objects from which data is collected), the range, the frequency of transmission, the volume of transmitted data, the time of the system operation.

In [12] authors summarize and systematize the physical and functional parameters of wireless technologies in the industrial scientific & medical band and short-range devices ranges, which are significant from the wireless smart systems radiofrequency cyber vulnerability perspective, including navigation technologies. The paper [13] proposes an improved method for choosing communication protocols depending on the type of interaction pattern and considering vulnerabilities of these protocols. In [14] authors show an example of the models' application for routing main and redundant UAVs of the fleet to cover the whole target MSs of the Zaporizhzhia NPP.

Study [15] describes an uplink NOMA resource allocation algorithm for a UAVIoT-based communication network serving a large number of IoT devices of different applications. The optimization problem is formulated under constraints with the objective to maximize the data rate and minimize the delay. The paper [16] provides a comprehensive survey on a set of relevant research issues and highlights the representative solutions and concepts that have been proposed thus far in the design and modeling of the logistics of drone delivery systems, with the purpose of discussing the respective performance levels reached by the various suggested approaches.

On the basis of the analysis of mentioned sources the authors of the current paper conclude that the reliability issues related to both UAVs and ABMSs utilization are not addressed in full, which forms a gap in the given topic.

The goal of this paper is to develop and research availability and reliability-models of a UAV fleet based NPP monitoring system (UAVMS) with battery recharging considering number of UAVs, vehicles and automation battery maintenance stations (ABMSs) reliability, parameters of routings and so on. The objectives are the following: to develop and analyse a general model for describing and assessing availability of the UAVMS by use of a queuing system model; to present and discuss a general UAVMS reliability model considering parameters of ABMSs and routings for monitoring; to research reliability models of UAVMS with ABMS (UAVMS-ABMS) considering reliability, structure and routing parameters.

Approach to research is the following: firstly, we develop and analyze the UAVMS structure, specify functions and structure of a fleet of UAV (FoUAV) as a part of the monitoring system; after that, develop a model of the UAVMS-ABMS considering that a set of UAVs can be interpreted as a queuing system in point of view different tasks of measurement, video data receiving and transmission, transmission of data from sensors with W-Fi interfaces, and finally, develop and investigate UAVMS-ABMS reliability models based on coverage of a set of routings by the FoUAV.

2 UAV Based System of Pre and Post Monitoring NPP

Structure of UAVMS presented in Fig. 1. This system is designed to monitor the reactor utility, the relevant NPP equipment, as well as the surrounding area of the station. It consists of two subsystems: the pre- and post-accident monitoring system (PAMS) which checks critical parameters of the reactor zone and the automation radiation monitoring system (ARMS) which measures and analyses radiation level at the NPP and near the station, in a multi-kilometer zone. Information is transmitted to the crisis center (CC) located in 1–2 km from the NPP. A fleet of UAV can be a part of the PAMS and the ARMS, and performs additional service functions [1, 2].

In general, drones of the FoUAV: measure environment and NPP zone parameters and transmit them to the CC; collect and transmit video data; perform functions of transmitter for wireless sensors of the PAMS and the ARMS for assurance of transmission to CC in post-accident conditions; form the additional Internet of Drones based system for redundant storing and presentation data for the private cloud infrastructure and so on [2].

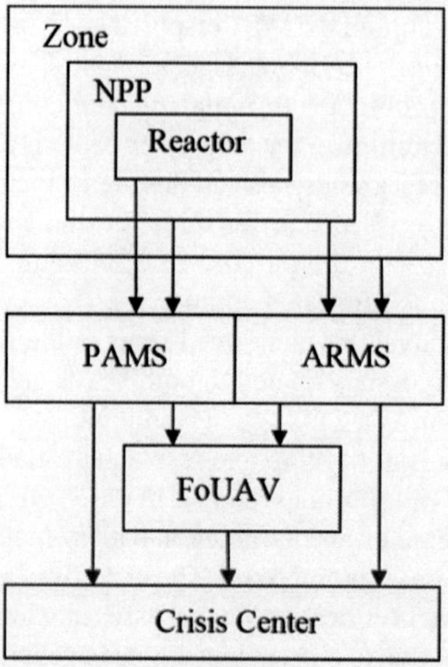

Fig. 1. General structure of UAVMS

Application of the FoUAV allows increasing reliability and survivability of the PAMS and the ARMS in pre- and post-accident conditions. In its turn, the FoUAV is a complex system consisting of three main parts (Fig. 2):

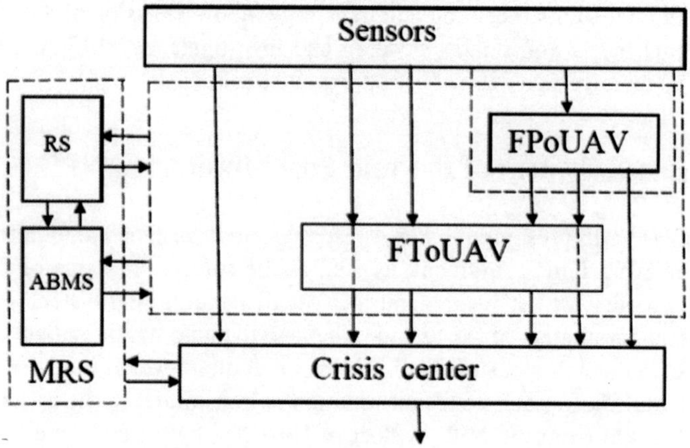

Fig. 2. Structure of UAV fleets for monitoring of critical objects

- the fleet of UAVs for transmission of data from sensors, videodata (FToUAV);
- the fleet of UAVs for preprocessing data from sensors (FToUAV);
- maintenance and recovery systems (MRS) consisting of the repairing and control subsystem (RS) and the ABMS.

3 Queuing System Model of UAVMS

Functioning UAVMS-ABMS can be described by model of queuing system (QS) consisting of three interconnected QSs (Fig. 3):

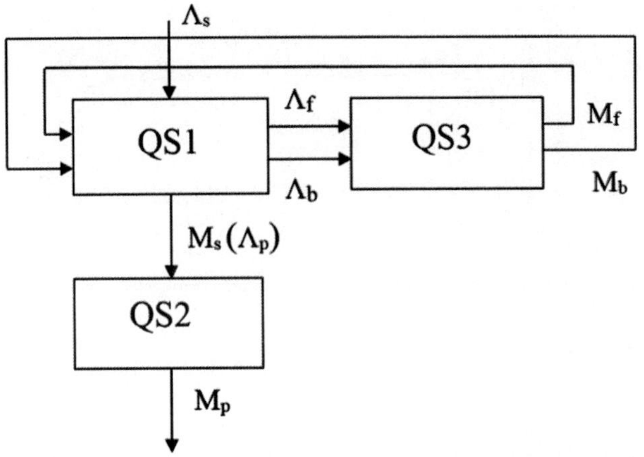

Fig. 3. General structure of UAV fleets for monitoring of critical objects

- QS1 is a system FToUAV maintaining stream of demands Λ_s from sensors and means of preprocessing (FPoUAV). The FToUAV can be described as a separate queuing subsytem as well. Output stream M_s is input stream of a system QS2. QS1 output streams of drone failures Λ_f and demands for drone recharging Λ_b are input streams of a system QS3;
- QS2 is a system maintaining demands formed by drons of FToUAV (and FPoUAV) and means of preprocessing (FToUAV). The FToUAV can be described as a separate queuing subsytem as well. Output stream of the system QS2 is output stream M_p of UAVMS-ABMS;
- QS3 is a system providing repair and recovery of drones (output stream M_f) and recharging drone batteries (output stream M_b).

Figure 4 presents graph of states UAVMS-ABMS as a queuing system. This model describes behavior of the systems in two modes: pre-accident and post-accident mnitoring (subrgaphs MM and EM correspindingly). Both subgraphs have the same structure consisting of chain "birth-death" (states a_m and a_e) describing processing demands by fleet of UAVs (rates of demands streams λ_{sm} and λ_{se} and maintenance streams μ_{sm} and μ_{se}) and similar chains FM (states a_{mf}) and EM (states a_{ef}) describing the same processes after drones failures (failure rates λ_{sf} and λ_{ef}) and recovery (recovery rates μ_{sf} and μ_{ef}). Transitions from states in mode "pre-accident"(MM) to states in mode "post-accident"(EM) are described by rate λ_{me}.

This model allows assessing and researching functions of UAVMS availability and performance in different modes depending on number of UAVs and required number

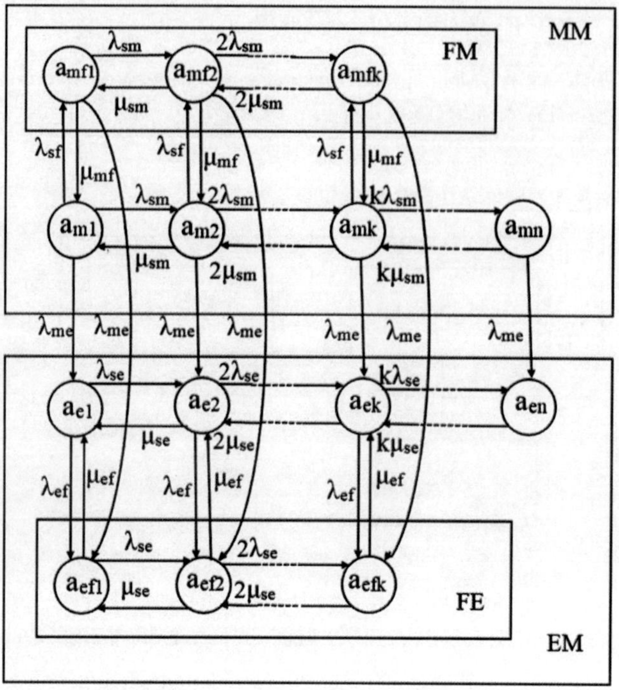

Fig. 4. Detailed structure of UAV fleets for monitoring of critical objects

of UAV which excludes losses of demands. To consider rates of failures and rates of demands for recharging separately the chains FM and FE should be decomposed. In this case task of assessing will be solved considering more complex graph of UAVMS-ABMS behavior.

4 UAVMS Reliability Models Considering Parameters of ABMS and Routings

Let us have *n* groups of monitoring stations (MSs). One UAV is used to visit all MSs of each group. All used UAVs is located at the same depot (QD). In general, UAV fleet can comprise both main UAVs (MDs) and redundant ones (RDs). Each of the last UAVs is ready to rich a point where a failed MD has stopped its monitoring mission, and to continue performing the mission instead of the failed UAV [11]. In this paper, we focus on application of MDs only. In order to perform a long-term NPP monitoring mission, UAVs need periodically recharging their batteries either at the QD or by using ABMSs deployed at certain points.

According to the first way of battery recharging, each UAV, after visiting the whole target MSs in its route, should return to the depot, recharge its battery, and repeat its previous route starting from the first visited MS. According to the second way of battery recharging, each UAV, after visiting the whole target MSs in its route, should rich ABMS, recharge its battery, and repeat its previous route starting from the last visited MS. A UAV fleet monitoring mission planning model with battery recharging is shown in Fig. 5 where MS_{i_ki} is MS k of MS group i where $i = 1,\ldots,n$, $fi = 1,\ldots,mi$; $L_{i\_(fi\text{-}1),fi}$ is the distance between points $(fi\text{-}1)$ and fi. For instace, $L_{i\_0,1}$ is the distance between the QD and MS_{i_1}, and $L_{i\_1,2}$ is the distance between MS_{i_1} to MS_{i_2}.

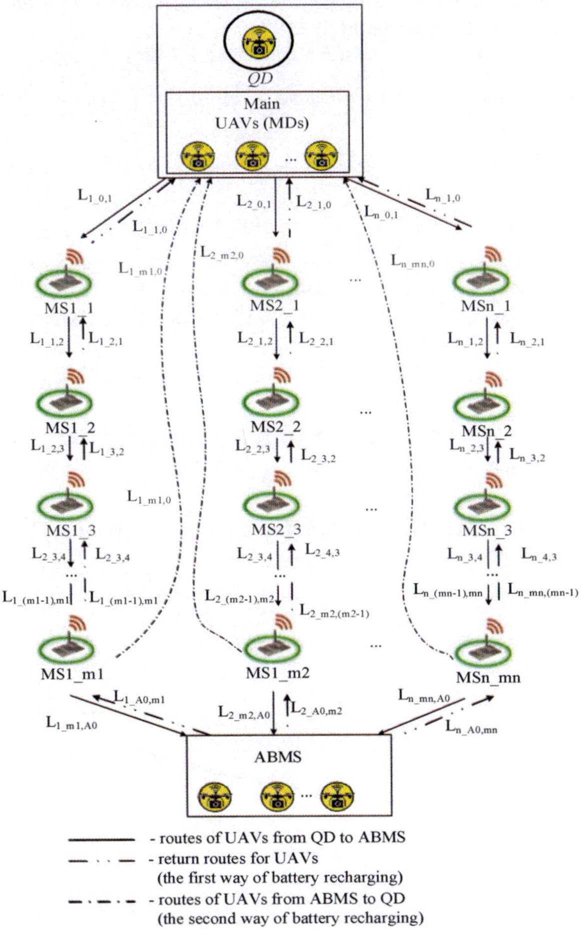

Fig. 5. General UAVMS reliability model considering parameters of ABMS and routings for monitoring

In order to consider failures of UAVs, it is reasonable to propose a set of reliability-based UAV fleet NPP monitoring mission planning models with battery recharging. First of all, let us give a classification of these models. To describe the proposed models, let us introduce data tuple S(n[m],k,b) where: n is the number of main routes; m is the number of route section; k is the number of redundant UAVs (NRD); b is the number of ABMSs.

The models' assumptions are the following. The models do not consider using RDs, but consider various ways for UAVs to follow their routes and recharge their batteries.

The NPP monitoring mission for each UAV of the fleet involves visiting the whole target MSs of the NPP twice. The probability of the successful plan fulfilment for the FoUAV to perform NPP monitoring mission (P_{SPF}) is used as an indicator. Redundant and non-redundant ABMSs can be used.

The P_{SPF} for models S(2[2],0,1) and S(2[2],0,2) considering that p_X is reliability function of the components X (MD, QD, ABMS) and p_{ret} is probability of successful UAV return after mission are calculated as:

$$P_{SPF}(S(2[2],0,1) = p_{MD}^{12} p_{ABMS} + \left(p_{MD}^{2} p_{ret}\right)^{4} p_{QD}(1 - p_{ABMS}), \tag{1}$$

$$\begin{aligned} P_{SPF}(S(2[2],0,2) &= p_{MD}^{12} p_{ABRAS}^{2} + 2p_{MD}^{12} p_{ABRAS}(1 - p_{ABRAS}) \\ &+ \left(p_{MD}^{4} p_{ret}^{2}\right)^{2} p_{QD}(1 - p_{ABRAS})^{2}. \end{aligned} \tag{2}$$

The reliability gain ΔP_{SPF} and the reliability gain as percentage $\Delta(\%)P_{SPF}$ due to the use of the redundant ABMS can be calculated via formulae (3) and (4).

$$\Delta P_{SPF} = P_{SPF}(S(2[2],0,2) - P_{SPF}(S(2[2],0,1). \tag{3}$$

$$\Delta(\%)P_{SPF} = \frac{P_{SPF}(S(2[2],0,2) - P_{SPF}(S(2[2],0,1)}{P_{SPF}(S(2[2],0,1)} 100\%. \tag{4}$$

Let us estimate the P_{SPF} for a UAV fleet visiting 4 intended MSs for the models S(2[2],0,1), and S(2[2],0,2). Assume that $L_{1\_0,1} = L_{1\_1,2} = L_{2\_0,1} = L_{2\_1,2}$, $p_{QD} = p_{ABMS} = p_{ret} = 0.9$, $p_{MD} = [0.5\ldots1]$. Results are presented in Figs. 6, 7, 8 and 9.

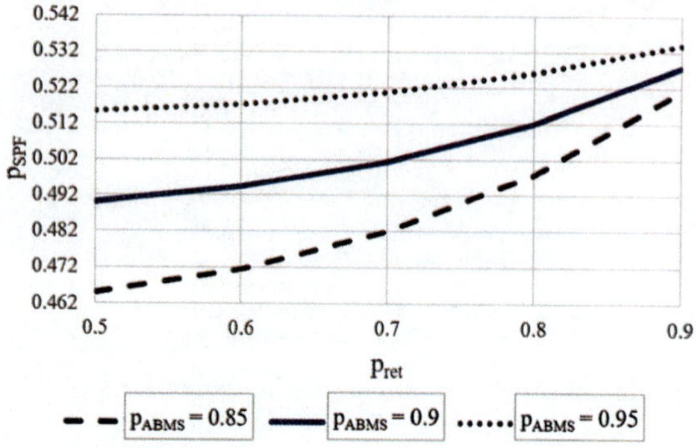

Fig. 6. Dependencies showing the relationship of the probability P_{SPF} to the probability p_{ret} when using the non-redundant ABMS

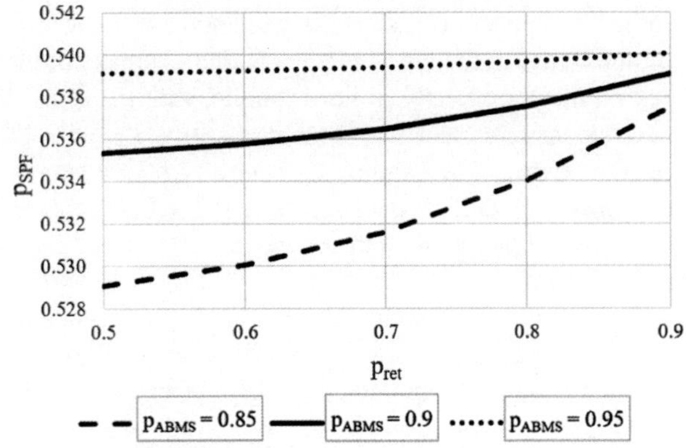

Fig. 7. Dependencies showing the relationship of the probability P_{SPF} to the probability p_{ret} when using the redundant ABMS

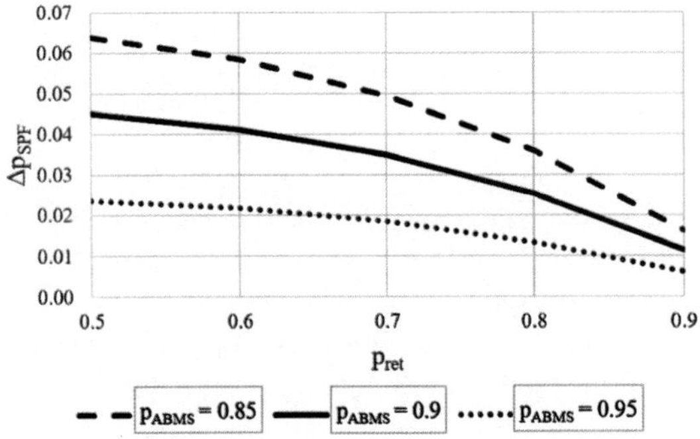

Fig. 8. Dependencies showing the reliability gain ΔP_{SPF} due to the use of a redundant ABMS

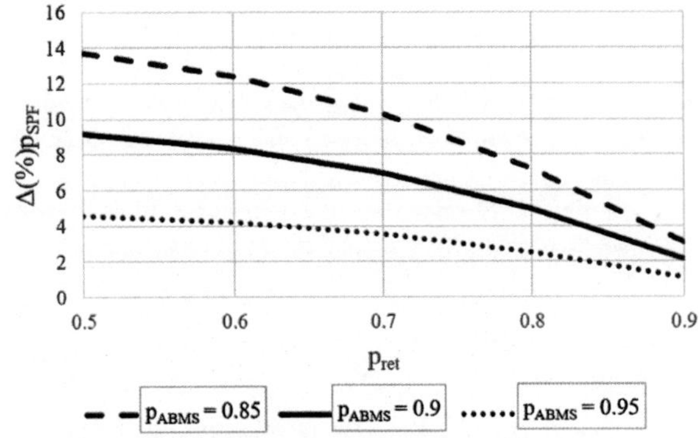

Fig. 9. Dependencies showing the reliability gain as a percentage $\Delta(\%)P_{SPF}$ due to the use of a redundant ABMS

The analysis of these dependencies shows that an increase in the reliability of ABMS (both a redundant and a non-redundant ones) leads to a decrease in the dependence of the parameter P_{SPF} on the parameter p_{ret}. For example, with $p_{ABMS} = 0.85$, an increase in p_{ret} from 0.5 to 0.9 leads to an increase in the probability P_{SPF} by 0.056117 (12.1%) when using a non-redundant ABMS and by 0.008418 (1.6%) when using a redundant ABMS, while, for $p_{ABMS} = 0.95$ increasing p_{ret} from 0.5 to 0.9 increases the probability P_{SPF} by 0.018706 (3.6%) when using non-redundant ABMS and by 0.000935 (0.2%) when using redundant ABMS.

It is also important to note that the increase in reliability is greater at lower values of probability 3, and this increase is more noticeable when using a non-redundant ABMS. Besides, the parameter P_{SPF} is more dependent on the parameter p_{ret} when using a non-redundant ABMS.

The results of the analysis show that the required value of the parameter P_{SPF} can be provided both when using non-redundant but expensive ABMSs, and when using inexpensive but redundant ABMSs. This gives rise to the problem of the optimal choice of the type and number of ABMS considering the cost and reliability indicators. The plots presented in Figs. 8 and 9 show reliability gain due to the use of a redundant ABMS.

5 Conclusion

The general model for describing and assessing availability of a UAV monitoring station (UAVMS) by use of a queuing system model is developed and analyzed in detail. Paper presents and discuss a general and partial UAVMS reliability models considering parameters of an ABMS and routings for monitoring.

The dependencies of the probability of the SFP for a UAV fleet visiting four intended MSs to the MD reliability function are obtained and explored for various cases. The analysis of the dependencies shows that the required value of the parameter P_{SPF} can be provided both when using non-redundant but expensive ABMSs, and when using inexpensive but redundant ABMSs. This gives rise to the problem of the optimal choice of the type and number of ABMS, considering the cost and reliability indicators.

Future research steps can be connected with simulating cyber-attacks on Internet of UAVs considering radio-frequency and protocols vulnerabilities [12, 13].

Acknowledgements. This work was supported by the ECHO project which has received funding from the European Union's Horizon 2020 research and innovation programme under the grant agreement no 830943.

The authors very appreciated to scientific society of consortium for invaluable inspiration, hardworking and creative analysis during the preparation of this paper.

References

1. Lüley, J., Vrban, B., Čerba, Š., Osuský, F., Nečas, V.: Unmanned radiation monitoring system. In: EPJ Web of Conferences, vol. 225 (2020)

2. Fesenko, H., Kharchenko, V., Sachenko, A., Hiromoto, R., Kochan, V.: An internet of drone-based multi-version post-severe accident monitoring system: structures and reliability. In: Kharchenko, V., Kor, A., Rucinski, A. (eds.) Dependable IoT for Human and Industry Modeling, Architecting, Implementation, pp. 197–217. River Publishers, Denmark, The Netherlands (2018)
3. Kliushnikov, I., Fesenko, H., Kharchenko, V.: Scheduling UAV fleets for the persistent operation of UAV-enabled wireless networks during NPP monitoring. Radioelectronic Comput. Syst. **1**(93), 29–36 (2020)
4. Sanchez-Aguero, V., Valera, F., Vidal, I., Tipantuña, C., Hesselbach, X.: Energy-aware management in multi-UAV deployments: modelling and strategies. Sensors **20**, 2791 (2020)
5. Galkin, B., Kibilda, J., DaSilva, L.A.: UAVs as mobile infrastructure: addressing battery lifetime. IEEE Commun. Mag. **57**, 132–137 (2019)
6. Yu, K., Budhiraja, A.K., Buebel, S., Tokekar, P.: Algorithms and experiments on routing of unmanned aerial vehicles with mobile recharging stations. J. Field Robot. **36**, 602–616 (2019)
7. Erdelj, D., Saif, O., Natalizio, E., Fantoni, I.: UAVs that fly forever: uninterrupted structural inspection through automatic UAV replacement. Ad Hoc Netw. **94**, 12–23 (2019)
8. Guimarães de Vargas, P., Kappel, K.S., Luis Marins, J., Milech Cabreira, T., Ferreira, P.R.: Patrolling strategy for multiple UAVs with recharging stations in unknown environments. In: Latin American Robotics Symposium (LARS), Brazilian Symposium on Robotics (SBR) and Workshop on Robotics in Education (WRE), pp. 346–351 (2019)
9. Ure, N.K., Chowdhary, G., Toksoz, T., How, J.P., Vavrina, M.A., Vian, J.: An automated battery management system to enable persistent missions with multiple aerial vehicles. IEEE/ASME Trans. Mechatron. **20**(1), 275–286 (2015)
10. Rohan, A., Rabah, M., Asghar, F., Talha, M., Kim, S.-H.: Advanced drone battery charging system. J. Electr. Eng. Technol. **14**(3), 1395–1405 (2019). https://doi.org/10.1007/s42835-019-00119-8
11. Kliushnikov, I., Fesenko, H., Kharchenko, V., Illiashenko, O., Morozova, O.: UAV fleet based accident monitoring systems with automatic battery replacement systems: algorithms for justifying composition and use planning. Int. J. Saf. Secur. Eng. **11**(4), 319–328 (2021)
12. Pevnev, V., Torinyk, V., Kharchenko, V.: Cyber security of wireless smart systems channels of intrusions and radio frequency vulnerabilities. Radioelectron. Comput. Syst. **4**(96), 74–91 (2020)
13. Kolisnyk, M.: Vulnerability analysis and method of selection of communication protocols for information transfer in internet of things systems. Radioelectron. Comput. Syst. **1**(97), 133–149 (2021)
14. Kliushnikov, I., Kharchenko, V., Fesenko, H., Zaitseva, E.: Multi-UAV routing for critical iinfrastructure monitoring considering failures of UAVs: reliability models, rerouting algorithms, industrial case. In: International Conference on Information and Digital Technologies 2021, IDT 2021, pp. 303–310 (2021)
15. Karem, R., Ahmed, M., Newagy, F.: Resource allocation in uplink NOMA-IoT based UAV for URLLC applications. Sensors **22**, 1566 (2022)
16. Benarbia, T., Kyamakya, K.: A literature review of drone-based package delivery logistics systems and their implementation feasibility. Sustainability **14**, 360 (2022)

Android Mobile Device as a 6DoF Motion Controller for Virtual Reality Applications

Maciej Kopczynski(✉)

Faculty of Computer Science, Bialystok University of Technology,
Wiejska 45A, 15-351 Białystok, Poland
m.kopczynski@pb.edu.pl

Abstract. This paper shows results and describes solution allowing for using an Android-based mobile device as a precise six degrees of freedom motion controller in **V**irtual **R**eality (**VR**) applications that are running on PC computer. Main goal of this approach is to create versatile system devoted to different types of applications, that are additionally accessible by wireless communication (no cables connecting PC and mobile device), therefore giving freedom of movements and allowing to use mobile device as a high-precision motion controller. Obtained results show possibility of achieving good compromise between low latency of tracked movements and high prediction rate, giving relatively small delay in movements tracking.

Keywords: Virtual Reality · Motion controller · Filtering · Prediction · Mobile device

1 Introduction

One of the main challenges of the VR industry is price of required devices, e.g. headset or controllers. Most high-quality applications, like games, are made for PC, because of their computing power and ways of interaction in virtual space with available peripherals. Because of the limited computing power and lack of basic keyboard or mouse interaction known from games, virtual reality mobile solutions are not used by most high-quality developers. Currently, PCs are most popular medium to create solutions for professional VR helmets like HTC Vive or Oculus Rift. Both of these solutions cost hundreds of euro. For most home users, these solutions are still not available due to the high price. The idea of developing technology described in this paper is to drastically reduce the price of entry into the VR for PC entertainment. It has to be noted, that IMU sensors in most mobile devices are not designed to provide high precision of motion tracking, what is crucial for virtual reality to achieve satisfactory levels of immersion.

The next problem in the mobile VR helmet and controller simulation is the excessive delay of the image in relation to the head and body movements (including hands) which results in motion sickness and prevents the comfortable gameplay. In order to overcome this effect, strong focus have to be put on optimization both the VR image transmission algorithms from the computer to the phone,

W. Zamojski et al. (Eds.): DepCoS-RELCOMEX 2022, LNNS 484, pp. 120–130, 2022.
https://doi.org/10.1007/978-3-031-06746-4_12

as well as motion tracking in the controllers to reduce any delay and instability. Motion sickness can be seen for most people, when delay is bigger than 50 ms [4].

At the moment there are not many ready-to-use solutions for transforming Android mobile device to high-precision motion controller using wireless network as transmission medium and focusing on low CPU usage, high prediction rate and low latency of movement tracking. Of course, there are many controller devices available on the market, but those are either expensive units or parts of VR helmet sets. Approach to remote motion controller with visual feedback on mobile devices was presented in [3]. Description of motion-based remote control device for interaction with 3D multimedia content can be found in [8]. Analysis of degrees of freedom using mobile phone for VR is described in [1]. 9-DOF customly designed motion controller based on MEMS chips is described in [5]. Optimizations for fast wireless image transfer using H.264 codec to Android mobile devices can be found in [7].

The paper is organized as follows. In Sect. 2 some information about the basic definitions are presented. The Sect. 3 focuses on description of implemented solution, while Sect. 4 is devoted to presentation of the experimental results.

2 Basic Definitions

For the purpose of measuring obtained results related to precision of motion tracking and movements prediction, some definitions have to be introduced. Selected formulas are presented below.

2.1 6 Degrees of Freedom

Six degrees of freedom in motion description (6DoF) refers to the axes of movement of a rigid body in three-dimensional space. The body is able to change position as forward/backward (surge), up/down (heave), left/right (sway) translation in three perpendicular axes, combined with changes in orientation through rotation in three perpendicular axes, often called yaw (normal axis), pitch (transverse axis), and roll (longitudinal axis). Basic IMU chip in mobile phone allows to track three degrees of freedom (3DoF) in rotational motion only: pitch, yaw, and roll. Solution described in this paper allows to translate 3DoF produced by mobile phone IMU sensor into 6DoF using inverted body model.

2.2 Kalman Filtering

Kalman filter is an algorithm proposed by R.E. Kalman. It is based on recursive determination of the minimal-variance estimate for state vector of a linear model, mostly discrete dynamic system, based on the measurements of the system's input and output. It is assumed that both the measurement and the processing inside the system are affected by an error with the Gaussian distribution. More details about Kalman filtering can be found in original article of R.E. Kalman [6] or general review like [2].

2.3 Quality Assesment

An important aspect of 3DoF to 6DoF conversion is smallest possible difference between perfect theoretical movement and the one that was created by using the algorithm. Additional aspect is the time delay required for data processing. Too big delay can cause nausea and motion sickness when using controllers in VR environment. Error in movement measurement can be summarized by using standard deviation on set of collected samples being a result of difference between real and ideal reference position:

$$\sigma = \sqrt{\frac{1}{N}\sum_{i=1}^{N}(x_i - \mu)^2} \qquad (1)$$

where N is number of collected samples, x_i is i-th sample difference, while μ is the average of the samples difference.

3 Solution Description

Test application was created in .NET technology using C# language and Microsoft Visual Studio 2017 on the PC side. Mobile device running Android operating system part was created in Java using Android Studio 3 IDE. Technologies that were used for data and graphics processing based extensively on low-level APIs for IMU raw data, as well as Direct3D 11 for visual part. For the precise and repeatable motion simulation, based on 2 different types of movements, which are described in details in Sect. 4, simple 1-axis robotic arm was used. Range of constant-speed rotational movement was limited to range 0–90°. Each tested mobile device was attached to the end of the arm.

Figure 1 presents general architecture of solution.

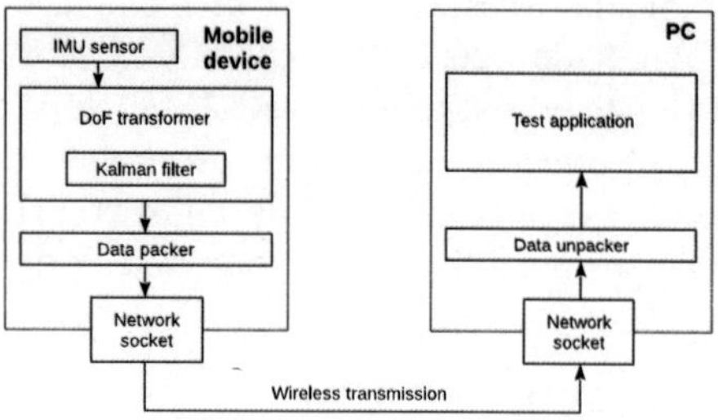

Fig. 1. Solution architecture in general

System consists of two parts - test application running on PC, and sensor data collecting and processing application on mobile device. Main functional parts of mobile device are:

- *IMU sensor* – main sensor of the mobile device,
- *DoF transformer* – transforms 3DoF data into 6DoF according to described algorithms,
- *Kalman filter* – Kalman filter for sensor data,
- *Data packer* – responsible for preparing collected and processed sensor data into MTU-sized frames.

Main functional parts of PC server are *Data unpacker* responsible for unpacking data from network stream and preparing it for processing and visualisation by *Test application.*

3.1 3DoF to 6DoF Transformation

The input for the algorithm is head and hand data. The hand is dependent on the initially unknown body position. For this reason, before estimation of hand to head relation, it is necessary to determine the approximate body direction. This process is divided into two stages - body position estimation and wrist position estimation.

Body Direction Estimation. The direction of the body is estimated basing on assumption, that wrist and head rotation is independent of the body itself, but with fixed angular restrictions. Head defines the direction of moving body, which is derived from spherical linear interpolation between the currently supposed body direction and the direction of the head. Additionally, intensity of the hand movements determine the body rotation speed. This approach prevents a false-estimated body rotation in a situation, where the player only looks to the side without directing his hand there, e.g. quickly looking back in response to an event in the played game. Because of space limitations, pseudocode of algorithm for body direction estimation is not presented here. Input to the algorithm is a transformation matrix M_H representing the head position and a transformation matrix M_C describing controller rotation in 3DoF. Output is the estimated body rotation quaternion Q_B, that should follow the head and hand, assuming human speed of movements. Q_B value is calculated as a spherical linear interpolation supported by the estimation function using current body direction Q_B and the current head direction with the fixed interpolation factor.

3DoF to 6DoF. After estimating the body position, it is possible to proceed to estimating the full position of the controller itself. For this purpose, a series of constants should be adopted in terms of body structure, e.g. the height from the elbow to the rotational center of the head, the horizontal distance from the center of the head to the elbow, or the length of the forearm with the wrist. Some constants are assigned to the controller (the structure of the controller), and others are assigned to a specific player (anatomical differences between players).

Assuming hand in the welcome position (e.g. for a handshake), it can be seen, that the wrist has relatively small rotation in the X and Z axes (pitch and roll).

In the Z axis (roll), the rotation is connected with the radius and ulna bones. X axis (pitch) has similar small rotation range. When making a gesture of aiming (e.g. with a weapon in the game) it is easy to notice, that when making changes in these axes, the entire forearm has to be moved. Only the Y axis (yaw) has a nearly 180° rotation field without changing the position of the forearm.

With these limitations in mind, it is assumed, that each wrist movement is a direct result of the forearm movement. Thus, the position of the elbow will be a direct result of wrist rotation.

The pseudocode describing the above process is described below.

INPUT: controller quaternion Q_C, body quaternion Q_B, vector V_A
OUTPUT: controller position matrix M_C

1: $Q_L \leftarrow Normalize(Inverse(Q_B) \cdot Q_C)$
2: $V_E, Q_E \leftarrow CalculateElbow(V_A, Q_L, Q_B)$
3: $V_W \leftarrow V_E + (Q_E \cdot WRIST\_OFFSET)$
4: $V_C \leftarrow Q_L \cdot CONTROLLER\_CENTER\_OFFSET$
5: $M_{CRot} \leftarrow RotationMatrix(Q_C)$
6: $M_{WPos} \leftarrow TranslationMatrix(V_W)$
7: $M_{CPos} \leftarrow TranslationMatrix(V_C)$
8: $M_C \leftarrow M_{CRot} \cdot M_{WPos} \cdot M_{CPos}$

Input to the algorithm is absolute controller rotation quaternion Q_C, estimated body rotation quaternion Q_B and vector $V_A = [-1, 1, 1]$ for the left controller and $[1, 1, 1]$ for the right controller. Output is the M_C matrix that represents controller position in three-dimensional space, including it's direction.

In line 1 quaternion Q_L representing local rotation of the controller with respect to the body, is computed. Then, in line 2, the elbow displacement vector V_E relative to the head along with the elbow quaternion Q_E is estimated on the basis of hand side multiplier V_A, controller rotation Q_L relative to the body, and estimated body rotation Q_B. Line 3 is responsible for calculating V_W vector, which describes head to wrist offset. This calculation is weighted by the anatomical user constant $WRIST\_OFFSET$. The goal of line 4 is to calculate the V_W vector offset from the wrist to the center of the controller rotation, which is constant for the controller and is based on its physical grip location. Given all the shift vectors and the quaternions of rotation, three transformation matrices are created on lines 5–7. M_{CRot} represents the absolute rotation of the controller, M_{Wpos} represents the offset to the wrist relative to the head, and M_{CPos} represents the offset to the center of the controller body relative to the wrist. In line 8, matrices are folded into a final transformation matrix M_C, which describes approximate 6DoF controller.

Kalman Filtering. Kalman filtering is used to make the tracking of the hand position smoother and with less perceptible delays, what included controller position data and its velocity. Results returned by filter allows to determine the predicted position for drawing the next frame. Next important advantage is the possibility of using existing measurements in the event of a delay in reading new data from stream and determining a new position at fixed time intervals.

Because of space limitations, pseudocode for Kalman filter prediction used for sensor data stream is not presented. According to the previous sections, input for the algorithm is M_C matrix that represents controller position in three-dimensional space (including it's direction) and single frame time F_T. Output of the algorithm is vector V_{PF} representing filtered and predicted value of controller position. Kalman filtering in this approach is based on 6-element state size and 3-element data size.

4 Experimental Results

Presented results were obtained using a PC equipped with an 8 GB RAM, 4-core Intel Core i5-6600k processor and Nvidia GeForce GTX 970 graphics card running Windows 10 operating system. Mobile devices used in tests were Google Pixel XL and Xperia Z1 Compact with the same version of Android operating system. Google Pixel XL is considered as mobile device with best quality IMU sensor and Xperia Z1 Compact has lowest grade sensor. Wireless network connection between PC and phone was established using TP-Link WDR3600 router configured to provide WiFi 802.11ac 5 GHz network.

Each of the controller movement measurements were 2 basic types:

- type 1 - swing of right hand over the left shoulder,
- type 2 - right hand rotation left to a ground parallel plane,

Each of the described movements were assuming 3DoF to 6DoF transformation using different mobile devices. Additionally, Kalman filtering was added in process of movement prediction. All collected samples were send using wireless network to specially written data processing application running on the PC. Purpose of the application was collecting IMU sensor data stream (raw and processed on the mobile device), as well as visualization of received data using 3 planes (XYZ). During rotational phase of the movement, each sample was acquired every 3° in the angle range of 0 to 90°, so 31 samples were collected for every single test phase.

Because of paper size limitations, tables with data contain every third sample, but images are built on basis of collected data.

Time needed by the mobile device to finish calculations on all implemented algorithms for each data sample was lower than 0.5 ms for each case.

Initial test set was performed for type 1 of the movement - swing of right hand over the left shoulder. First mobile device tested was Xperia Z1 Compact. Table 1 presents results acquired by test application.

Each presented table shows rotation in degrees for 3 axes - pitch, yaw and roll in column *Rotation*. *Reference* is the column describing ideal single point description using XYZ coordinates after transformation from 3DoF to 6DoF. Following columns have the same transformation. *Raw data* column shows non-filtered data, while *Kalman* column represents points after Kalman filtering. It should be additionally noted, that reference data was collected every time after arm or mobile device reconfiguration took place.

Table 1. Results acquired for type 1 movement for Xperia Z1 Compact

Sample	Rotation			Reference			Raw data			Kalman		
	pitch	yaw	roll	X	Y	Z	X	Y	Z	X	Y	Z
1	0.013	0.008	0.002	0.262	−0.399	−0.438	0.282	−0.392	−0.430	0.282	−0.392	−0.430
4	6.351	6.398	0.364	0.208	−0.358	−0.447	0.211	−0.350	−0.439	0.211	−0.341	−0.443
7	12.629	12.948	1.441	0.158	−0.317	−0.448	0.172	−0.294	−0.445	0.160	−0.285	−0.447
10	18.732	19.822	3.302	0.110	−0.281	−0.439	0.125	−0.269	−0.424	0.117	−0.246	−0.438
13	24.555	27.200	6.030	0.062	−0.248	−0.423	0.062	−0.248	−0.405	0.069	−0.231	−0.419
16	29.998	35.271	9.740	0.023	−0.216	−0.403	0.041	−0.207	−0.401	0.033	−0.202	−0.401
19	34.893	44.232	14.569	−0.013	−0.189	−0.371	−0.005	−0.176	−0.349	−0.005	−0.170	−0.369
22	39.051	54.229	20.596	−0.042	−0.167	−0.338	−0.021	−0.161	−0.333	−0.033	−0.149	−0.336
25	42.261	65.336	27.846	−0.068	−0.148	−0.296	−0.065	−0.129	−0.282	−0.061	−0.134	−0.291
28	44.292	77.373	36.120	−0.083	−0.136	−0.252	−0.071	−0.126	−0.245	−0.083	−0.116	−0.251
31	44.998	90.011	45.010	−0.092	−0.131	−0.203	−0.088	−0.128	−0.192	−0.094	−0.113	−0.201

Standard deviation for type 1 movement for Xperia Z1 Compact based on all originally collected samples (31) and compared on basis of errors to reference data is equal to:

- raw sensor data - $\sigma_{X_R} = 0.00632, \sigma_{Y_R} = 0.00677, \sigma_{Z_R} = 0.00671$,
- filtered sensor data - $\sigma_{X_F} = 0.00489, \sigma_{Y_F} = 0.00606, \sigma_{Z_F} = 0.00234$.

Table 2 presents results acquired by test application for Google Pixel XL mobile device.

Table 2. Results acquired for type 1 movement for Google Pixel XL

Sample	Rotation			Reference			Raw data			Kalman		
	pitch	yaw	roll	X	Y	Z	X	Y	Z	X	Y	Z
1	0.013	0.001	0.004	0.262	−0.396	−0.437	0.261	−0.398	−0.438	0.261	−0.398	−0.438
4	6.355	6.396	0.361	0.208	−0.356	−0.447	0.207	−0.357	−0.447	0.208	−0.357	−0.448
7	12.627	12.949	1.447	0.158	−0.319	−0.448	0.157	−0.319	−0.448	0.157	−0.318	−0.455
10	18.733	19.826	3.306	0.109	−0.282	−0.440	0.108	−0.282	−0.440	0.107	−0.280	−0.452
13	24.553	27.205	6.041	0.064	−0.246	−0.423	0.063	−0.247	−0.424	0.060	−0.245	−0.436
16	30.006	35.274	9.752	0.022	−0.215	−0.400	0.021	−0.216	−0.401	0.017	−0.212	−0.412
19	34.897	44.232	14.561	−0.012	−0.190	−0.371	−0.013	−0.190	−0.372	−0.019	−0.185	−0.382
22	39.059	54.237	20.596	−0.043	−0.165	−0.336	−0.044	−0.166	−0.338	−0.051	−0.161	−0.346
25	42.252	65.332	27.842	−0.067	−0.148	−0.296	−0.068	−0.149	−0.297	−0.076	−0.144	−0.305
28	44.294	77.381	36.126	−0.083	−0.135	−0.251	−0.084	−0.136	−0.252	−0.093	−0.131	−0.259
31	44.998	90.015	45.012	−0.092	−0.128	−0.204	−0.093	−0.130	−0.204	−0.102	−0.123	−0.209

Standard deviation for type 1 movement for Google Pixel XL based on all originally collected samples (31) and compared on basis of errors to reference data is equal to:

- raw sensor data - $\sigma_{X_R} = 0.00043, \sigma_{Y_R} = 0.00053, \sigma_{Z_R} = 0.00042$,
- filtered sensor data - $\sigma_{X_F} = 0.00034, \sigma_{Y_F} = 0.00025, \sigma_{Z_F} = 0.00037$.

Second test set was performed for type 2 of the movement - right hand rotation left to a ground parallel plane. First mobile device tested was Xperia Z1 Compact. Table 3 presents results acquired by test application.

Table 3. Results acquired for type 2 movement for Xperia Z1 Compact

Sample	Rotation			Reference			Raw data			Kalman		
	pitch	yaw	roll	X	Y	Z	X	Y	Z	X	Y	Z
1	0.014	0.013	0.006	0.263	−0.397	−0.438	0.284	−0.382	−0.424	0.284	−0.382	−0.424
4	0.008	9.005	0.000	0.204	−0.398	−0.433	0.221	−0.391	−0.415	0.210	−0.380	−0.426
7	0.003	18.009	0.009	0.146	−0.396	−0.423	0.149	−0.375	−0.421	0.139	−0.372	−0.421
10	0.008	27.002	0.016	0.093	−0.399	−0.403	0.100	−0.395	−0.397	0.081	−0.372	−0.409
13	0.007	36.009	0.010	0.044	−0.399	−0.373	0.061	−0.398	−0.351	0.041	−0.382	−0.376
16	0.006	45.001	0.015	−0.000	−0.399	−0.340	0.010	−0.395	−0.321	0.002	−0.385	−0.334
19	0.003	54.000	0.012	−0.039	−0.397	−0.304	−0.023	−0.380	−0.298	−0.037	−0.384	−0.298
22	0.005	63.005	0.016	−0.071	−0.398	−0.260	−0.050	−0.390	−0.255	−0.060	−0.390	−0.253
25	0.006	72.009	0.009	−0.098	−0.399	−0.212	−0.092	−0.391	−0.200	−0.093	−0.387	−0.208
28	359.993	81.000	0.012	−0.115	−0.398	−0.162	−0.102	−0.384	−0.141	−0.114	−0.390	−0.152
31	0.003	90.007	0.0131	−0.124	−0.397	−0.113	−0.106	−0.382	−0.108	−0.124	−0.386	−0.106

Standard deviation for type 2 movement for Xperia Z1 Compact based on all originally collected samples (31) and compared on basis of errors to reference data is equal to:

- raw sensor data - $\sigma_{X_R} = 0.00617, \sigma_{Y_R} = 0.00667, \sigma_{Z_R} = 0.00641$,
- filtered sensor data - $\sigma_{X_F} = 0.00523, \sigma_{Y_F} = 0.00629, \sigma_{Z_F} = 0.00529$.

Table 4 presents results acquired by test application for Google Pixel XL mobile device.

Table 4. Results acquired for type 2 movement for Google Pixel XL

Sample	Rotation			Reference			Raw data			Kalman		
	pitch	yaw	roll	X	Y	Z	X	Y	Z	X	Y	Z
1	0.000	0.010	0.006	0.262	−0.397	−0.436	0.261	−0.398	−0.438	0.261	−0.398	−0.438
4	0.005	9.012	0.006	0.204	−0.399	−0.433	0.203	−0.399	−0.434	0.202	−0.399	−0.436
7	0.002	18.003	0.009	0.148	−0.399	−0.422	0.146	−0.399	−0.423	0.143	−0.400	−0.431
10	0.002	27.008	0.008	0.094	−0.398	−0.401	0.093	−0.399	−0.402	0.088	−0.400	−0.416
13	0.001	36.000	0.014	0.043	−0.399	−0.376	0.043	−0.399	−0.376	0.037	−0.399	−0.389
16	0.004	45.004	0.008	−0.001	−0.398	−0.342	−0.002	−0.399	−0.343	−0.007	−0.399	−0.353
19	0.001	54.000	0.011	−0.039	−0.399	−0.301	−0.040	−0.399	−0.303	−0.047	−0.399	−0.311
22	0.002	63.001	0.016	−0.072	−0.399	−0.258	−0.073	−0.399	−0.259	−0.080	−0.399	−0.266
25	0.003	72.009	0.003	−0.097	−0.396	−0.211	−0.098	−0.398	−0.212	−0.106	−0.398	−0.218
28	0.005	80.992	0.004	−0.115	−0.397	−0.163	−0.116	−0.398	−0.164	−0.125	−0.398	−0.168
31	−0.001	89.998	0.012	−0.126	−0.398	−0.112	−0.126	−0.399	−0.113	−0.136	−0.399	−0.116

Standard deviation for type 2 movement for Google Pixel XL based on all originally collected samples (31) and compared on basis of errors to reference data is equal to:

- raw sensor data - $\sigma_{X_R} = 0.00045, \sigma_{Y_R} = 0.00049, \sigma_{Z_R} = 0.00048$,
- filtered sensor data - $\sigma_{X_F} = 0.00029, \sigma_{Y_F} = 0.00009, \sigma_{Z_F} = 0.00041$.

Figure 2 and Fig. 3 shows visualization of acquired data points for two types of movements for Xperia Z1 Compact and Google Pixel XL mobile devices respectively. Green line is data points for type 1 movement, while blue line is for type 2 movement. Part a) on each image presents reference set of data points, part b) is for raw and non-filtered data, while part c) is for data after Kalman filtering process.

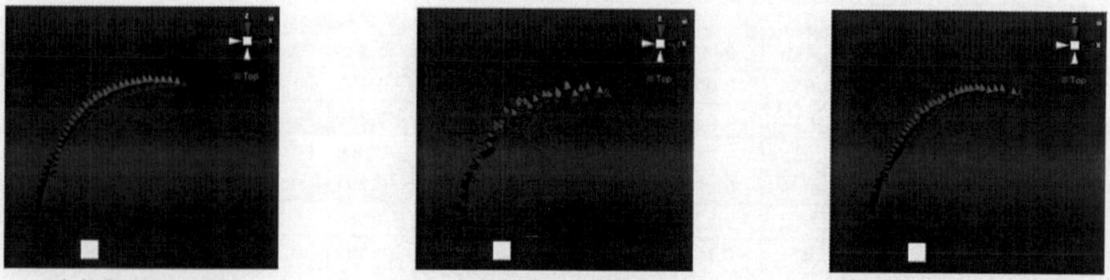

(a) Reference (b) Raw data (c) Kalman

Fig. 2. Sensor data points visualization for 2 types of movement for Xperia Z1 Compact

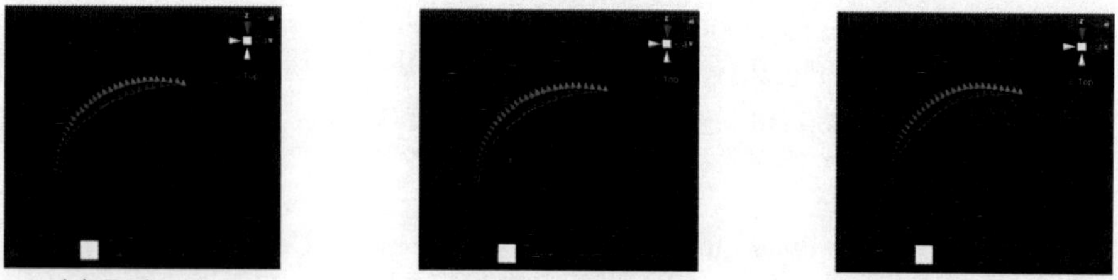

(a) Reference (b) Raw data (c) Kalman

Fig. 3. Sensor data points visualization for 2 types of movement for Google Pixel XL

According to obtained results, both in tables and calculated standard deviation, it can be clearly seen, that 3DoF to 6DoF algorithm works properly, but quality of the IMU sensor used by the mobile device has big impact on the output stability of 6DoF conversion. Google Pixel XL equipped with very precise sensor needs no converted data filtering, in opposite to Xperia Z1 Compact mobile device. Kalman filtering allows removing noise, but adds small delay to the output stream in terms of real position of controller.

5 Conclusions

Performed research shows, that creating efficient and versatile methods of using 3DoF mobile devices as 6DoF controllers is possible, when specified limitations are assumed and by using properly designed 3DoF to 6DoF transformation algorithm. Additionally, Kalman filtering approach allows to improve quality of predicted movements by removing mobile device sensor noise. Computational cost has very small effect on player immersion in VR world and allows using mobile device as a fully functional 6DoF controller. Only one limitation coming from Kalman filtering used for processing sensor data is a small added delay related to nature of Kalman filter algorithm. This can be especially problematic in very dynamic FPP games.

Further research will focus on developing methods, that could be useful for even better movement prediction and noise filtering and which could overcome the lag-effect of Kalman filtering. One of the paths will be related to changing and optimizing body model limitations that are used in 3DoF to 6DoF conversion. Another path of research will be related to testing more mobile devices to check, how IMU sensor quality impacts final results and how to change algorithms parameters on the basis of observations.

Acknowledgment. The work was supported by the grant WZ/WIIIT/2/2020 from Bialystok University of Technology. Research results are based on the project "Development of new, innovative tools and interaction mechanisms in VRidge technology" financed by National Center of Research and Development.

References

1. Benzina, A., Toennis, M., Klinker, G., Ashry, M.: Phone-based motion control in VR: analysis of degrees of freedom. In: Proceeding of CHI EA 2011: CHI 2011 Extended Abstracts on Human Factors in Computing Systems, pp. 1519–1524. ACM (2011)
2. Campestrini, C., Heil, T., Kosch, S., Jossen, A.: A comparative study and review of different Kalman filters by applying an enhanced validation method. J. Energy Storage **8**, 142–159 (2016)
3. Sravan, M.S., Moolam, S., Sreepathi, S., Kokil, P.: Implementation of remote motion controller with visual feedback. In: 2017 8th International Conference on Computing, Communication and Networking Technologies (ICCCNT), pp. 1–7. IEEE (2017)
4. Hell, S., Argyriou, V.: Machine learning architectures to predict motion sickness using a virtual reality rollercoaster simulation tool. In: 2018 IEEE International Conference on Artificial Intelligence and Virtual Reality (AIVR), Taichung, Taiwan, pp. 153–156 (2018)
5. Jayaraj, L., Wood, J., Gibson, M.: Engineering a mobile VR experience with MEMS 9DOF motion controller. In: 2018 IEEE Games, Entertainment, Media Conference (GEM), pp. 1–9 (2018)
6. Kalman, R.E.: A new approach to linear filtering and prediction problems. Trans. ASME-J. Basic Eng. **82**, 35–45 (1960)

7. Kopczynski, M.: Optimizations for fast wireless image transfer using H.264 codec to Android mobile devices for virtual reality applications. In: Zamojski, W., Mazurkiewicz, J., Sugier, J., Walkowiak, T., Kacprzyk, J. (eds.) DepCoS-RELCOMEX 2021. AISC, vol. 1389, pp. 203–212. Springer, Cham (2021). https://doi.org/10.1007/978-3-030-76773-0_20
8. Santos, R.A., Rasteiro, M.A., Castro, H.F., Bento, L.B., Barata, M., Assunção, P.A.: Motion-based remote control device for enhanced interaction with 3D multimedia content. In: Proceedings of the Conference on Telecommunications - ConfTele, Aveiro, Portugal, pp. 1–4 (2015)

Performance Analysis and Comparison of Acceleration Methods in JavaScript Environments Based on Simplified Standard Hough Transform Algorithm

Damian Koper and Marek Woda(✉)

Faculty of Information and Communication Technology,
Wroclaw University of Science and Technology, Wrocław, Poland
marek.woda@pwr.edu.pl

Abstract. In this paper, we present an analysis of popular acceleration methods in JavaScript execution environments including Chrome, Firefox, Node, and Deno. We focus evenly on adopting the same codebase to take advantage of every method, benchmarking our solutions and caveats of building libraries compatible with multiple environments. To compare performance, we use a simplified standard Hough transform algorithm. As reference points of our benchmarks, we use a sequential version of the algorithm written in both JavaScript and C++. Our study shows that Chrome is the fastest JS environment in every benchmark and Firefox is the slowest in which we identified optimization problems. WebGL appears as the fastest acceleration method. Without parallel execution native C++ addon in Node is the most performant. This analysis will help to find the most efficient way to speed up execution making JavaScript a more robust environment for CPU-intensive computations.

Keywords: JavaScript · Acceleration · SHT · Standard Hough transform · Node · Browser · Deno · WebGL · Webpack · WASM · SIMD · Workers

1 Introduction

JavaScript (JS) has become one of the most widely used programming languages in the world. However, it has not been widely adopted to perform scientific computing and data analysis by a community and maintainers for a long time. This includes the development of libraries and features of the language itself. JavaScript is a dynamically typed scripting language, evaluated in the runtime and has only a single thread available to the developer in general. Released in 1995, it was executed only in an isolated sandbox environment of a web browser and was used mostly to add dynamic content to websites.

With the release of Node in 2009, we were able to run server-side code using the V8 engine from the Chromium project. At this point, there were at least two

W. Zamojski et al. (Eds.): DepCoS-RELCOMEX 2022, LNNS 484, pp. 131–142, 2022.
https://doi.org/10.1007/978-3-031-06746-4_13

ways of building libraries - one for the browser environment (AMD) and one for the CommonJS (CJS) format used by Node. With the release of ECMAScript 2015, better known as the ES6 standard, the modularity of JS code was standardized as ESModules. They are now supported by the vast majority of environments but with an ecosystem that big (1851301 npm packages as of 2022-01-19) it is still common to build and publish libraries in legacy formats.

The reason why CPU-intensive computing in JS is not a popular solution is complex. First of all, slow evolution, the predefined and narrow purpose of the language at first has slowed down the efforts to adapt the language to the task of heavy computation. Secondly, not unified module format and differences between environments in the implementation of corresponding features were not encouraging the community to develop multi-platform libraries. Finally, in the meantime when JS was evolving, other and more popular solutions like MatLab, Python, and R language were already around with their ecosystems.

In this paper, we check the state of the CPU-intensive computing potential of JS. We try to find how the same algorithm performs when implemented using multiple acceleration methods in popular JS environments with as few adjustments as possible for each method. We also want to find if all of these methods can be built into a single output format with the same library interface for maximum compatibility. We hope to find interesting facts and caveats about each analyzed environment and acceleration method.

2 Related Work

Analyzing the problem we have to look at it from different perspectives. The first one is the performance of the native JS, specifically the performance of JS engines. We analyze the performance of V8 and SpiderMonkey engines focusing on the first one. It was shown that simple optimizations, even changing one line of code, can increase performance due to the JIT optimizations in the runtime [7,12, 13,18]. Thread-level speculation could also be beneficial to the performance [11] but it is not implemented in common JS engines.

The second perspective is the performance of each environment. We analyze performance in web browsers - Chrome and Firefox with V8 and SpiderMonkey as JS engines respectively. We also analyze the same code on server-side environments - Node and Deno, both using V8 as JS engine. Node, written in C++, has proven to be great environment for I/O heavy tasks outperforming other environments [3,10]. Deno, written in Rust, is a Node's younger brother providing a modern ecosystem with built-in support for ESModules. It keeps up the performance with Node with insignificant differences [4].

The last aspect is the performance of acceleration methods in each environment. Analyzing CPU-heavy algorithm we want to adjust it and compare its performance against available acceleration methods in each environment. Multithreading is available in each environment in some form of "Worker". Parallel execution increases performance as expected [5] but this performance may differ between environments.

Another way of increasing performance, this time in server-side environments, is to use native plugins which were developed primarily to allow usage of existing codebase written in other languages. Node and Deno use plugins that can be written in C++ and Rust respectively. The same reason for development stands behind WebAssembly (WASM) [15]. It allows to compile and make any code portable across environments performing better than native JS. The study shows, that depending on the case we can get mixed performance results [9,19].

The last analyzed method is the usage of GPGPU (General-purpose computing on graphics processing units). It involves tricky usage of the WebGL's graphics pipeline not to generate graphics, but using its advantages, executing algorithms that can be massively parallelized [17]. The caveats of this method are slow data transfer between CPU and GPU, and lack of shared memory which is the characteristic of the pipeline. There were attempts of trying to standardize API specific to GPGPU (e.g. WebCL) and the most promising today is WebGPU [2].

It is important to mention the building ecosystem which provides high compatibility and the possibility of transforming assets. Some optimization or transpilation techniques may interfere with implementation specific to a particular acceleration method. We use Webpack as a module bundler with ESModule [1] as the desired output for each library. We built one library per acceleration method.

3 Benchmarking

We tested the performance of mentioned acceleration methods in different environments. Implementation status and reason if not implemented is shown in Table 1. Tested JS environments and their versions are described in Table 2.

As an algorithm to benchmark, we chose a simplified standard variant of Hough transform (SHT) [14]. Choosing a single algorithm over the whole benchmark suite gives us granular control over implementation, building process, and adaptation for each acceleration method. Hough transform, in the standard variant, is used to detect lines in binary images. It maps points to values in an accumulator space, called parameter space. Unlike the original [8], modern version maps points to curves (x, y) using polar coordinates (θ, ρ) according to (1) [6]. Peaks in the parameter space can be mapped back to line candidates.

$$f(x, y) = \rho(\theta) = x \cos\theta + y \sin\theta \tag{1}$$

The resolution of the accumulator determines the precision of line detection, the size of the computational problem, and the required memory. The computational complexity of the sequential algorithm equals $O(wh)$ where w and h are dimensions of an input image. It could be also expressed as $O(S_\theta S_\rho)$ with constant input dimensions where S_θ and S_ρ denotes angular and pixel sampling respectively. We benchmark each method for various problem sizes keeping everything constant but S_θ. We implemented a version simplified from commonly seen ones. Our implementation defines the anchor point of polar coordinates in the upper left corner of the image instead of its center. It also bases the voting process on a simple threshold instead of analyzing the image space [16].

We believe that the algorithm being a representation of CPU-intensive task is sufficient for performing valuable benchmarks across different environments since it requires many iterations to fill the accumulator and enforces intensified memory usage for input data and the accumulator. Relying heavily on sin and cos functions, also allows us to test their performance. Because of that, we implemented two variants of each algorithm - *non-LUT* and *LUT*. The first one uses standard sin and cos functions and the second one caches their results in a lookup table. As shown in Fig. 1, we also use the image after threshold operation instead of commonly used edge-detection. It requires more pixels to be processed thus increases problem size.

Table 1. Acceleration methods and environments.

	Chrome	Firefox	Node	Deno
Sequential	✓	✓	✓	✓
Native addon	✗[a]	✗[a]	✓	✗[b]
asm.js	✓	✓	✓	✗[b]
WASM	✓	✓	✓	✗[b]
WASM+SIMD	✓	✓	✓	✗[b]
Workers	✓	✓	✓	✓
WebGL	✓	✓	✗[b]	✗[a]
WebGPU	✗[c]	✗[c]	✗[a]	✗[c]

[a]Not available in environment
[b]Requires external package or non-C++ codebase
[c]Unstable or under a flag

Each benchmark lasts 5 s minimum and 30 s maximum or 50 runs. We want to rely on the most likely execution scenarios which are influenced by JS engine optimizations, resulting in a shorter execution time. To resolve this cold start problem we use the coefficient of variance metric (c_v). We start the actual benchmark after the window of 5 executions where $c_v \leq 1‰$.

Table 2. Versions of environments.

Env	Version	Engine version
Chrome	97.0.4692.71	V8 9.7.106.18
Firefox	96.0	SpiderMonkey 96.0
Node	16.13.2	V8 9.4.146.24-node.14
Deno	1.18.0	V8 9.8.177.6

(a) Input image (b) Accumulator (c) Detected lines

Fig. 1. Input image (419 × 421), accumulator and visualized result of sequential SHT algorithm in *non-LUT* variant ($S_\theta = 1, S_\rho = 1$).

Benchmarks were performed on a platform equipped with Intel® Core™ i7-12700KF CPU and Nvidia 970 GTX GPU using Ubuntu 20.04.1. The CPU had frequency scaling turned off and due to its hybrid architecture only 4 P-Cores were enabled for a benchmark using `taskset` utility.

4 Results and Details

In this section, we present execution times for each method depending on angular sampling S_θ which should result in linear computational complexity. Section 4.1 shows times for sequential execution which is then marked as a grey area for comparison. Every chart contains native C++ times.

4.1 Sequential

Analyzing benchmark times shown in Fig. 2 we can see the advantage of Chrome over Firefox being 1.61× faster in the *non-LUT* variant and 2.87× in the *LUT* one. It is worth noticing the optimization in Firefox for the $S_\theta = 5$ and subsequent sampling values in the *LUT* variant which could be an object of further research. The performance of server-side environments, Node and Deno, since they are very similar environments, has only insignificant differences, yet still being $(1.47, 2.06)\times$ slower than Chrome for both variants.

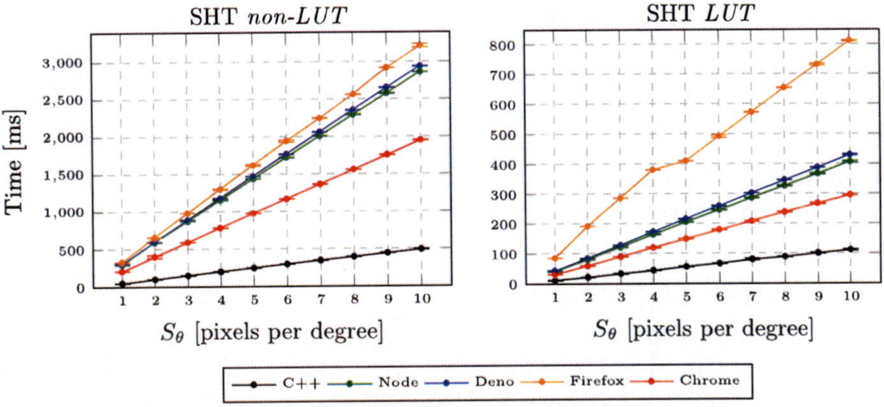

Fig. 2. Sequential SHT execution benchmark results.

We detected one-pixel difference in generated accumulator between variants as shown in Fig. 3a (upper right corner). We implemented lookup table for *LUT* variants using `Float32Array`. JS internally, without optimizations, represents numbers in double-precision and reduced precision of cached values can have a significant impact on detection results.

(a) Sequential LUT (b) WASM (c) WebGL

Fig. 3. Normalized absolute accumulator difference from sequential *non-*LUT variant. Note the two-pixel difference in the upper right corner for Sequential LUT.

4.2 Node C++ Addon

C++ addon for Node was built using the same shared library as the C++ version. Thus the difference in performance between native C++ and the addon arises mostly from handling data transfer between C++ – JS boundary since the data needs to be copied and transformed to corresponding C++ structures. Results are shown in Fig. 4.

Compilation with optimization of trigonometric functions in *non-*LUT variant allowed to gain more performance (2.24×) than compilation of the LUT variant (1.32×) relative to their sequential variants. This case allows us to draw a conclusion that if an algorithm has trigonometric functions and output which cannot be cached beforehand, the usage of the C++ addon in Node is beneficial.

4.3 WebAssembly and asm.js

Benchmark results for asm.js and WASM were shown on Figs. 5 and 6 respectively. In our case asm.js - a highly optimizable subset of JS instructions, operating only on numeric types and using heap memory - is actually slower in all environments than sequential execution. We suspect that it is caused by the building process. Webpack adds its own module resolution mechanisms that prevent part of the bundle with asm.js code from being recognized and compiled ahead-of-time. Performance flame chart from Chrome DevTools tools shows a lack of `Compile Code` blocks, unlike any other isolated asm.js sample.

WASM on the other hand improves performance for *non-*LUT variant and has no effect on LUT variant besides preventing optimization mentioned in Sect. 4.1 in Firefox. Again, it is beneficial to use this method if the output of trigonometric functions cannot be cached.

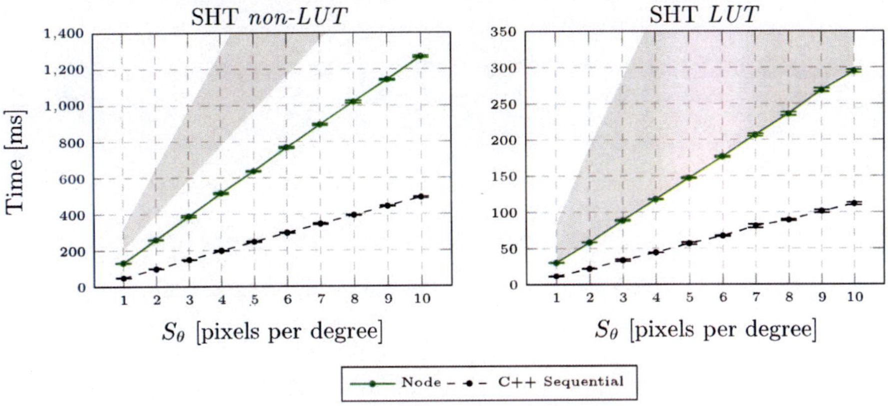

Fig. 4. Node C++ addon SHT execution benchmark results.

In our C++ implementation we use single precision floating point variables. This results in accumulator differences shown in Fig. 3b since WASM distinguishes between `f32` and `f64` types.

4.4 WebAssembly SIMD

SIMD instructions in WASM are available from Chrome 91 and Firefox 89 for all users. The usage of SIMD instructions can be done implicitly by letting the compiler (commonly LLVM) perform the auto-vectorization process or explicitly by using vector instructions in code. We tested both solutions resulting in no difference from sequential benchmarks for the first one. Because of that, we present only an explicit usage attempt. In benchmarks shown in Fig. 7 we can see that the performance difference between Chrome and Firefox decreased compared to sequential execution and Chrome is only 1.16× faster than Firefox. Moreover, Firefox overtakes Node in performance, which was not as prone to SIMD optimization as other environments.

4.5 Workers

All worker benchmarks used concurrency $n = 4$. Results are shown in Fig. 8. Because of the simplified implementation described in Sect. 3, precisely the center of the polar coordinate system in image space, the 3rd worker is redundant. The 3rd vertical quarter of the accumulator will always be empty as shown on example accumulator visualization in Fig. 1b. This was not optimized in our implementation.

The Table 3 shows the speedup and its efficiency for environments and variants. The big difference in speedup efficiency between variants again shows us

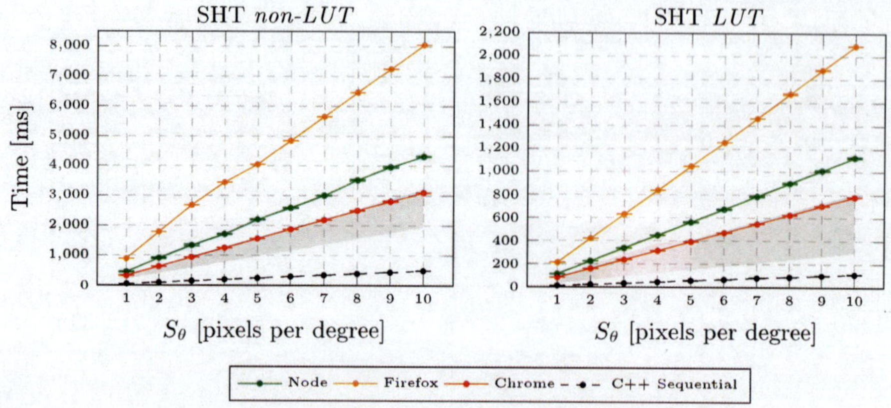

Fig. 5. asm.js SHT execution benchmark results.

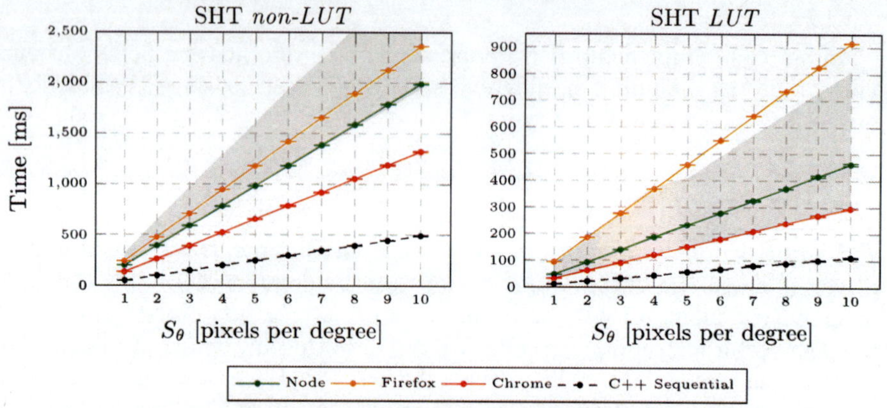

Fig. 6. WASM SHT execution benchmark results.

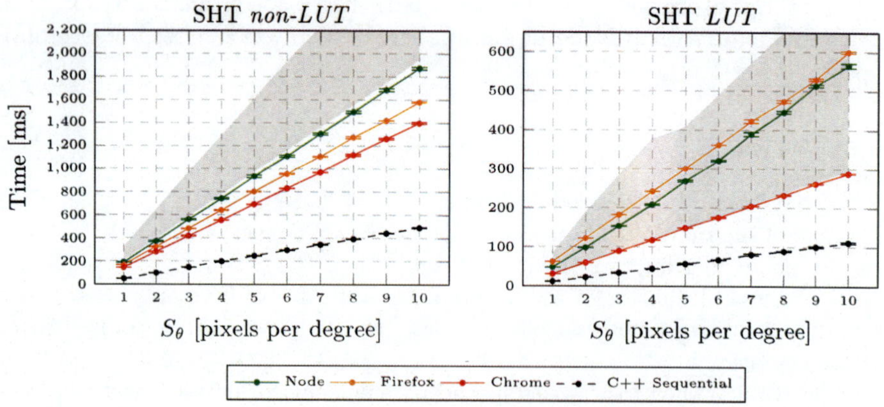

Fig. 7. WASM SIMD (explicit) SHT execution benchmark results.

how demanding calculations of trigonometric functions are. Only the accumulator filling process was parallelized thus the speedup difference between environments is expected since the worker calculations take less time due to the lookup tables.

Our implementation can be improved to achieve better performance because the voting process is not parallelized. Even though, the current state of implementation still allows us to compare this method across environments.

We share the accumulator array between workers and it is important to mention that our implementation does not use `Atomics` since every worker operates on a different part of the array. According to our benchmarks, usage of `Atomics` tends to slow down performance and was not necessary in this case.

Table 3. Speedup metrics for worker acceleration method ($S_\theta = 1, p = 4$).

Env.	Speedup	Efficiency
Chrome	2.99	0.75
Firefox	2.82	0.70
Node	3.18	0.80
Deno	3.22	0.81
Chrome *LUT*	1.81	0.45
Firefox *LUT*	1.87	0.47
Node *LUT*	1.64	0.41
Deno *LUT*	2.05	0.51

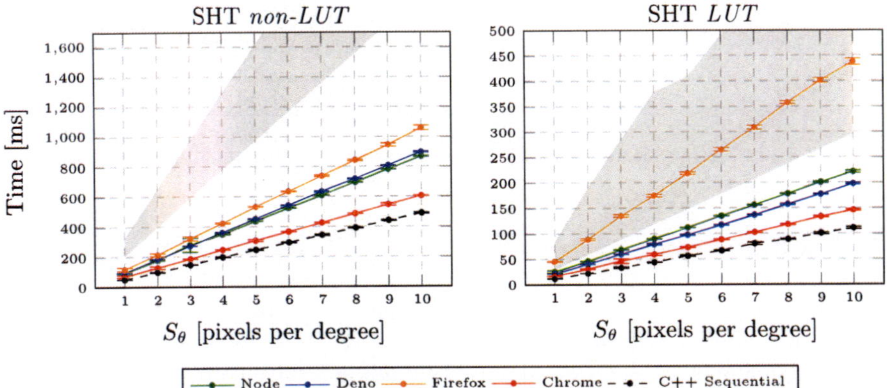

Fig. 8. Workers SHT execution benchmark results with concurrency $n = 4$.

4.6 WebGL

Our last acceleration method uses a GPGPU to fill the accumulator array. With help of the WebGL and the gpu.js library, we implemented kernel functions calculating every pixel separately. It is the only possible solution since the WebGL pipeline does not provide shared memory. This results in a bigger accumulator difference shown in Fig. 3c. First of all, the pipeline provides only single-precision operations. Secondly, for every accumulator value - pair (θ, ρ), we had to sum image pixels laying on a possible line. This operation is prone to rounding errors. Additionally, the minification of the output bundle provided by Webpack was interfering with the way the gpu.js library transpiles code to a GLSL language. We had to construct the function from string to prevent minification of the kernel function – `new Function('return function (testImage)...')()`.

This method tends to have the biggest result variance which comes directly from communication between CPU and GPU. It has also the biggest cold start times since the kernel has to be compiled by an environment on the first run. There is no big difference between both variants because in the *non-LUT* variant each thread on the GPU has to calculate *sin* and *cos* functions once which is not a significant overhead (Fig. 9).

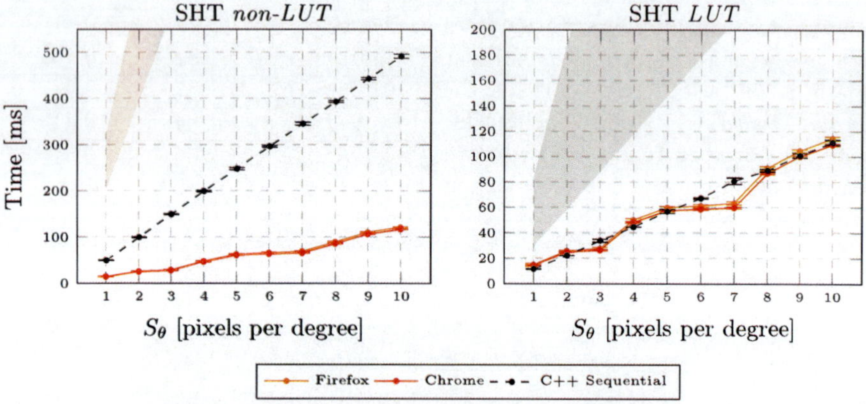

Fig. 9. WebGL SHT execution benchmark results.

5 Conclusions

We performed various benchmarks of the same algorithm in Chrome, Firefox, Node, and Deno environments listed in Table 2. In each one, we tested available and popular acceleration methods including a native addon, WASM alone, WASM with SIMD instructions, multi-threading with workers, and GPGPU using WebGL graphics pipeline. We did not test every method on every environment because of some being unavailable, unstable, under a flag, or based on non-C++ codebase as shown in Table 1. Summarized results for the same problem size are shown in Table 4. In every benchmark Chrome appears as the fastest environment with Firefox being 2.40×, Node 1.45×, and Deno 1.46× slower in general. As expected, in the case without involving parallel execution, the Node C++ native addon brings the best results across all environments. Server-side environments performed similarly with a slight predominance of Node over Deno.

According to our results, the performance of *LUT* variant was always better than its *non-LUT* counterpart. Trigonometric functions are demanding but our study shows that using native addon or compiling code to WASM can prevent

Table 4. Comparison of implemented methods in analyzed environments.

LUT	Method	Execution time [ms]			
		Chrome	Firefox	Node	Deno
	JS Sequential	200 (1.00)	324 (1.62)	296 (1.48)	303 (1.51)
	C++ addon	– (–)	– (–)	132 (–)	– (–)
	asm.js	317 (1.00)	887 (2.80)	451 (1.42)	– (–)
	WASM	131 (1.00)	238 (1.82)	199 (1.52)	– (–)
	WASM SIMD (impl.)	130 (1.00)	239 (1.84)	200 (1.54)	– (–)
	WASM SIMD (expl.)	141 (1.00)	164 (1.16)	188 (1.33)	– (–)
	Workers	67 (1.00)	115 (1.72)	93 (1.39)	94 (1.40)
	WebGL	13 (1.00)	14 (1.08)	– (–)	– (–)
✓	JS Sequential	29 (1.00)	84 (2.90)	41 (1.41)	43 (1.48)
✓	C++ addon	– (–)	– (–)	31 (–)	– (–)
✓	asm.js	79 (1.00)	210 (2.66)	114 (1.44)	– (–)
✓	WASM	30 (1.00)	93 (3.10)	47 (1.57)	– (–)
✓	WASM SIMD (impl.)	30 (1.00)	93 (3.10)	47 (1.57)	– (–)
✓	WASM SIMD (expl.)	30 (1.00)	61 (2.03)	48 (1.60)	– (–)
✓	Workers	16 (1.00)	45 (2.81)	25 (1.56)	21 (1.31)
✓	WebGL	14 (1.00)	15 (1.07)	– (–)	– (–)
	Geometric mean (non-LUT)	(1.00)	(1.64)	(1.45)	(1.46)
	Geometric mean (LUT)	(1.00)	(2.40)	(1.52)	(1.40)
	Geometric mean (all)	(1.00)	(1.99)	(1.48)	(1.43)

significant performance loss, especially in the Firefox environment. We can see that Firefox is not able to optimize code as well as other environments, where using vector instructions explicitly actually lowers performance, increasing it in Firefox. When using lookup tables results may be different. Looking at Chrome performance of all WASM methods with *LUT* variant we can see that performance is roughly the same with the sequential. Data exchanged between JS and WASM must be transformed and copied, which takes time, so it is not safe to assume performance benefit with intensive memory input and output when adopting WASM. We also identified a problem with asm.js. which wasn't compiled ahead-of-time. We suspect that the bundling system and minification process prevented environments from recognizing asm.js specific code.

To sum up, all environments serve similar cases but differ in terms of the performance of various acceleration methods. It is important to analyze which method best suits our needs depending on requirements. With all this, it is important to remember that no acceleration method can increase the performance of an algorithm like improvement of the computational complexity of the algorithm itself.

References

1. Paltoglou, K., Zafeiris, V.E., Diamantidis, N.A., Giakoumakis, E.A.: Automated refactoring of legacy JavaScript code to ES6 modules. J. Syst. Softw. **181**, 111049 (2021)
2. January 2022. https://gpuweb.github.io/gpuweb/
3. Chitra, L.P., Satapathy, R.: Performance comparison and evaluation of Node.js and traditional web server (IIS). In: 2017 International Conference on Algorithms, Methodology, Models and Applications in Emerging Technologies (ICAMMAET), pp. 1–4. IEEE (2017)
4. Choubey, M.: Deno vs node performance comparison: hello world, October 2021. https://medium.com/deno-the-complete-reference/deno-vs-node-performance-comparison-hello-world-774eda0b9c31
5. Djärv Karltorp, J., Skoglund, E.: Performance of multi-threaded web applications using web workers in client-side JavaScript (2020)
6. Duda, R.O., Hart, P.E.: Use of the Hough transformation to detect lines and curves in pictures. Commun. ACM **15**(1), 11–15 (1972)
7. Gong, L., Pradel, M., Sen, K.: JITProf: pinpointing JIT-unfriendly JavaScript code. In: Proceedings of the 2015 10th Joint Meeting on Foundations of Software Engineering, pp. 357–368 (2015)
8. Hough, P.V.: Method and means for recognizing complex patterns, US Patent 3,069,654, 18 December 1962
9. Jangda, A., Powers, B., Berger, E.D., Guha, A.: Not so fast: analyzing the performance of webassembly vs. native code. In: 2019 USENIX Annual Technical Conference (USENIX ATC 2019), pp. 107–120 (2019)
10. Lei, K., Ma, Y., Tan, Z.: Performance comparison and evaluation of web development technologies in PHP, Python, and Node.js. In: 2014 IEEE 17th International Conference on Computational Science and Engineering, pp. 661–668. IEEE (2014)
11. Martinsen, J.K., Grahn, H., Isberg, A.: Combining thread-level speculation and just-in-time compilation in Google's V8 JavaScript engine. Concurr. Comput. Pract. Exp. **29**(1), e3826 (2017)
12. Meurer, B.: An introduction to speculative optimization in V8, November 2017. https://ponyfoo.com/articles/an-introduction-to-speculative-optimization-in-v8
13. Meurer, B.: JavaScript performance pitfalls in V8, March 2019. https://ponyfoo.com/articles/javascript-performance-pitfalls-v8
14. Mukhopadhyay, P., Chaudhuri, B.B.: A survey of Hough transform. Pattern Recogn. **48**(3), 993–1010 (2015)
15. Nießen, T.: WebAssembly in Node.js. Ph.D. thesis, University of New Brunswick (2020)
16. Palmer, P.L., Kittler, J., Petrou, M.: An optimizing line finder using a Hough transform algorithm. Comput. Vis. Image Underst. **67**(1), 1–23 (1997)
17. Sapuan, F., Saw, M., Cheah, E.: General-purpose computation on GPUs in the browser using GPU.js. Comput. Sci. Eng. **20**(1), 33–42 (2018)
18. Selakovic, M., Pradel, M.: Performance issues and optimizations in JavaScript: an empirical study. In: Proceedings of the 38th International Conference on Software Engineering, pp. 61–72 (2016)
19. Yan, Y., Tu, T., Zhao, L., Zhou, Y., Wang, W.: Understanding the performance of webassembly applications. In: Proceedings of the 21st ACM Internet Measurement Conference, pp. 533–549 (2021)

Clustering Algorithms for Efficient Neighbourhood Identification in Session-Based Recommender Systems

Urszula Kużelewska(✉)

Faculty of Computer Science, Bialystok University of Technology,
Wiejska 45a, 15-351 Bialystok, Poland
u.kuzelewska@pb.edu.pl

Abstract. Recommender systems are applications to support users in searching relevant information or items on the Internet. Such examples are/news, articles, products in e-stores, books in libraries, music in broadcasting media. Recently, a new approach to recommendation generation emerged that concentrates on short-term users' activities organized in sessions. Such an approach is effective in real-world solutions due to a predominant number of one-time users and a limited time of items' availability. Session-based recommender systems are algorithms that focus particularly on users' ongoing sessions with the aim to predict their next actions. This article presents results of experiments with session-based recommenders which rely on an active user's neighbourhood identification. The algorithms were used in a standard form, where the neighbourhood was calculated with k-nearest neighbours and in a form of clusters generated in advance by clustering methods. The cluster-based approach was more efficient in terms of accuracy and diversity of recommendation lists.

Keywords: Session-based recommender systems · Clustering · Nearest neighbours

1 Introduction

Recommender systems (RSs) emerged as an answer to information overload. They are electronic, the most often internet, applications that help users to reach the data (e.g. news, products, items, media content), which is the most relevant for them. Collaborative filtering methods (CF) are the most popular type of RSs [4,9]. They are based on users' past behaviour data: navigating to the particular web sites, buying items, listening to the music, rating items. The data is used for similarity searching, with an assumption that users with corresponding interests prefer the same items. As a result, they predict the level of interest of those users on new, never seen items [4,21]. The collaborative filtering approach has been very successful due to its precise prediction ability [6,20].

W. Zamojski et al. (Eds.): DepCoS-RELCOMEX 2022, LNNS 484, pp. 143–152, 2022.
https://doi.org/10.1007/978-3-031-06746-4_14

The user-item matrix in the original CF approach is composed of the interactions that users performed over a long period of time (that is the users are required to register and use their accounts). In real-world applications, such an assumption might not be reasonable. First of all, one-time or anonymous visitors predominate in the internet applications (e.g. shops or media services). No long-term preference is registered for them. Next, many items (products to buy, songs or movies) are available for a limited-time. Furthermore, only a high level of their novelty might be important [11].

Session-based recommenders focus on a prediction of the next interaction in the given user's session, which can be anonymous. Most of them consider the time order of the items in the available data. Then, the information of the ongoing session and the sessions of other users collected in past, are analysed to generate recommendations.

The goal of this paper is to verify whether the neighbourhood identification based on clustering or multi-clustering is advantageous in session-based recommender systems. The following are the major contributions:

1. The neighbourhood of an active user might be modelled by clustering-based methods more precisely than k-nearest-neighbor kNN approach since during the clustering, all similarities (not only the nearest neighbours) among objects in the dataset are considered.
2. The session-based (item-based k-nearest-neighbor - kNN) methods' performance may be improved when the procedure of the neighbourhood identification in is replaced with clustering-based techniques.

The algorithms selected to use in the experiments described below were: SKNN (Session-based kNN) and V-SKNN (Vector Multiplication Session-Based kNN) as outperforming other, sometime complex, solutions: recurrent neural networks and factorization-based methods [15].

The article is organised as follows: the first section presents the background of session-based recommender systems including a description of the selected kNN methods as well as the clustering-based approach. The following section describes metrics of session-based techniques' performance evaluation. The next section contains the results of the performed experiments. The last section concludes the paper.

2 Background and Related Work

Session-based methods work on vectors (usually binary), which store users' sessions - that is, registered users' interactions with items. The data are derived from internet web sites or services, e.g. shops or media streaming ones and contains: items' views, purchasing, listening, coded into the session vectors. The aim of session-based systems is to predict the next activity of users. It is obtained using similarity values among an active user (a user to whom recommendations are generated) vector and the vectors from their neighbourhood. In the case of kNN approach, the neighbourhood area is determined by the most similar k

objects from the dataset [15,22]. Below, the methods used in the experiments as well as a baseline algorithm are described.

A baseline *kNN* method is IKNN (Item-based *kNN*) described in [7]. It examines only the last item in an active user's session and searches for the most similar to it in terms of their co-occurrence in other sessions. User vectors are encoded by binary values, that is if a particular item is present in the particular session, its value is 1. The similarity between two session vectors is calculated by employing the metrics, which are widely-used in *RS* domain, e.g. Pearson correlation or cosine similarity [2].

The SKNN algorithm [15], despite the most recent item, considers the entire session of an active user and analyses its similarity to the other sessions of full-length, as well. The search space is limited to k most similar vectors. Then the item that is present in many neighbourhood sessions, which are the most similar to the active user, is indicated as the predicted next activity.

The V-SKNN technique [15] considers not only the entire sessions but also the order (or partial sequences) of the items in them. Such an approach is sometimes separated as session-aware recommenders [16]. The characteristic of this method is vectors' encoding that prioritizes the more recent activity. It uses real values that depend on the position of the item in the session: the earlier action is performed, the lower value it has. Only the last element obtains the value 1. A special linear decay function is applied for this purpose [15].

Despite *kNN* approach to the session-based recommendation, many other, more sophisticated techniques were presented. Examples are Bayesian Personalized Ranking [17] or Factorized Personalized Markov Chains, proposed by [18]. Recently, the approaches based on Recurrent Neural Networks emerged, with their most important implementation GRU4REC [8].

2.1 Clustering-Based Methods Used in RS Domain

Clustering is a part of the Machine Learning domain. The aim of clustering methods is to organize data into separate groups without any external information about their membership, such as class labels. They analyse only the relationship among the data, therefore clustering belongs to Unsupervised Learning techniques [13].

Due to the independent *á priori* clusters identification, clustering algorithms are an efficient solution to the problem of RSs scalability, providing for the recommendation process a predefined neighbourhood [19]. The efficiency of clustering techniques is related to the fact that a cluster is a neighbourhood that is shared by all cluster members, in contrast to *kNN* approach determining neighbours for every object separately [2]. The disadvantage of this approach is usually the loss of prediction accuracy.

The multi-clustering approach includes a broad range of methods that are based on widely understood multiple runs of clustering algorithms or multiple applications of a clustering process on different input data [3]. In [14] $M-CCF$ method was examined, which, instead of one clustering scheme, works on a set of partitions, therefore it selects dynamically for an active user the most appropriate one that models the neighbourhood precisely. Thus, it solves two major problems related to the quality of clustering, which occurs in some cases.

3 Evaluation Metrics

The following metrics were used for the recommenders' performance evaluation. Three indices: Hit Rate (HR) and Mean Reciprocal Rank (MRR) and Normalized Discounted Cumulative Gain (NDCG) were calculated to measure recommendation accuracy [7]. They are standard and classification statistics. Additionally, two other factors, Coverage and Popularity, were used to assess diversity as well as resistance to the tendency of an algorithm to recommend mostly popular items [15].

- **HR** and **MRR:** The procedure of both HR and MRR calculation is based on the evaluation of the content of recommendation lists when adding incrementally the consecutive items to the test sessions. It was adopted from [7] and other works. Then, after propositions are generated, the list is cut off at the particular position and the content is examined in terms of the presence of the items from the test vector. MRR, additionally, calculates weights that are inversely related to the order of the predicted items in the test sessions.
- **NDCG:** The Normalized Discounted Cumulative Gain is a widely used metric in the field of information retrieval that considers additionally relevance score, that is the position of items from generated recommendation lists, that was predicted properly [12].

Besides the accuracy, the ability to generate recommendation lists of diversified content as well as handle a long-tail phenomenon is encouraged factors of RS techniques. They are measured by Coverage and Popularity metrics.

- **Coverage:** Coverage [1] reports the frequency of items to appear in the recommendation lists. Its high values denote different sets of propositions for different users, its low level relates to the tendency to recommend the same set to many users, even though they have different preferences.
- **Popularity:** Popularity index [7] measures the algorithm's bias to put only popular items in the recommendation lists. Its low values are advantageous, related to methods' ability to tackle the long-tail problem. This index is calculated by average popularity score for the top-k elements in the recommendation lists. This final score is an average of the individual popularity scores of each recommended item. It is estimated by counting the items' occurrences in one of the training sessions and by then applying min-max normalization to obtain a score between 0 and 1.

4 Experiments

This section presents the results of performance evaluation of the selected session-based algorithms. The experiments were executed on ACM RecSys 2015 Challenge dataset [26]. The data was constructed by the YOOCHOOSE company and contains a collection of sessions from a retailer. The sessions are based on the click events that the user performed on the website, as well as occasional buy events at the end of the sessions. There are 17793 sessions and 2932 items. The following recommender systems were taken into consideration: SKNN (Session-based kNN), V-SKNN (Vector Multiplication Session-Based kNN) since they are reference kNN methods. SKNN is a baseline neighbourhood-based algorithm confirmed as an accurate one in [10], whereas V-SKNN improves the performance of SKNN [15]. It underlines more recent items in a session in a similarity value and works on real-valued vectors. Both algorithms had the part of a neighbourhood calculation replaced with the clusters, obtained by clustering or multi-clustering methods. Finally, the following additional techniques were formed:

- SCCF-S-B, as development of SKNN, where the clustering method was Birch [23],
- SCCF-S-E, as development of SKNN, where the clustering method was EM [5],
- SCCF-V-B, as development of V-SKNN, where the clustering method was Birch,
- MCCF-V, as development of V-SKNN, where the clusters were formed by M-CCF - a multi-clustering implementation of Birch,

The code of both V-SKNN and SKNN as well as evaluation framework was taken from GitHub repository Session-Rec [24], whereas the both EM and Birch clustering methods come from the Scikit Learn package [25]. The multi-clustering $M - CCF$ is the author's implementation in the framework Session-Rec. The implementation of SKNN as well as V-SKNN is optimized over time efficiency by introducing s factor, that is a number of the most recent sessions for neighbours searching. The other coefficient is k, that is a number of items in the neighbourhood. The data for the experiments were prepared in the following way. It was applied a procedure of one single training-test split [15]. The training sessions were composed of users' activities of the entire period except the very last day, which was used for the testing sessions' formation.

The results presented in this section were generated using the following configuration of the methods. All recommenders were executed with different values of the nearest neighbours k (100, 200 and 500) as well as different values of the latest sessions s (500 and 5000). All SCCF algorithms worked on 3 clusters obtained from Birch or EM, however, Birch clustered the sessions coded by integer values, whereas EM clustered binary-coded sessions. The sessions coded by binary values were formed in the standard way described in Sect. 2. Integer values in the sessions were related to a number of item's occurrences in the particular session, without no upper limit in the value. Other values of the algorithms, as well as the parameters, were examined in the experiments, however, only selected the best ones are presented in this paper.

In the case of the M-CCF method, two Birch clustering schemes were used (with a number of clusters equal to 2 and 3), which were obtained on integer-coded sessions. M-CCF works as follows. Instead of one clustering scheme, it works on many partitionings generated by one or different clustering algorithms. During recommendations generation for an active user, the first step is to select for them the most appropriate cluster from all clusters in the set of partitionings. Then the recommendations generation phase is corresponding to one from the SCCF version (see [14] for more information about the M-CCF algorithm).

Table 1 contains HitRate values obtained during experiments. The following list's cut off thresholds (CT) were examined: 3, 5, 10, 15 and 20. Different algorithms' configuration (in terms of k and s) is reported by their name, in brackets. Note, that the range of values' variation is very slight with a small advantage of V-SKNN over SKNN, however, the combination of M-CCF with V-SKNN led to the best results in the case of $CT = 5, 10, 15$. In the other instances, V-SKNN with the neighbourhood based on kNN (for $k = 500$ and $s = 500$) generated better outcomes. It is worth mentioning that although the combination of SCCF and SKNN did not produce the best values, it outperforms the original SKNN version. However, the analogous relationship does not occur with the configuration V-SKNN and SCCF.

Table 1. Evaluation of recommender systems' performance - HitRate.

Algorithm	Metrics				
	HR@3	HR@5	HR@10	HR@15	HR@20
SKNN (k = 200, s = 500)	0.3848	0.4779	0.5768	0.6273	0.6563
SKNN (k = 500, s = 500)	0.3833	0.4761	0.5766	0.6264	0.6575
SKNN (k = 500, s = 5000)	0.3831	0.4757	0.5767	0.6263	0.6576
V-SKNN (k = 500, s = 500)	**0.4311**	0.5211	0.6188	**0.6628**	**0.6883**
V-SKNN (k = 500, s = 5000)	0.4294	0.5199	0.6183	0.6616	0.6861
SCCF-S-B (k = 100, s = 500)	0.3837	0.4772	0.5738	0.6209	0.6439
SCCF-S-B (k = 200, s = 500)	0.3844	0.4778	0.5766	0.6278	0.6554
SCCF-S-B (k = 500, s = 500)	0.3841	0.4768	0.5780	0.6272	0.6583
SCCF-S-E (k = 100, s = 500)	0.3824	0.4755	0.5723	0.6177	0.6428
SCCF-S-E (k = 200, s = 500)	0.3849	0.4783	0.5764	0.6271	0.6544
SCCF-S-E (k = 500, s = 500)	0.3841	0.4773	0.5754	0.6257	0.6553
SCCF-V-B (k = 500, s = 5000)	0.4308	**0.5213**	0.6178	0.6621	0.6877
MCCF-V (k = 500, s = 5000)	0.4303	**0.5213**	**0.6191**	**0.6628**	0.6877

The results of another metric - MRR - are shown in Table 2. The list's cut off thresholds were the same: 3, 5, 10, 15 and 20. In this case, the methods combined with the clustering-based technique was leading. The best results generated SCCF-V-B, however, both SCCF-S-B and SCCF-S-E outperformed the original SKNN.

NCDG values are presented in Table 3. Still, the list's cut off thresholds were the same. The original V-SKNN opens the ranking of the methods in this case. It is followed by SCCF-V-B and MCCF-V, however, the difference is not significant.

In the case of remaining metrics (see Table 4), Coverage and Popularity, both combinations of V-SKNN with clustering-based neighbourhood resulted

Table 2. Evaluation of recommender systems' performance - MRR.

Algorithm	Metrics				
	MRR@3	MRR@5	MRR@10	MRR@15	MRR@20
SKNN (k = 200, s = 500)	0.2817	0.3027	0.3161	0.3202	0.3218
SKNN (k = 500, s = 500)	0.2807	0.3018	0.3154	0.3194	0.3212
SKNN (k = 500, s = 5000)	0.2805	0.3016	0.3153	0.3192	0.3210
V-SKNN (k = 500, s = 500)	0.3281	0.3487	0.3619	0.3654	**0.3669**
V-SKNN (k = 500, s = 5000)	0.3270	0.3478	0.3611	0.3645	0.3659
SCCF-S-B (k = 100, s = 500)	0.2826	0.3038	0.3167	0.3205	0.3218
SCCF-S-B (k = 200, s = 500)	0.2821	0.3033	0.3167	0.3207	0.3222
SCCF-S-B (k = 500, s = 500)	0.2817	0.3028	0.3165	0.3204	0.3223
SCCF-S-E (k = 100, s = 500)	0.2810	0.3022	0.3152	0.3188	0.3202
SCCF-S-E (k = 200, s = 500)	0.2820	0.3032	0.3164	0.3205	0.3220
SCCF-S-E (k = 500, s = 500)	0.2820	0.3032	0.3164	0.3204	0.3221
SCCF-V-B (k = 500, s = 5000)	**0.3282**	**0.3489**	**0.3620**	**0.3655**	**0.3669**
MCCF-V (k = 500, s = 5000)	0.3276	0.3485	0.3616	0.3651	0.3665

Table 3. Evaluation of recommender systems' performance - NDCG.

Algorithm	Metrics				
	NDCG@3	NDCG@5	NDCG@10	NDCG@15	NDCG@20
SKNN (k = 200, s = 500)	0.3596	0.4031	0.4374	0.4513	0.4582
SKNN (k = 500, s = 500)	0.3588	0.4022	0.4372	0.4508	0.4583
SKNN (k = 500, s = 5000)	0.3585	0.4019	0.4369	0.4505	0.4580
V-SKNN (k = 500, s = 500)	**0.4059**	0.4482	**0.4821**	**0.4941**	**0.5002**
V-SKNN (k = 500, s = 5000)	0.4042	0.4469	0.4809	0.4928	0.4987
SCCF-S-B (k = 100, s = 500)	0.3592	0.4031	0.4363	0.4492	0.4548
SCCF-S-B (k = 200, s = 500)	0.3599	0.4035	0.4378	0.4518	0.4585
SCCF-S-B (k = 500, s = 500)	0.3596	0.4031	0.4382	0.4516	0.4591
SCCF-S-E (k = 100, s = 500)	0.3579	0.4016	0.4350	0.4474	0.4535
SCCF-S-E (k = 200, s = 500)	0.3596	0.4033	0.4372	0.4511	0.4577
SCCF-S-E (k = 500, s = 500)	0.3593	0.4030	0.4370	0.4507	0.4578
SCCF-V-B (k = 500, s = 5000)	0.4057	**0.4483**	0.4817	0.4939	0.5001
MCCF-V (k = 500, s = 5000)	0.4051	0.4480	0.4818	0.4938	0.4978

in outstanding performance outperformed the original V-SKNN in both cases. Note, that V-SKNN is particularly efficient at generating recommendation lists that are complete and has the lowest tendency to recommend popular items. Furthermore, when it is equipped with a clustering-based neighbourhood the efficacy increases.

Table 4. Evaluation of recommender systems' performance - Coverage and Popularity.

Algorithm	Metrics	
	Coverage@20	Popularity@20
SKNN (k = 200, s = 500)	0.9062	0.2002
SKNN (k = 500, s = 500)	0.9011	0.2115
SKNN (k = 500, s = 5000)	0.9011	0.2113
V-SKNN (k = 500, s = 500)	0.9423	0.1530
V-SKNN (k = 500, s = 5000)	0.9441	0.1533
SCCF-S-B (k = 100, s = 500)	0.9199	0.1810
SCCF-S-B (k = 200, s = 500)	0.9083	0.1977
SCCF-S-B (k = 500, s = 500)	0.9035	0.2069
SCCF-S-E (k = 100, s = 500)	0.9134	0.1751
SCCF-S-E (k = 200, s = 500)	0.9032	0.1912
SCCF-S-E (k = 500, s = 500)	0.8998	0.2001
SCCF-V-B (k = 500, s = 5000)	**0.9533**	**0.1523**
MCCF-V (k = 500, s = 5000)	0.9529	0.1533

Slight differences in values were observed in the results presented in this section. Moreover, the results' relationship does not correspond to different metrics. However, the algorithms with the nearest neighbours identification procedure replaced by clustering schemes outperformed their original variants. It is also noted in the case of the SKNN algorithm, which generates less accurate outcomes than V-SKNN, however, the introduction of clusters as the neighbourhood of users was advantageous, as well.

5 Conclusions

Recent research directions in the domain of recommender systems led towards session-based algorithms, which are more related to real users' behaviour. The implementations based on the nearest neighbours have validated their efficacy to generate accurate, complete and unbiased recommendations. This article presents an improvement to two kNN-based methods, in which the neighbourhood identification procedure includes clustering schemes. The partitionings were generated by 2 clustering algorithms: Birch and EM with the number of clusters set to 3 in all cases. The multi-clustering method, M-CCF, was examined, as well. The clustering- and multi-clustering-based neighbourhoods improved recommendations performance in terms of the following metrics: HitRate, MRR, Coverage and Popularity. However, future researches are focused on the development of methods based on clustering as well, but with the purpose to concentrate on Coverage and Popularity metrics, since these features were benefited the most.

Acknowledgments. The work was supported by the grant from Bialystok University of Technology WZ/WI-IIT/2/2020 and funded with resources for research by the Ministry of Education and Science in Poland.

I would also like to thank Prof. Dietmar Jannach for a consultation and his support and Sara Latifi for evaluation framework advice.

References

1. Adomavicius, G., Kwon, Y.O.: Improving aggregate recommendation diversity using ranking-based techniques. IEEE Trans. Knowl. Data Eng. **24**(5), 896–911 (2012)
2. Aggrawal, C.C.: Recommender Systems. The Textbook. Springer, Cham (2016). https://doi.org/10.1007/978-3-319-29659-3
3. Bailey, J.: Alternative Clustering Analysis: A Review. Intelligent Decision Technologies: Data Clustering: Algorithms and Applications, pp. 533–548. Chapman and Hall/CRC (2014)
4. Bobadilla, J., Ortega, F., Hernando, A., Gutiérrez, A.: Recommender systems survey. Knowl.-Based Syst. **46**, 109–132 (2013)
5. Dempster, A.P., Laird, N.M., Rubin, D.B.: Maximum likelihood from incomplete data via the EM algorithm. J. Roy. Stat. Soc. **39**, 1–38 (1977)
6. Gorgoglione, M., Pannielloa, U., Tuzhilin, A.: Recommendation strategies in personalization applications. Inf. Manag. **56**(6), 103143 (2019)
7. Hidasi, B., Karatzoglou, A., Baltrunas, L., Tikk, D.: Session-based recommendations with recurrent neural networks. In: Proceedings of ICLR 2016 (2016)
8. Hidasi, B., Karatzoglou, A.: Recurrent neural networks with top-k gains for session-based recommendations. CoRR abs/1706.03847 (2017). arXiv:1706.03847
9. Jannach, D.: Recommender Systems: An Introduction. Cambridge University Press, Cambridge (2010)
10. Jannach, D., Malte, L.: When recurrent neural networks meet the neighborhood for session-based recommendation. In: Proceedings of the Eleventh ACM Conference on Recommender Systems (RecSys 2017), pp. 306–310 (2017)
11. Jannach, D., Mobasher, B., Berkovsky, S.: Research directions in session-based and sequential recommendation. User Model. User-Adap. Interact. **30**, 609–616 (2020). https://doi.org/10.1007/s11257-020-09274-4
12. Jarvelin, K., Kekalainen, J.: Cumulated gain-based evaluation of IR techniques. ACM Trans. Inf. Syst. (TOIS) **20**(4), 422–446 (2002)
13. Kaufman, L.: Finding Groups in Data: An Introduction to Cluster Analysis. Wiley, Hoboken (2009)
14. Kużelewska, U.: Dynamic neighbourhood identification based on multi-clustering in collaborative filtering recommender systems. In: Zamojski, W., Mazurkiewicz, J., Sugier, J., Walkowiak, T., Kacprzyk, J. (eds.) DepCoS-RELCOMEX 2020. AISC, vol. 1173, pp. 410–419. Springer, Cham (2020). https://doi.org/10.1007/978-3-030-48256-5_40
15. Ludewig, M., Jannach, D.: Evaluation of session-based recommendation algorithms. User Model. User-Adap. Interact. **28**(4–5), 331–390 (2018). https://doi.org/10.1007/s11257-018-9209-6
16. Quadrana, M., Cremonesi, P., Jannach, D.: Sequence-aware recommender systems. ACM Comput. Surv. **51**, 1–36 (2018)

17. Rendle, S., Freudenthaler, C., Gantner, Z., Schmidt-Thieme, L.: BPR: Bayesian personalized ranking from implicit feedback. In: Proceedings of the 25th Conference on Uncertainty in Artificial Intelligence, pp. 452–461 (2009)
18. Rendle, S., Freudenthaler, C., Schmidt-Thieme, L.: Factorizing personalized Markov chains for next-basket recommendation. In: Proceedings of the 19th International Conference on World Wide Web, pp. 811–820 (2010)
19. Sarwar, B.: Recommender systems for large-scale E-commerce: scalable neighborhood formation using clustering. In: Proceedings of the 5th International Conference on Computer and Information Technology (2002)
20. Schafer, J.B., Frankowski, D., Herlocker, J., Sen, S.: Collaborative filtering recommender systems. In: Brusilovsky, P., Kobsa, A., Nejdl, W. (eds.) The Adaptive Web. LNCS, vol. 4321, pp. 291–324. Springer, Heidelberg (2007). https://doi.org/10.1007/978-3-540-72079-9_9
21. Singh, M.: Scalability and sparsity issues in recommender datasets: a survey. Knowl. Inf. Syst. **62**, 1–43 (2018). https://doi.org/10.1007/s10115-018-1254-2
22. Verstrepen, K., Goethals, G.: Unifying nearest neighbors collaborative filtering. In: Proceedings of RecSys 2014, pp. 177–184 (2014)
23. Zhang, T., Ramakrishnan, R., Livny, M.: BIRCH: a new data clustering algorithm and its applications. Data Min. Knowl. Discov. **1**(2), 141–182 (1997). https://doi.org/10.1023/A:1009783824328
24. Implementation of session-based framework. https://github.com/rn5l/session-rec/. Accessed 10 Dec 2021
25. Scikit Learn Clustering Algorithms. https://scikit-learn.org/stable/modules/model_evaluation.html. Accessed 15 Dec 2021
26. RecSys Challenge Dataset. https://www.kaggle.com/chadgostopp/recsys-challenge-2015. Accessed 12 Dec 2021

Reliability- and Availability-Aware Mapping of Service Function Chains in Softwarized 5G Networks

Jerzy Martyna(✉)

Institute of Computer Science, Faculty of Mathematics and Computer Science, Jagiellonian University, ul. Prof. S. Lojasiewicza 6, 30-348 Cracow, Poland
jerzy.martyna@uj.edu.pl

Abstract. For improving the efficiency and flexibility of 5G networks, a technique called network function virtualization (NFV) is commonly used. The service function chains (SFCs) are used in such networks to represent various network services. Mapping them can reduce hardware costs and increase the reliability of the entire network. In this paper, reliable SFC placement in virtualized and softwarized platform is formulated. To solve this problem, a new algorithm for reliability- and availability-aware SFC mapping is presented. The physical resources are then used with greater reliability than usual. The simulation results show that the proposed method can assign virtual network functions effectively and efficiency when the number of targets is large.

Keywords: Virtual network function · Service function chain · Resource allocation · 5G wireless networks

1 Introduction

Softwarization of 5G networks is essential to deliver advanced services such as enhanced mobile broadband or the very high transmission reliability required inter alia in industrial networks. This is a challenge faced by the designers of these networks, as confirmed in the requirements document [1]. Useful technologies are Software-Defined Networking (SDN) and service function chains (SFCs), which allow the traditional hardware limitations of 5G networks to be overcome [2]. Moreover, to achieve network softwarization and virtualization, new design and implementation is needed in different wireless network segments such as transport networks, core networks, mobile-edge networks, and radio network clouds. In other words, softwarization is an essential tool in the design and management of all network functions and services in 5G mobile networks.

The SFC concept was introduced in [3] and [4]. It assumes that the functions providing a service (e.g., firewalls, traffic monitors, load balancer, etc.) that are arranged in a linear chain are inserted. Then, end-to-end services made available in a virtualized and layered network structure determine the processing properties of the entire system. As the requirements for 5G networks increase, including

W. Zamojski et al. (Eds.): DepCoS-RELCOMEX 2022, LNNS 484, pp. 153–162, 2022.
https://doi.org/10.1007/978-3-031-06746-4_15

the creation of radio clouds or enabling mobility, SFC placement becomes a big challenge for designers of these networks. In general, efficient SFC placement into 5G mobile networks can minimize the costs of computing resources and link resources.

Finding the optimal solution for SFC placement is an NP-hard problem. Obtaining an accurate solution requires a large amount of computational time, hence it is an impractical method for practical applications. A number of approaches have been used to find a solution to this problem. The most important of these are heuristic methods. Among others, a heuristic algorithm to coordinate the composition of SFCs and their embedding into the substrate network was presented by Beck *at al.* [5]. A genetic-based algorithm as well as a bee colony-based algorithm to solve the SFC placement and routing problem to find a pareto optimal solution considering heterogenous physical nodes was proposed by Khoshkholghi *et al.* [6]. The optimization model was investigated by the authors using a pareto optimal solution so that it optimizes multiple objectives as much as possible.

The inclusion of reliability of SFC mapping into softwarized networks has also resulted in a significant number of papers. Among others, heuristic algorithms and resource allocation with the adaptive mapping method, which jointly optimizes the processes of designing an SFC and allocating resources requested by the chain was presented by Jalalitabar *et al.* [7]. Reliability-aware SFCs in carrier-grade softwarized networks ere given by Qu *et al.* [8]. This paper addressed the problem of reliable provisioning of service function chains with resource sharing. The dynamic reliability-aware SFC placement as an infinite horizon Markov decision process (MDP) was formulated by Karimzadeh [9]. In this approach, to minimize the placement cost and maximize the number of admitted services the number of active services in the network was considered to be the state of the system.

To efficiently utilize the available resources and enhance the reliability of service chains without assigning backups, an Integer Linear Programming (ILP) method was proposed by Kaliyammal Thiruvasagam *et al.* [10]. In this paper, the authors additionally presented a modified stable matching algorithm to provide a near-optimal solution in polynomial time.

The main goal of this paper is to present a new method that allows for simultaneous mapping of the reliability and availability of SFCs in softwarized 5G networks. This method makes it possible to find a near-optimal solution to this problem. A simulation of the performance of the proposed solution was carried out, which confirmed the validity of this method. Compared to other methods, it is a better solution and a more effective solution to this problem.

The rest of the paper is as follows. Section 2 presents a system model and problem formulation. Section 3 defines the problem of reliability- and availability-aware SFC mapping into a physical network. Section 4 introduces an approximate algorithm for efficient, reliable, and available SFC design placement. The performance of the proposed algorithm is provided in Sect. 5. The paper is concluded in Sect. 6.

2 Network Model and Problem Formulation

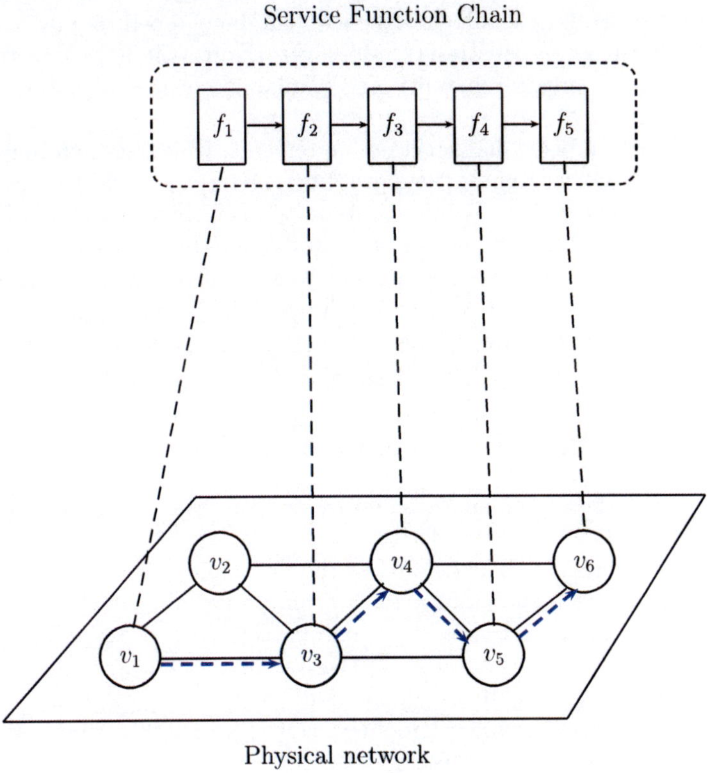

Fig. 1. An example of a service request

In this section, a network model of a 5G architectural framework is presented. Additionally, the problem formulation for the proposed study is provided.

The physical network, also called substrate network, is defined as an undirected physical graph $G = (V, E)$, where V is a set of N edge (cloud) physical nodes expressed as $V = \{v_1, ..., v_N\}$, and E is a set of L physical links between nodes denoted by $E = \{e_1, ..., e_L\}$. Every physical node $v, v \in V$, is to be characterized by processing power, memory size, etc. Let the set of service chains be given by S. A single service chain consists of a list of VNFs and is represented by a directed graph $SF = (F, M)$, where F is the set of virtual functions and M is the set of virtual links, respectively. The set of virtual functions consists of requests for processing power, memory size, etc. In turn, each virtual link $m, m \in M$, consists of bandwidth requests, data rates, etc.

Thus, an ordered set of VNFs allows an SFC to be constructed that is connected by VNF links. Figure 1 shows an example of the constructed SFC, which consists of five VNFs chained by VNF links.

The mapping rules are as follows. Individual requests from service chains s must be allocated one or more VNFs f, provided that the following requirements are met. Each VNF f can be placed on only one physical node v, which meets the defined constraints. Analogously, each l can be mapped to only one physical path that satisfies the constraints. Moreover, due to the approach taken here, requests such as certain reliability and availability values must be taken into account.

Each type of industry has service requirements, which have an impact on the applications they use. Hence, there are different service requests, each of which may have different specific requirements in terms of high data rate, bandwidth demand, allowable delay, reliability, and availability. Each service request s must be mapped onto a VNF to satisfy all its requirements. Moreover, it should also be considered whether VNF redundancy will be applied, which significantly affects the values of the mapped SFCs. Additionally, the virtual links in the SFC are mapped to the physical path of the NFV.

3 Network Availability and Reliability

In this section, the end-to-end availability and reliability models of VFNs with potential redundancy are considered.

The network reliability is defined as the probability that a system performs its intended functions successfully for a given period of time [11]. Otherwise, if the network operates successfully at time t_0, the network reliability yields the probability that in the interval 0 to t_0 there were no failures. It is assumed that the probability of successful communication between a source and a target node is k-terminal reliability, which is defined as the probability that a path exists and connects k nodes in a network.

The k-terminal reliability for the k nodes $\{n_1, \ldots, n_k\} \subset V(G,p)$ for the graph (G,p), where p is the link failure probability, can be expressed as [12]:

$$R_C^{n_1,\ldots,n_k}(G,p) = \sum_{i=w_{n_1},\ldots,n_k}^{\omega} T_i^{n_1,\ldots,n_k}(G)p^{\omega-i}(1-p)^i \tag{1}$$

where $\omega =\mid E(G) \mid$ is the size of the graph, $T_i^{n_1,\ldots,n_k}(G)$ denotes the number of subgraphs connecting the nodes $n_1, \ldots, n_k$ with i-th edges. and $w_{n_1}, \ldots, w_{n_k}$ is the size of the minimum tieset connecting the nodes $n_1, \ldots, n_k$.

Furthermore, the k-terminal reliability can be also expressed as follows:

$$R_C^{n_1,\ldots,n_k}(G,p) = 1 - \sum_{i=\beta(G)}^{\omega} C_i^{n_1,\ldots,n_k}(G)p^i(1-p)^{\omega-1} \tag{2}$$

where $C_i^{n_1,\ldots,n_k}(G)$ denotes the number of edge cutsets of cardinality i and $\beta(G)$ is the cohesion.

In turn, the network availability is defined as the probability that any instant time t the network is up and available, i.e. no disconnections. Otherwise, it is the portion of the time the network is operational.

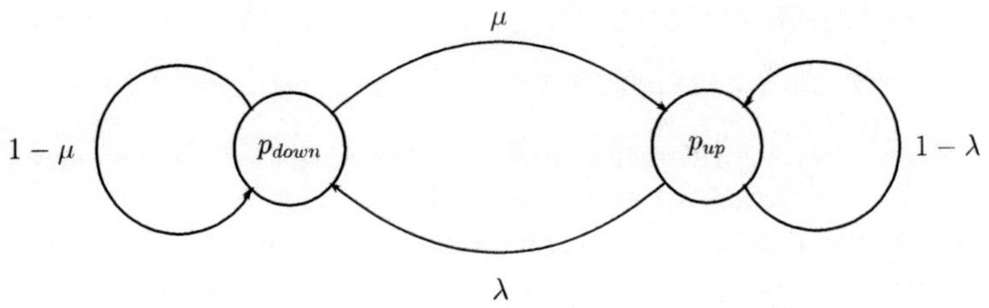

Fig. 2. A two state Markov model of the reliability of a link.

Given network in steady state ($t \to \infty$), the steady-state availability can be expressed as:

$$A = \frac{MTBF}{MTBF + MTTR} \tag{3}$$

where $MTBF$ is *Mean Time Between Failures* and $MTTR$ is *Mean Time To Repair*, respectively.

In the case where a link in the network can be repaired after failure, it can be modeled using a two-stage Markov diagram. In this model, one state represents the repaired link and the other state represents the broken link. Let the operational link failure e_{ij} be an exponential distribution with a rate parameter μ - see Fig. 2.

For the link being up (p_{up}) or down (p_{down}) the state probabilities at the steady-state are as follows:

$$p_{up} = \frac{\mu}{\mu + \lambda} \tag{4}$$

$$p_{down} = \frac{\lambda}{\mu + \lambda} \tag{5}$$

It is assumed that the network can be represented as a random graph G, in which there can be $k - 1$ different nodes, namely $d_i \in D$, where $D = \{d_1, \ldots, d_{k-1}\}$. It is also assumed that there is one node r which is the root of the tree. The network is fully operational if there is an operational path between the root node r and each of the other nodes in the network. Then availability can be expressed for k nodes as k-terminal availability, namely:

$$A_C^{r,d_1,\ldots,d_{k-1}}(G, p_{down}) = 1 - \sum_{i=\beta(G)}^{\omega} C_i^{r,d_1,\ldots,d_{k-1}} p_{down}^i (1 - p_{down})^{\omega - i} \tag{6}$$

where $C_i^{r,d_1,\ldots,d_{k-1}}(G, p_{down})$ denotes the number of edge cutsets of cardinality i.

4 Algorithm for Reliability- and Availability-Aware SFC Mapping

In this section, the algorithm for finding paths with the desired reliability and availability will be presented and explained.

The algorithm designed must provide, regardless of the physical topology of the 5G network, all elements of this network, the provision of which guarantees the fulfillment of all SFCs of requests. It also takes into account the redundancy possibilities of individual components of this network, including links and nodes. Moreover, in the case of an SFC request concerning, for example, the most demanding data flows with time constraints, the time constraints and the required flow rates have been taken into account. The algorithm is designed so that each of the constraints can be applied separately or all treated together.

The pseudo-code for the aforementioned algorithm is given in Fig. 3. Its operation is as follows. For a given 5G physical topology, there is a number of shortest paths between the given tree roots and leaves. These make available n in N paths. Then, given the time constraints, some of them satisfy the reliability and availability requirements, or both. For this purpose, values for the reliability of the k-terminal and the availability of the k-th terminal are used. In the event

Algorithm 1 Algorithm for reliable and available SFC mapping

```
procedure RESOURCE ALLOCATION
Require: R_con, A_con, delay_con;
Initialisation:
    [buffer] ← ∅;
    for ∀f ∈ VNF do
        sort the_paths(f);
        put buffer[sorted_paths];
    end for
    for ∀s ∈ S do
        for ∀f ∈ VNF do
            compute A_f, R_f for k-terminal
            while (A_f > A_con and R_f > R_con) do
                compute end_to_end_delay_f
                if end_to_end_delay_f < delay_con then
                    map_s_in_buffer[sorted_paths];
                else
                    find_the_redundant_path_s'
                end if
                map_s_and_s'_in_buffer[sorted_paths];
            end while
        end for
    end for
end procedure
```

Fig. 3. Pseudo-code of algorithm for reliable and available SFC mapping.

that such paths are missing, the possibility of redundancy of individual elements is checked. Thanks to the redundancy of these elements, it is possible to find paths with the desired parameters. The paths found in this way are assigned to the given SFC requests.

5 Performance Evaluation

In this section, the performance evaluation of the proposed solution presented in the previous sections is provided.

The physical network is here represented by a 5G network with a two-tier configuration with three nodes. Each node of the network has a single data center with a capacity equal to 2000 units. Two traffics are assumed in the model: URLLC and eMBB. The URLLC traffic demands extremely reliable and low latency radio transmission, i.e., one way radio transmission with a latency of 1 ms [13]. The eMBB traffic provides massive connectivity solutions for various Internet of Things. Each of the data centers can provide six to eight virtual network functions. The availability of each VNF is randomly distributed within [0.9, 0.98] and the reliability with a packet error rate (PER) of 10^{-3}. [14].

For each of the service chain requests in the network only 4 VNFs have been accepted. Each of them requires two resources (power capacity, memory size) and each virtual link has a bandwidth requirement equal to 200, 300, 400 Gb/s with equal probability. The value of the processing delay of the VNF is equal here to 60–120 μs [15].

Using a written simulation program, the proposed algorithm was assessed for performance evaluation. The first scenario examined the reliability of the requirement for the provided request mapping algorithm from service chains. The availability for both data traffics was then examined with the adopted mapping algorithm and its absence.

Figure 4 shows the reliability requirement of the service versus the admission ratio. The admission ratio ρ used here is defined as the number of accepted request services with the required reliability to the number of incoming services. The graph indicates that the used algorithm can significantly improve the reliability with the increase of the admission ratio.

Figure 5 shows the availability versus the probability of virtual link failure for a varying number of redundant paths. It is evident that increasing the number of redundant paths will increase availability. This improvement is particularly noticeable with relatively low probabilities of link failures.

Figure 6 shows the availability versus average number of backup VNFs for both data traffics: eMBB and URLLC, respectively. It can be seen in the figure that the eMBB traffic requires more back-up VNFs than URLLC traffic for the same admission ratio $\rho = 0.95$.

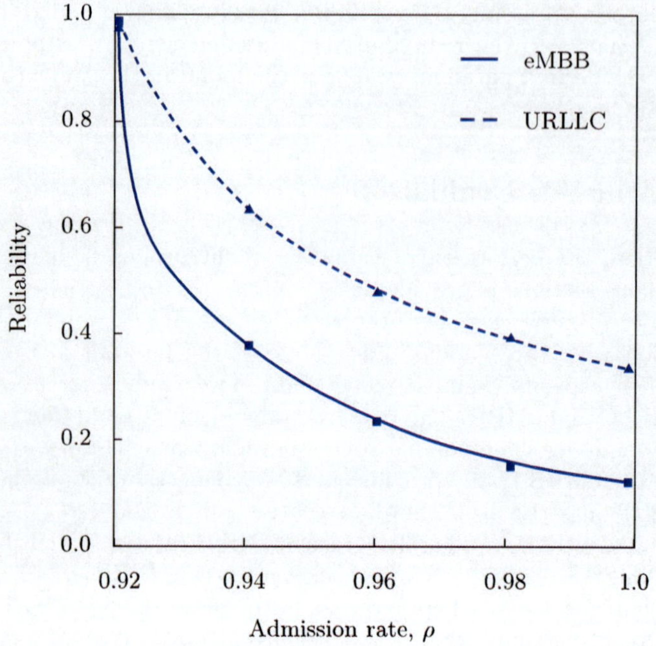

Fig. 4. The reliability requirement of service versus the admission ratio.

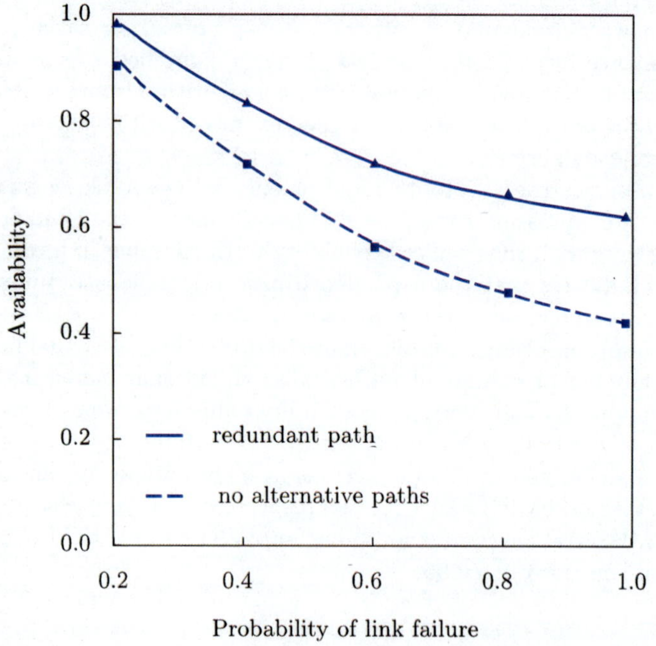

Fig. 5. The availability versus the probability of virtual link failure.

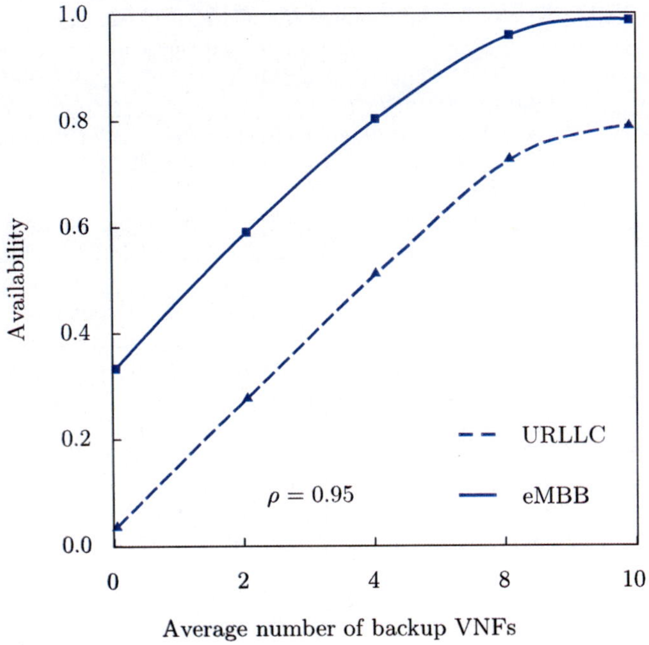

Fig. 6. The availability versus average number of backup VNFs.

6 Conclusion

Mapping of service function chains over virtualized 5G networks allows such paths to be found which provide the required reliability and availability, or only certain reliability or availability. The algorithm presented in the paper can be applied online in real time. n addition, a possible backup of VNF, is included, which allows for an additional increase in the efficiency of the virtualized network. Furthermore, the proposed algorithm was tested through simulation, and the results obtained indicate the efficiency in achieving its key design goals with regard to placing VNFs into physical 5G network.

References

1. NGMN, 5G Extreme Requirements: End-to-End Considerations, August 2018. https://ngmn.org/wp-content/uploads/Publications/2018/180819_NGMN_5G_Ext_Req_TF_D2_2_v2.0.pdf
2. Yousaf, F.Z., Bredel, M., Schaller, S., Schneider, F.: NFV and SDN - key technology enablers for 5G networks. IEEE J. Sel. Areas Commun. **35**(11), 2468–2478 (2017). https://doi.org/10.1109/JSAC.2017.2760418
3. Quinn, P., Nadeau, T.: Problem Statement for Service Function Chaining, RFC 7498, Technical report, 7498, April 2015. https://rfc-editor.org/rfc/rfc7498.txt

4. Medhat, A.M., Taleb, T., Elmangoush, A., Carella, G.A., Covaci, S., Magedanz, T.: Service function chaining in next generation networks: state of the art and research challenges. IEEE Commun. Mag. **55**(2), 216–223 (2017). https://doi.org/10.1109/MCOM.2016.1600219RP
5. Beck, M.T., Botero, J.F.: Coordinated allocation of service function chains. In: 2015 IEEE Global Communications Conference (GLOBECOM), pp. 1–6 (2015). https://doi.org/10.1109/GLOCOM.2015.7417401
6. Khoshkholghi, M.A., et al.: Service function chain placement for joint cost and latency optimization. Mob. Netw. Appl. **25**(6), 2191–2205 (2020). https://doi.org/10.1007/s11036-020-01661-w
7. Jalalitabar, M., Guler, E., Luo, G., Tian, L., Cao, X.: Dependence-aware service function chain design and mapping. In: IEEE Global Communications Conference (GLOBECOM), pp. 1–6 (2017). https://doi.org/10.1109/GLOCOM.2017.8254485
8. Qu, L., Khabbaz, M., Assi, C.: Reliability-aware service chaining in carrier-grade softwarized networks. IEEE J. Sel. Areas Commun. **36**(3), 558–573 (2018). https://doi.org/10.1109/JSAC.2018.2815338
9. Karimzadeh-Farshbafan, M., Shah-Mansouri, V., Niyato, D.: A dynamic reliability-aware service placement for network function virtualization (NFV). IEEE J. Sel. Areas Commun. **38**(2), 318–333 (2020). https://doi.org/10.1109/JSAC.2019.2959196
10. Kaliyammal Thiruvasagam, P., Kotagi, V.J., Murthy, S.R.: A reliability-aware, delay guaranteed, and resource efficient placement of service function chains in softwarized 5G networks. IEEE Trans. Cloud Comput. (2020). https://doi.org/10.1109/TCC.2020.3020269
11. Sahner, K., Trivedi, K.S., Puliafito, A.: Performance and Reliability Analysis of Computer Systems: An Example-Based Approach Using the SHARPE Software Package. Kluwer Academic Publishers, Boston (1996)
12. Egeland, G., Engelstad, P.E.: The reliability of wireless backhaul mesh networks. In: Proceedings of the 2008 IEEE International Symposium on Wireless Communication Systems, pp. 178–183 (2008). http://citeseerx.ist.psu.edu/viewdoc/download?doi=10.1.1.586.3676&rep=rep1&type=pdf
13. Dahlman, E., et al.: 5G wireless access: requirements and realization. IEEE Commun. Mag. **52**(12), 42–47 (2014). https://doi.org/10.1109/MCOM.2014.6979985
14. Popovski, P., Trillingsgaard, K.F., Simeone, O.: 5G wireless network slicing for eMBB, URLLC, and mMTC: a communication-theoretic view. IEEE Access **6**, 55765–55779 (2018). https://doi.org/10.1109/ACCESS.2018.2872781
15. Basta, A., Kellerer, W., Hoffmann, M., Morper, H.J., Hoffmann, K.: Applying NFV and SDN to LTE mobile core gateways, the functions placement problem. In: Proceedings of the 4th Workshop on All Things Cellular: Operations, Applications, and Challenges, pp. 33–38. ACM (2014). https://dl.acm.org/doi/pdf/10.1145/2627585.2627592. https://doi.org/10.1145/2627585.2627592

Softcomputing Approach to Virus Diseases Classification Based on CXR Lung Imaging

Jacek Mazurkiewicz[1] (✉) and Kamil Nawrot[2]

[1] Faculty of Information and Communication Technology, Wrocław University of Science and Technology, ul. Wybrzeże Wyspiańskiego 27, 50-370 Wrocław, Poland
jacek.mazurkiewicz@pwr.edu.pl

[2] Comarch SA, ul. Długosza 2-6, 51-162 Wrocław, Poland

Abstract. The goal of the paper is to create a system for automated classification of virus diseases based on CXR lung imaging. The system designed and implemented as softcomputing engine in the form of a convolutional neural network, is able to perform multiclass classification of CXR images. The learning process is using created dataset and chosen hyperparameters - model is capable of achieving almost 93% accuracy of classification. The implemented solutions give very satisfactory results. Research conducted in this study shall be treated as a discussion of chances and potential prospects for future integration of machine learning techniques in COVID-19 detection. As such, it is not intended for being used as a testing solution offered to actual patients. It is a theoretical work that was not verified in real-life scenarios. Nevertheless, conclusions pointed out as a result of various experiments may push forward further research on implementations of tools based on the described concept in a real environment.

Keywords: Virus disease classification · ANN · CNN · CXR image

1 Introduction

The problem of lung disease is a vital component of healthcare around the world. Over the years mankind has learned to recognize and heal the majority of lung illnesses. However, from time to time new diseases appear that require the fastest possible response in the form of quick classification and appropriate medical treatment. Early detection of the symptoms is vital to apply adequate treatment to patients and prevent the illness from dealing the most harmful damage. Extensive and accurate testing is the key to control the spread of the virus, as confirmed cases can be isolated from other people before they will be able to infect anyone. Various tests and techniques, differing in complexity and accessibility, are available for almost everyone. Laboratory or clinical detection using swabs is the most common approach. Even though the procedure seems straightforward, and such tests are cheap and easily available, they still require several steps and involve multiple people to analyze samples and give a final prediction. Clinical testing methods most commonly have turnaround time equal to several hours or even days. Their sensitivity in the clinical environment varies from 86 to almost 100 per

W. Zamojski et al. (Eds.): DepCoS-RELCOMEX 2022, LNNS 484, pp. 163–175, 2022.
https://doi.org/10.1007/978-3-031-06746-4_16

cent [4]. An important aspect is, that tests tailored specifically to detect the SARS-CoV-2 virus are performed mostly on people that are suspected of being potentially infected: exhibit typical symptoms or were recently exposed to increased risk. Taking into consideration the fact that COVID-19 disease in numerous cases has an asymptomatic course, a considerable amount of infected patients may be basically omitted. These reasons yield the need for the development of alternative methods for effective COVID-19 detection. Not to replace existing solutions, but to broaden the variety of tools in the fight with the pandemic. In many cases distinguishing its symptoms from other types of viral pneumonia becomes a truly demanding task. However, subtle differences always exist, and if they cannot be spotted with a bare eye, machine learning methods should be incorporated as they are capable of learning patterns imperceptible for humans. The only requirement is to provide them with enough input data in a standardized format, which will allow processing - medical images seem to be ideal candidates. The rest of the paper is organized in the following way: the second chapter contains medical background of the work - short elaboration of viral lung diseases (in particular viral pneumonia and COVID-19) and lung image overview, healthy and diseased one. Section 3 presents the datasets. The fourth section aim is to guide the reader through the process of all experiments conducted on the developed solution and extensive analysis of the obtained results carried out. This chapter describes the methodology and outcomes of research conducted in order to formulate conclusions on the subject. The last chapter contains a discussion of all conclusions drew from the research and results' analysis, including assessment of the solution, attempt to generalize the outcomes and predict possible further work in the field.

2 Viral Respiratory Diseases

About 200 viruses are known, that can cause disease in humans [5]. Viruses are the smallest microorganisms that can cause illness. There are specific types of viruses that human organisms are most susceptible to. The most common virus infection occurs via droplets. The microscopic droplets are produced by the sick person and spread by sneezing, coughing or talking. If another person comes within their reach, an infection may occur through the dropping of these droplets in the mouth and nose [2]. It is also important that the viruses may live on the surface within several hours, and the infection may occur also by touching the surface and then by touching space around the face, eyes, nose or mouth. Viruses attacking the respiratory tract have a similar structure. They also occur seasonally, which means that diseases caused by them appear at certain times of the year. They cause different symptoms depending on the part of the respiratory ways they are attacking. The upper respiratory tract infection will manifest itself differently than the lower respiratory tract infection. In the case of the lungs, the most common symptoms are shortness of breath, difficulty breathing and wheezing. Many viruses are known to cause viral infections [2].

2.1 Viral Pneumonia

Pneumonia is a pretty common disease. It is a threat especially for young children in developing countries and for seniors in developed countries. There is still speculation about the causes of pneumonia in children, especially after vaccination programs, e.g. against pneumococci have been introduced. Viruses are mentioned here as the major cause of pneumonia. The role of respiratory viruses appears to be significant following the emergence of severe acute respiratory syndrome (SARS), avian influenza A (H5N1) virus, and the 2009 pandemic influenza A (H1N1). All of the above have shown that at any time a virus can emerge, capable of spreading around the world and killing more people than previously known diseases. It should be noted that thanks to the wide access to tests (e.g. PCR tests), the ability to quickly and accurately classify respiratory viruses has significantly increased [14]. Pneumonia is the largest infectious cause of death in children worldwide. It killed 808694 children under the age of 5 in 2017, accounting for 15% of all deaths of children under five years old. Pneumonia affects people all over the world, but most severely in South Asia and sub-Saharan Africa [2]. The diagnosis of viral pneumonia consisted in taking a sample from the lower or upper respiratory tract and searching for the virus or its antigen. Thanks to the large scale of use PCR-based methods the detectability of viruses has increased significantly. This is particularly evident in case of adults. In general, PCR-based methods has between two and five more sensitivity in compare with conventional virus diagnosis methods [1]. The American Thoracic Society recommends that the diagnosis of pneumonia be made on the basis of a chest radiography. When analyzing the image, the radiologist searches for specific changes that indicate the cause of the disease. Interstitial infiltrates on chest radiographs points to a viral cause of pneumonia and alveolar infiltrates points to a bacterial cause [6]. On the other hand, bacteria and viruses alone or together can cause a number of radiological changes. Changes are only helpful in specific cases when it is necessary to confirm whether the cause of the pneumonia is microbial. Correct diagnosis of the bacterial or viral substrate is important when choosing a further treatment. Viral pneumonia requires a different approach than bacterial pneumonia.

2.2 COVID-19 Influence on Lungs

Chest imaging using radiographic projection, commonly referred to as chest x-ray (CXR) is the most popular and widely used diagnosis method from the medical imaging branch. Around 72% of all medical images acquired using any screening method are pictures of chest [9, 11]. In the case of traditional radiography, this proportion is even higher. Reasons for such a major share of lung imaging in the statistics partially overlap with the aforementioned causes of radiography commonness. Besides that, imaging is the optimal method of diagnosing numerous pulmonary disorders. In a typical scenario, valid for detecting lung abnormalities, the image obtained as a result should be a frontal picture presenting the whole chest with internal organs - lungs and heart together with vessels - visible. Such format allows convenient analysis of the lungs' internal structure which is crucial for detecting abnormalities.

Chest x-ray images are used widely on a daily basis, for detecting pneumonia, tuberculosis, lung fibrosis, cancer, sarcoidosis, and more. Experts responsible for analyzing and interpreting CXR images are radiologists - highly specialized professionals with expertise and experience that allow them to spot even the most subtle differences and abnormalities. As was mentioned earlier, lungs are the major target for the SARS-CoV-2 virus and organ frequently receiving the most damage, peculiarly in more severe cases of

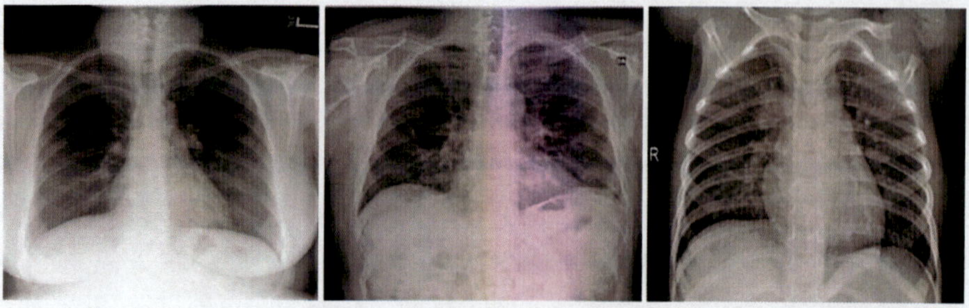

(a) healthy lungs (b) COVID-19 (c) viral pneumonia

Fig. 1. Chest radiographs comparison of patients: healthy lungs, COVID-19, viral pneumonia

disease course. Main lung-related pathologies discovered in bodies of patients infected with coronavirus include: minor exudation of serum and fibrin, inflammation, diffuse alveolar damage, pulmonary fibrosis. These are mostly symptoms that may be perceived as mild or moderate without more detailed evaluation and easily confused with other pulmonary diseases. However, in more severe cases, acute respiratory distress syndrome may occur, causing the reduced size of patient's lungs and as a result - a need for artificial ventilation. Reflections of those issues can be potentially discovered after examination of CXR images. When the course of the disease is severe, external symptoms can be clearly observed without image diagnostic, but in more light cases or early stages of the illness, symptoms is less specific and the importance of correct diagnosis increases. Influenza and COVID-19 shares many similarities: they both cause respiratory distress syndrome with mostly overlapping symptoms, transmission happens in the same manner through the droplets in the air. Nonetheless, differences exist as well. From standard symptoms occurrence rate comparison [10], presented in Table 1, we can conclude that the typical course of the illness is similar, but not the same. In such a situation, the most important aspect seems to be an effective way of distinguishing between those two disease entities. Subtle divergences can be also matched with differences visible on CXR images. Comparison of chest x-rays images from patients without any type of pneumonia diagnosed, COVID-19 cases and other types of viral pneumonia, was shown in Fig. 1. Small speckles and blurred entities can be spotted on infected lungs pictures, and their density and arrangement may vary with regard to the nature of the infection. However, as one may conclude from observation of those images, correct interpretation based on radiographic projections is not trivial and requires high-level expertise in the field of radiography. An alternative solution may be an automated approach utilizing machine learning.

Table 1. Symptoms comparison of pneumonia caused by COVID-19 and Influenza virus

Symptoms or clinical features	COVID-19 [%]	Influenza [%]
Fever	78.5	89.2
Cough	69.7	86.1
Chest pain	13.7	27.8
Sore throat	9.5	37.3
Fatigue	63.6	39.0
Anosmia	7.0	0
Diabetes	16.4	11.1
Asthma	8.4	16.1
Hypertension	28.8	14.1
Oxygen therapy	65.0	42.3
Hospitalization duration (days)	8.6	3.0
Hospital mortality	21	3.8

3 Datasets

Choosing chest X-rays was dictated by a couple of factors. First of all rich resources of crafted datasets containing lots of CXR images were easily available. Secondly - popularity and commonness of traditional radiography could increase potential interest in the developed solution. Relatively easy access to CXR imaging may help patients in regions where laboratory testing is not available or impeded. Moreover the area of recognizing COVID-19 on CXR images was not fully researched and require improvements. Finally X-rays characterize with a much higher variance of the images than CT scans which can be an additional challenge for the classifier. The dataset selected for the purpose of this project is called COVID-19 Radiography Database [3, 12, 13], and was selected a winner of the COVID-19 Dataset Award by Kaggle Community. In its most recent version, used in the project, it has the size of 342 MB and consists of: 3616 images of patients diagnosed with COVID-19, 4598 images of healthy patients, 1345 images of patients diagnosed with other types of viral pneumonia. The total number of examples was equal to 9559 - it can be considered a sufficient amount of data for a given problem. Many pieces of research proved, that effective training of convolutional neural network is possible even with a much smaller dataset available. The number of samples presenting viral pneumonia is visibly lower than the representation of other classes, however, the discrepancy between them is not that big to cause serious issues related to class imbalance. All images have unified dimensions: 299 × 299 pixels. They were saved as PNG files in greyscale even though radiographs operate only on a single color channel it was ensured that any potential anomalies on the images causing single pixels from outside the greyscale domain will not be passed into the classifier. According to the study focused on the impact of the image resolution on convolutional neural network algorithms performance in the area of radiography [15], for most of the classified diseases several network architectures reached the optimal level of performance when the image resolution was equal to around 320 × 320 px. Therefore, images present in the selected dataset can be treated as valid and fully useful in terms of their size. Even though all examples were retrieved using exactly the same medical procedure, they significantly vary in shape and quality. Factors that can influence the look of the image include angle, patient's pose, brightness, ribs opacity level, labels and various markings added by devices or doctors, minor distortions. Such diversity lays in the nature of CXR imaging, and it is impossible to unify all examples automatically. However, according to the research on the impact of various image enhancements techniques on the CNN performance in chest X-rays classification [13], some methods of preprocessing can be implemented in order to increase the overall accuracy of the model. Two of them, that fine-tuned networks examined in the aforementioned study in a relatively significant way, were: gamma correction and histogram equalization. Hence, these two enhancements were decided to be verified also on the example of the network being the topic of this work. Gamma correction - the first applied preprocessing method, sometimes abbreviated as gamma, is a non-linear operation that can be performed on any image to effectively change its luminance. Histogram equalization - the second action in preprocessing - is a process of distributing grey levels with the whole image. It aims to equalize the number of pixels representing every possible brightness. After this transformation, the histogram shall be stretched over the whole range of pixel intensity values. The equalization process lightens dark images and darkens bright ones, thus increases the contrast in most cases. Initially, the gamma value was set to 1.5 on images with previously equalized histograms. It was based on similar enhancements presented in the literature [13] and preliminary tests. Effects of utilizing these two described enhancement methods with various parameter values will be examined during the experiments stage. Besides them, no additional image transformations were applied to the input chest X-rays. Linear transformations posed a great risk of failing because of the diversity of the input data - in such circumstances unified, automatic approach to preprocessing, even if useful for some samples, might be a drawback for another subset. Before starting any training procedure, the dataset was divided into three subsets [8] intended to be used at different stages and various purposes: training subset (60%) - samples with labels presented to the network during training, validation subset (20%) - images used for intermediate model parameters validation and tuning during the training procedure, test subset (20%) - separated set of images reserved for final model testing after the learning process is finished.

4 Virus Neural Classifier

4.1 Structure

The main and the most important part of the developed softcomputing engine, was the convolutional neural network - classifier making decisions. CNN built for this project was intended to be a shallow network, with straightforward and compatible architecture to ensure a relatively fast time of computation and ease of introducing modifications. In its final shape, it consists of: five convolutional layers, varying in terms of number of filters five pooling layers, chained with each convolution, implementing maximum pooling strategy, two dropout layers, inserted before and after the last convolution to differentiate weights and prevent overfitting, flattening layer after the last convolution and pooling, to transform input matrices into a vector that can serve as input layer of the fully-connected network, hidden dense layer, operating in a similar manner to its equivalent in multilayer perceptron - fully connected with both input and output, final output layer consisting of three neurons representing possible classes. The exact structure was developed based on various experiments conducted on different architectures. Convolutional layers are always chained with non-linear activations. In the case of the described network, rectifiers were selected to play this role. ReLU performed the best among other widely accepted activation functions. The number of output filters in the convolution was increasing in subsequent layers as the size of the output was reduced. Kernel size of the convolution window had the constant size of 3×3. The total number of trainable parameters was equal to 3637635 - significant but acceptable in terms of computational complexity. The correct definition of a fully connected section at the end of CNN required deciding on two parameters: the number of hidden layers and the number of neurons in them. Rules typically applied to multilayer perceptron models remains valid and suitable for the convolutional network as well. Therefore, it was decided to include a single hidden layer, which is sufficient for most problems as it allows to perform any continuous mapping between finite spaces. In the case of any convolutional network, the majority of the work is performed by preceding layers, thus the hidden layer does not have to be too complex. The number of hidden neurons was set to 512, which was the best-tested compromise between performance and precision. The only layer with an activation function other than ReLU was the output layer of the fully connected network. In its case, the softmax curve was incorporated. This is a standard technique in multi-class classification, allowing to achieve results in the form of normalized probabilities of recognizing every possible class.

4.2 Training Procedure

Learning procedure was conducted on image batches with batch size parameter set to 16, according to performance tests that were performed for different values of this hyperparameter. Initial weights were generated randomly. Classes had equal weights because the dataset was assessed as differentiated enough to not cause a class imbalance problem. Typically, the net is trained for a specified number of epochs, with all subset of training samples being presented in each epoch. However, this approach requires knowledge about the training progress and speed, and can be misleading as the procedure may proceed in different manners during separate runs. Therefore, there is a possibility to define stop criterion - condition defined on metrics calculated after each batch or epoch, that needs to be fulfilled to finish the training. It shall rely on the validation subset measures, as they provide unbiased feedback regarding intermediate performance. As far as the particular example of the net that is the core of this work is concerned, the early stopping function was formed to monitor accuracy achieved on the validation set. Important detail related to such criterion is the patience parameter - it is a number defining the number of epochs that needs to pass without monitored metric improvements to definitely finish the training. Its responsibility is to prevent individual cases in which the network did not make any progress during a single epoch. Patience was set to 5 in the present case. Before training can be started, neural network models require specifying two essential parameters: optimizer and loss function. They define the goal of the learning process and the way of achieving it. The loss function of the classifier became cross-entropy - a popular and well-proven solution for multi-class classification tasks. On the other hand, Adaptive Moment Estimation (Adam) was set to be an optimizer for efficient minimizing the loss value during the learning procedure. Adam, since its introduction a few years ago [7], become a standard, especially for a problem with a large number of parameters, because of many improvements in comparison with older stochastic gradient descent methods (Table 2).

Table 2. List of convolutional neural network layers

Layer	Output shape	Number of parameters
conv2d_1	(297, 297, 32)	896
max_pooling_2d_1	(148, 148, 32)	–
conv2d_2	(146, 146, 64)	18496
max_pooling_2d_2	(73, 73, 64)	–
conv2d_3	(71, 71, 64)	36928
max_pooling_2d_3	(35, 35, 64)	–
conv2d_4	(33, 33, 128)	73856
max_pooling_2d_4	(16, 16, 128)	–
dropout_1	(16, 16, 128)	–
conv2d_4	(14, 14, 256)	295168
max_pooling_2d_4	(7, 7, 256)	–
dropout_2	(7, 7, 256)	–
flatten	(12544, 1)	–
hidden_dense	(256, 1)	3211520
output	(3, 1)	771

Depending on the hyperparameters settings and current performance of the machine, the whole learning procedure took from 60 up to 80 min to finish. During each epoch, lasting around 150 s, around 6000 images from the training subset was flowing through the network. Accuracy stabilizes typically between 25 and 32 epoch of training. To that point, progress in terms of accuracy increase and loss decrease was smooth and constant.

4.3 Pretrained Models

Two pre-trained deep convolutional neural network models were implemented: VGG16 and Inception V3. VGG16 - network composed of 13 convolutional layers divided into five groups, each ended by maximal pooling layers. It accepts images of resolution 224 × 224 as the input. All non-linearity is introduced using ReLU. Used convolution filter size is 3 × 3 in most cases, but even 1 × 1 linear filters are present in the model. In some aspects, architecture is similar to the structure of the custom model presented in this paper. Inception V3 - the most complex model among all tested - it is composed of several layers together with maximum pooling, average pooling, dropouts and additional mechanisms intended to boost the performance. Input images must be of size 299 × 299.

5 Results Analysis

Basic metrics computed for validation subset after training procedure were presented in Table 3. Overall accuracy was equal to 92.26% for validation subset, whereas for training data - 93.69%. Small divergence between these two measures may lead to a conclusion that any overfitting did not occur during the training. A satisfactory compromise between precision and recall may be observed for all three classes. Worst scores are observed for viral pneumonia classification. This difference may come from a smaller subset of samples labelled as pneumonia in the training dataset. Useful in observing how well particular classes are predicted is the confusion matrix, shown in Fig. 2. One interesting conclusion one can draw from this confusion matrix is that, despite doubts, the model misclassified COVID-19 with pneumonia and pneumonia with COVID-19 case very rarely. This was the most possible potential weakness of the solutions, however, actual results did not confirm that. Besides that, a good level of accuracy was achieved based on those results. Except for taking into consideration correct classifications, one significant aspect of the model's performance is the certainty of this classification. To assess it, the average probability of the selected class returned by the classifier was calculated. Only correct recognition was taken into account. Computed values of these probabilities, named average certainty, were presented in Table 3. As one can notice, the model performs classification with certainty over 90%.

Additionally, the change of accuracy during subsequent learning epochs was compared with the same measure calculated for two pretrained models mentioned in previous sections. The last section of the architecture, a fully connected network, was common for all three models. Early stopping criterion was present in every case as well, thus every model finished training after a different number of epochs. As can be observed from the chart shown in Fig. 3, the custom model outperformed pre-trained CNN architectures in terms of accuracy. VGG16 was not able to learn anything and finished after five epochs without any improvements. The training process of the Inception network was much more interesting - it started from a similar level of accuracy as the custom model. But at the end of the training, Inception V3 architecture was not able to achieve stabilized results, thus its performance was also worse than the custom model. Considering solely the complexity of particular models, results achieved during learning may be perceived as counterintuitive. However, a couple of possible reasons for such a situation may be outlined. The complexness of the model does not always provide better outcomes, as the

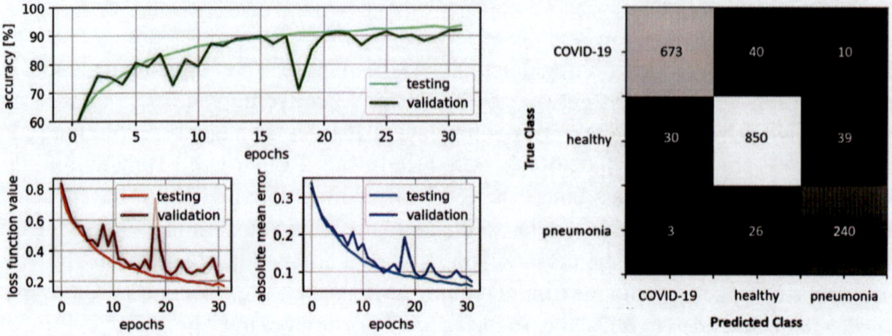

Fig. 2. Training metrics, confusion matrix calculated for validation set on optimal model

Table 3. Basic metrics calculated for model classifying validation set of examples layers

	Precision	Recall	F1-score	Samples	Certainty
COVID-19	0.95	0.93	0.94	723	0.9590
Healthy	0.93	0.92	0.93	919	0.9134
Pneumonia	0.83	0.89	0.86	269	0.9436

key element may be tailoring the architecture to the given problem. Pretrained models are universal and may be applied for numerous kinds of problems. Architecture designed for a specific task and optimized for maximal performance in it may perform with higher accuracy despite simpler architecture. Another problem may lay in the fully connected network, which was not suited well for preceding layers in the case of external models.

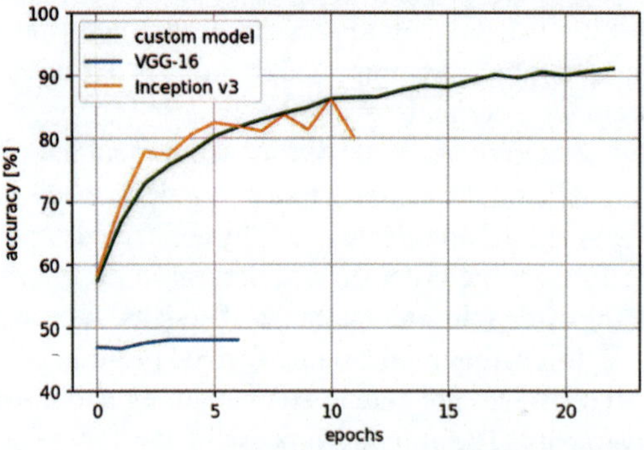

Fig. 3. Training metrics, confusion matrix calculated for validation set on optimal models

The process of system tuning starts early, at the stage of data preprocessing, when optimal transformations, increasing the model's performance, has to be decided and applied to the set of images. In the described case, based on external research in the field, gamma correction and histogram equalization were chosen to be utilized as the most promising transformations from the broad set of considered operations. Different combinations were tested, all results were presented in Table 4.

Table 4. Overall accuracy scores depending on implemented image transformations

	With histogram equalization	Without histogram equalization
Gamma = 0.5	90.22%	89.44%
Gamma = 1.0	91.32%	90.65%
Gamma = 1.5	92.26%	90.78%

Similar procedure was conducted for each hyperparameter to choose its best value. Optimal batch size of input samples during learning process was decided to be equal to 32. optimization functions that prove itself the best among others, was Adam with a learning rate equal to 0.001 and decay rates for the first and second moment - 0.900 and 0.999, respectively. Considering the assessment of a neural network, or any machine learning algorithm, just an evaluation of their results may not be sufficient. Feature maps, generated by convolutional layers, is a set of images representing regions or objects present on the input image that the CNN engine recognized and distinguished from other elements. Visual analysis of feature maps may be used to verify, whether the model is recognizing the intended object and no other, random or negligible patterns. The output of the first convolutional layer, was presented in Fig. 4. Most of the maps are focusing on the lungs, and many red regions (denoting the highest importance) cover exactly the elements where patterns specific to COVID-19 disease are appearing in the image. A similar process may be repeated for activation maps, presented as a heatmap, describing how particular regions of the input image influence the final decision made by the model. If the most activated parts contain objects or patterns that are being detected, it may be proof that CNN works correctly. In Fig. 4, activations of the last convolutional layer in the chain, for samples from each class, were shown. Even though drawing straight, explicit conclusions are not possible in that case, activation heatmaps present classification from the model perspective. And as such, it can be assessed as valid, because the main factor in each example is the lungs, especially their lower parts. This is also the region, where differences between healthy lungs, COVID-19 and viral pneumonia cases are condensed and the most visible.

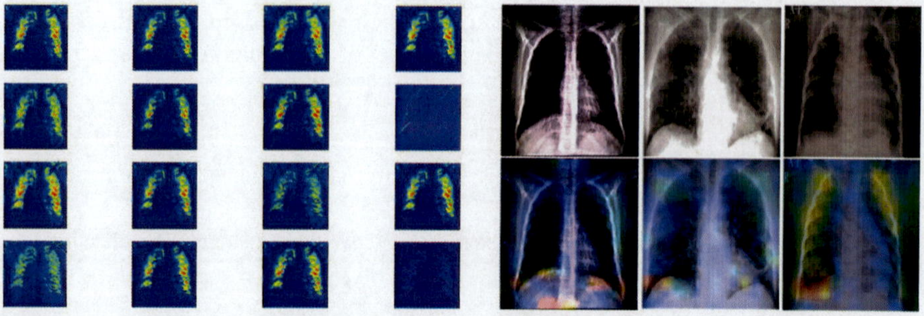

Fig. 4. Set of feature maps generated by first convolutional layer (left), activation regions in last convolution layer: healthy/COVID-19/pneumonia (right)

6 Conclusions

The model depicted in this paper proved that a simple convolutional network structure is capable of classifying COVID-19 based on lung imaging with satisfactory results. Overall accuracy - 93% - is a precision level of some widely-used types of laboratory tests, which are characterized by accuracy from range 86–100% [4]. At the same time, the sensitivity of the classifier is two percentage points higher. In medical fields, where, in many cases, human life depends on the correct diagnosis of the disease, is the most important measure. On the basis of internal network processes analysis, using visualization methods, one can assume, that the model recognized intended patterns, focusing on minor differences in lungs structure. This is a prospect that the model will work correctly on any set of chest X-rays images. In general, the recognition process was valid and in accordance with expectations. In conclusion, outcomes from the designed model are satisfactory and competitive to other solutions in that field. This may be proof of the high effectiveness of machine learning techniques in recognizing diseases from medical images.

References

1. Arens, M.Q., et al.: Comparison of the eragen multi-code respiratory virus panel with conventional viral testing and real-time multiplex PCR assays for detection of respiratory viruses. J. Clin. Microbiol. **48**(7), 2387–2395 (2010)
2. Caceres, V.: Types of Respiratory Viral Infections (2020). https://health.usnews.com/conditions/articles/types-of-respiratory-viral-infections
3. Chowdhury, M.E.H., et al.: Can AI help in screening viral and COVID-19 pneumonia? IEEE Access **8**, 132665–132676 (2020)
4. Giri, B., Pandey, S., Shrestha, R., Pokharel, K., Ligler, F.S., Neupane, B.B.: Review of analytical performance of COVID-19 detection methods. Anal. Bioanal. Chem. **413**(1), 35–48 (2021)
5. Hansen-Flaschen, J., Bates, D.V.: Respiratory disease (2019). https://www.britannica.com/science/respiratory-disease. Accessed 15 Mar 2022
6. Jennings, L.C., et al.: Incidence and characteristics of viral community-acquired pneumonia in adults. Thorax **63**(1), 42–48 (2008)

7. Kingma, D.P., Ba, J.: Adam: a method for stochastic optimization. arXiv preprint arXiv:1412.6980 (2014)
8. Kohavi, R., Provost, F.: Glossary of terms. Mach. Learn. **30**(1), 5–6 (1998)
9. National Health Service. Diagnostic Imaging Dataset 2012–2021 (2021). Accessed 15 Mar 2022
10. Osman, M., Klopfenstein, T., Belfeki, N., Gendrin, V., Zayet, S.: A comparative systematic review of COVID-19 and Influenza. Viruses **13**(3), 452 (2021)
11. Picano, E.: Sustainability of medical imaging. BMJ **328**(7439), 578–580 (2004)
12. Rahman, T., Chowdhury, M., Khandakar, A.: COVID-19 Radiography Database, ver. 4 (2021). https://www.kaggle.com/tawsifurrahman/covid19-radiography-database. Accessed 20 Mar 2022
13. Rahman, T., et al.: Exploring the effect of image enhancement techniques on COVID-19 detection using chest X-ray images. Comput. Biol. Med. **132**, 104319 (2021)
14. Ruuskanen, O., Lahti, E., Jennings, L.C., Murdoch, D.R.: Viral pneumonia. Lancet **377**(9773), 1264–1275 (2011)
15. Sabottke, C.F., Spieler, B.M.: The effect of image resolution on deep learning in radiography. Radiol. Artif. Intell. **2**(1), e190015, (2020). PMID 33937810

Single and Series of Multi-valued Decision Diagrams in Representation of Structure Function

Michal Mrena[1], Miroslav Kvassay[1(✉)], and Stanislaw Czapp[2]

[1] Faculty of Management Science and Informatics, University of Zilina, Zilina, Slovakia
{michal.mrena,miroslav.kvassay}@fri.uniza.sk

[2] Faculty of Electrical and Control Engineering, Gdańsk University of Technology, Gdańsk, Poland
stanislaw.czapp@pg.edu.pl

Abstract. Structure function, which defines dependency of performance of the system on performance of its components, is a key part of system description in reliability analysis. In this paper, we compare two approaches for representation of the structure function. The first one is based on use of a single Multi-valued Decision Diagram (MDD) and the second on use of a series of MDDs. The obtained results indicate that the series of MDDs can be more efficient than a single MDD in case of series-parallel systems, which belong to the most fundamental types of topologies studied by reliability engineers.

Keywords: Binary decision diagram · Dynamic creation of decision diagram · Multi-valued decision diagram · Reliability · Structure function

1 Introduction

The first step in analyzing system reliability is creation of a mathematical model of the analyzed system. The properties of the model and its elements depend on the properties of the system and the main goals of the analysis. However, regardless of these specifics, the model usually contains one specific map known as a structure function. This function defines how performance or operability of the system depends on performance or operability of the components that the system is composed of. Performance or operability of the system as well as its components is usually defined by a set of m integers $\{0, 1, \ldots, m-1\}$, where number 0 means that the system/component is completely failed, while number $m-1$ implies that it is perfectly functioning [1]. These integers are also known as states of the system [1]. If a system and all its components can operate only at two possible states (states 0 and 1), then the system is recognized as a Binary-State System (BSS) [2], and its structure function can be viewed as a Boolean function [2, 3]. If $m > 2$, then the system is labeled as a Multi-State System (MSS) [1]. In this case, the structure function can be viewed as a Multiple-Valued Logic (MVL) function [1, 4]. However, in both cases, the structure function can be viewed as a special case of integer function, which can be represented by a decision diagram [5].

W. Zamojski et al. (Eds.): DepCoS-RELCOMEX 2022, LNNS 484, pp. 176–185, 2022.
https://doi.org/10.1007/978-3-031-06746-4_17

A decision diagram is a graph data structure that has been developed originally for representation of Boolean functions [6]. In this case, the decision diagram is known as a Binary Decision Diagram (BDD) and it can be used in reliability analysis to represent the structure function of a BSS [7].

A decision diagram can also be used for representation of MVL [5] and integer functions [5, 8]. In this case, we say about Multi-valued Decision Diagrams (MDDs). This kind of decision diagrams is also very useful in reliability analysis because it allows us to represent efficiently the structure function of an MSS [4, 9]. However, the structure function of an MSS can also be represented by a series of MDDs [10, 11]. This is based on a transformation of the structure function into a series of integer functions with Boolean-valued output [12].

In reliability analysis of MSSs, a single MDD as well as a series of MDDs can be used to compute basic characteristics, such as the system state probabilities [10, 13], the system availability [11, 12], or to evaluate importance of the system components [13]. However, the key question is which of these two representations is more efficient in terms of computing. Answering this question is the main goal of the paper.

2 Basic Concepts

Real systems can be considered as BSSs or as MSSs. The BSS and its components can be in one of two possible states that are functioning and failed. This approach is suitable for systems that are binary in their nature or in case of systems where even a slightest degradation of their performance can cause some catastrophic failure. Many real-world systems can, however, operate on multiple performance levels. Therefore, it is suitable to consider them as MSSs. Each component of the MSS and the system itself can be in one of m possible states denoted by integer numbers 0 (for completely failed state), $1, \ldots, m-1$ (for perfectly functioning state).

2.1 Structure Function

After defining number of system states m, it is necessary to create a mathematical description of the system, which describes the relationship between the state of individual system components and the state of the entire system. Such a description is called a structure function and has the next form for a system of n components [1, 4]:

$$\phi(\boldsymbol{x}) : \{0, 1, \ldots m-1\}^n \rightarrow \{0, 1, \ldots, m-1\}, \tag{1}$$

where $\boldsymbol{x} = (x_1, x_2, \ldots x_n)$ is a vector that contains states of all components, i.e., state vector, while variable x_i represents state of the i-th component for $i = 1, 2, \ldots, n$.

Structure function of a system with a large number of components can be considerably extensive and complicated. It is therefore necessary to represent such a function in an efficient way. Definition (1) agrees with the definition of an MVL function and, in the special case where $m = 2$, it agrees with the definition of a Boolean function. This allows us to use the approaches of MVL and Boolean logic in reliability analysis of the systems described in this way [3, 4, 7]. One of them is the decision diagram.

2.2 Decision Diagrams

A decision diagram is a graph data structure that can be used to efficiently represent integer function. It consists of two types of nodes, that are arranged in levels. The first type are internal nodes, which correspond to variables of the function it represents. Each internal node is associated with exactly one variable with index i and has m outgoing edges (in the case of BDD, $m = 2$), which represent the possible values of the i-th variable. These edges are oriented and can only enter nodes that are at lower levels of the diagram. This implies that the graph is oriented and acyclic. The second types of nodes are terminal nodes. These nodes represent possible values of the function ranging from 0 to $m - 1$. All nodes that are at the same level of the diagram must either be associated with the same variable i or be terminal nodes while the terminal nodes are always at the last (lowest) level of the diagram. At the first (highest) level of the diagram, there is always exactly one node, which we denote as the root of the diagram.

The efficiency of the diagrams is ensured by the fact that each node of the diagram is unique. This means that there are no two internal nodes associated with the same variable i with edges that lead to the same nodes in the diagram. Also, there are no two terminal nodes representing the same value of the function described by the diagram. In Fig. 1, we can see examples of both BDD and MDD.

When working with a structure function, the size of its representation is very important. In terms of size, we are usually interested in its relationship to the number of components n. A fairly well-known representation is the truth table. The table is very simple and easy to work with, but its size grows exponentially with increasing number of variables. Working with it is therefore problematic even for dozens of variables [14]. The great advantage of decision diagrams is that the relationship of their size (number of nodes) to the number of variables is in many practical cases significantly better than exponential. Another advantage of decision diagram is that it is an orthogonal representation of the structure function, which makes it easier for us to work with probabilities during calculation of basic reliability measures [13].

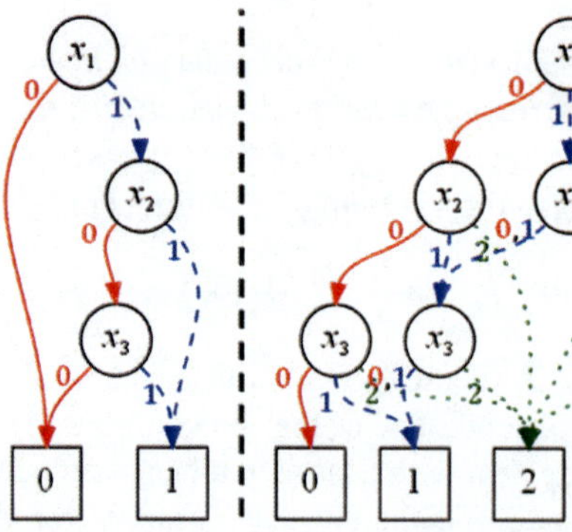

Fig. 1. A BDD of Boolean function $f(\boldsymbol{x}) = x_1(x_2 \vee x_3)$ on the left and an MDD of MVL function $g(\boldsymbol{x}) = \max(x_1, x_2, x_3)$ on the right.

3 Decision Diagrams in Reliability Analysis

There are two basic approaches to represent the structure function using decision diagrams. The most straightforward one is to use a single diagram to represent the entire structure function. Another one is to use a series of diagrams representing series of functions [11, 12] describing different availability levels of the system.

3.1 Single Diagram

Since the definition (1) of the structure function agrees with the definition of MVL function, it is possible to use a single MDD to represent the entire structure function. An example of such a diagram can be seen in Fig. 2.

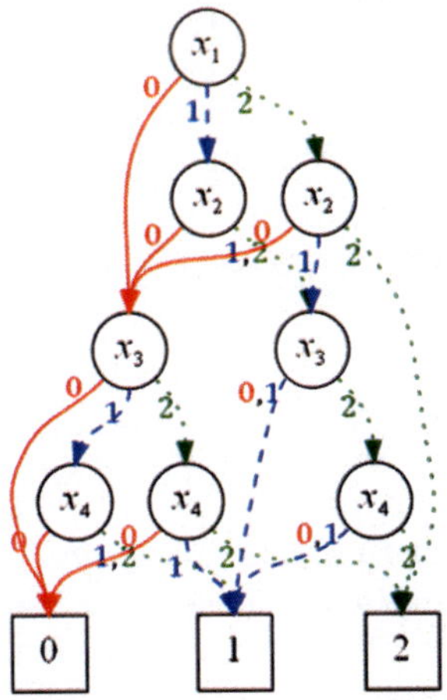

Fig. 2. An MDD of function $\phi(\boldsymbol{x}) = \max(\min(x_1, x_2), \min(x_3, x_4))$.

Obtaining the value of the structure function for a given state vector is simple in this case. It requires finding a path from the root of the diagram to a terminal node. Decision, hence, the name of the diagram, is made in each visited internal node and an edge is selected based on the value of the variable that the given node is associated with. The value of the structure function is then given by the value at the terminal node at which the path is terminated.

3.2 Multiple Diagrams

An alternative to using one MDD is to use a series of MDDs of functions $\phi(\boldsymbol{x}) \geq j$ for $j = 1, 2, \ldots, m-1$ to represent the structure function of a system with m states. Each diagram represents integer function with Boolean-valued output containing at most n m-valued variables. An example can be viewed in Fig. 3.

In contrast to using a single diagram, obtaining a system state for a particular state vector $\boldsymbol{x}$ is more complicated. It is necessary to find the value j for which the diagram of the function $\phi(x) \geq j$ evaluates to the value 1 and the diagram of the function $\phi(x) \geq j+1$ evaluates to the value 0. The value j is then the system state for the state vector $\boldsymbol{x}$. If we cannot find the value of j that satisfies the above rules, then the system is either in state 0 (if it holds that $\{\phi(x) \geq 1\} = 0$) or in state $m - 1$ (if $\{\phi(x) \geq m - 1\} = 1$).

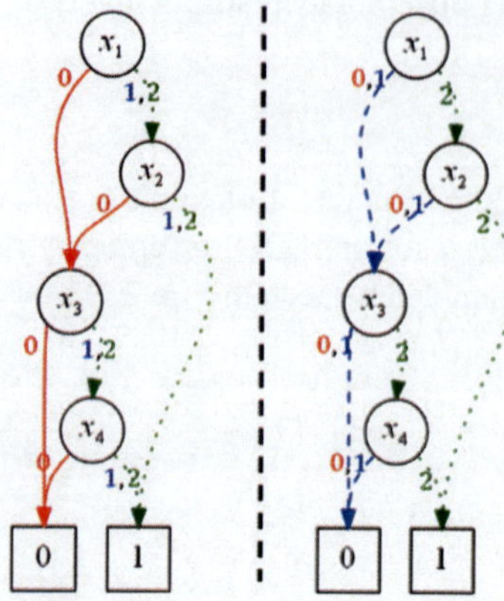

Fig. 3. An MDD of function $\phi(x) \geq 1$ on the left and an MDD of function $\phi(x) \geq 2$ on the right.

3.3 Different System Types

System reliability can be examined using either a single diagram that represents the entire structure function or using a series of diagrams describing particular states of the system. An interesting question is whether one of these approaches is more advantageous in practice than the other. The advantage can be understood, for example, as memory and time requirements for the calculation of various system characteristics. To answer this question, we compared these two ways of representing a structure function in case of series, parallel, and series-parallel systems.

Series and parallel systems [2] are well-known systems with simple topologies. The series BSS is functioning if all of its components are functioning. The structure function of such a system can be expressed as logical conjunction of n Boolean variables. One of the ways to generalize the series BSS to an MSS is to use the min function [15]. Parallel system that are functioning if at least one of their components is operational can be generalized in a similar way by using the max function [15]. By combining the two aforementioned kinds of system we obtain series-parallel systems.

4 Experiments

Experiments described in this section were performed using the Templated Decision Diagram library (TeDDy) [16]. TeDDy can create and manipulate several types of decision diagrams, including BDDs and MDDs. The library is written in the C++ programming language and, in addition to general manipulation of decision diagrams, it also contains a module that implements several algorithms aimed at reliability analysis of systems represented by decision diagrams. All experiments were conducted on a PC with Intel Core i9-10900KF (5.3 GHz, 20 MB Cache) processor, 128 GB DDR4 RAM running Ubuntu 20.04.03 LTS operating system. Code for the experiments was compiled using the G++ compiler in version 11.1.0.

4.1 Diagram Creation

The aim of the experiments was to compare two approaches to the representation of the structure function. We made a comparison on three common system topologies that are series, parallel and series-parallel. To create decision diagrams for the structure functions of these systems, we used a procedure called dynamic decision diagram creation [14]. The essence of this procedure is the gradual merging of diagrams representing parts of the function into the resulting diagram representing the entire function. Diagram creation for series and parallel systems was done in two steps. In the first step, simple diagrams representing the individual variables were created, and in the second step, they were merged by the min(series) and max(parallel) functions. Fold procedure [14] was used to merge the diagrams in the second step. Illustration of this procedure can be seen in Fig. 4, where the initial diagrams are in the left and the resulting diagram is on the right. All series (parallel) systems composed of n components have the same structure function and thus the same diagram. Therefore, there was no need to specifically generate different system configurations.

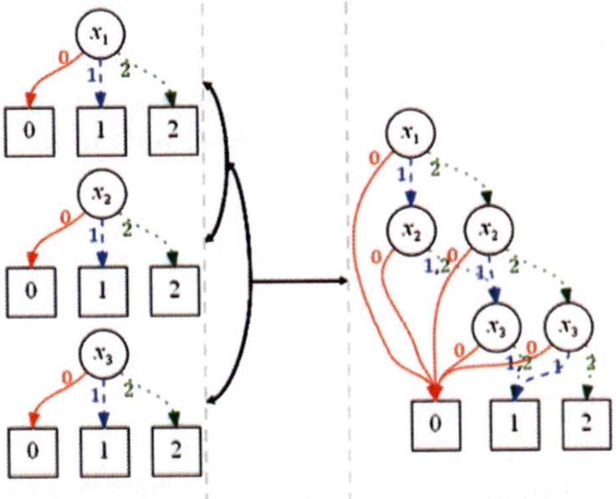

Fig. 4. Merger of diagrams of single variables into a diagram of function $\phi(\boldsymbol{x}) = \min(x_1, x_2, x_3)$.

On the other hand, a series-parallel system composed of n components can exist in many different configurations. Therefore, in the experiment, we needed to randomly generate these configurations. Again, we did this in two steps. In the first step, we created an Abstract Syntax Tree (AST) in which the min and max functions represented serial and parallel parts of the system. We started to generate the tree at its root, in which we randomly selected min or max function and the number of components n that the tree was to contain was randomly divided into two parts. We then used these parts to recursively generate the left and right son of the root. We stopped the generation when the number of components that the given subtree was to contain reached 1. In this case we created a leaf containing the next variable. It is worth noting that we started with creation of leaf containing variable x_1. It follows from this procedure, that each variable was present in the tree exactly once. An example of such an AST can be seen in Fig. 5. A decision diagram was then created by post-order traversal of the AST, in which the min/max function corresponded to a merger of two decision diagrams using given function. Variable ordering of the diagram was given by the order in which we visited leaves of the AST in the post-order traversal, i.e., by indices of the variables.

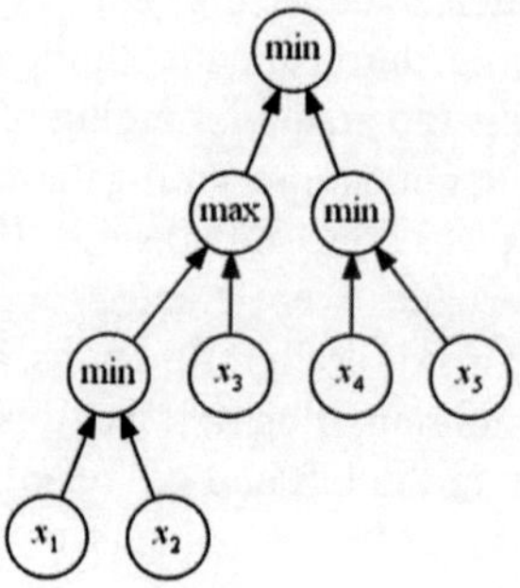

Fig. 5. Abstract syntax tree representing function $\phi(x) = \min(\max(\min(x_1, x_2), x_3), \min(x_4, x_5))$.

So far, we have described how we created diagrams for the structure function represented by a single diagram. We obtained a series of decision diagrams used in the other approach by merging this diagram with a diagram representing constant function by operation $\geq$ for values $1, 2, \ldots, m-1$ of the constant function. This merging can be realized using the *apply* algorithm mentioned in [14].

4.2 Properties Compared

We compared the two approaches to the representation of the structure function by observing certain properties of diagrams and the speed of the algorithms that worked with them. The property of the diagram we observed was the number of its nodes. In the case of the series of diagrams, the number of nodes was calculated as the sum of the number of unique nodes of all diagrams in the series. From the point of view of algorithms, we monitored the average time required to obtain the system state for an arbitrary state vector. Since obtaining the state of the system can be very fast due to the properties of the diagram (the edge can lead from the root of the diagram to the terminal node), we calculated the state of the system for 100, 000 random state vectors in one replication. Average times were obtained from 1, 000 replications for series and parallel systems and from 5, 000 replications for series-parallel systems. One repetition involved creating a diagram for the system with the given parameters and measuring the observed properties.

4.3 Experiment Results

Firstly, we compared the two approaches to the representation of the structure function in case of series and parallel systems. The number of unique nodes needed to represent the structure function of a series system with different numbers of components and different numbers of system/component states {3, 4, 5} is shown in Table 1. As we can see, the number of required nodes is the same in both cases. Diagrams representing the structure function of a parallel system show the same property. Other results showed that in addition to the number of nodes, the time required to calculate system state for an arbitrary state vector is also the same. In further experiments, we therefore focused on comparing the two approaches in series-parallel systems.

Table 1. Average number of nodes in a single MDD and in a series of MDDs depending on the number of system components in case of series systems

n	Single MDD			Series of MDDs		
	3	4	5	3	4	5
500	1,002	1,502	2,002	1,002	1,502	2,002
1,000	2,002	3,002	4,002	2,002	3,002	4,002
1,500	3,002	4,502	6,002	3,002	4,502	6,002
2,000	4,002	6,002	8,002	4,002	6,002	8,002
2,500	5,002	7,502	10,002	5,002	7,502	10,002

In Table 2, we can see a comparison of the number of nodes in the diagrams representing series-parallel systems. An interesting observation is that when using one diagram, the number of nodes depends on the topology of the generated system. In the case of a series of diagrams, we can see that the number of nodes is given by the number of components, regardless of the topology of the system. This number was smaller in all cases than when using one diagram. These results imply that approach that uses series of diagrams is more efficient in terms of diagram size.

To compare the two approaches in terms of reliability algorithms, we monitored the time required to calculate the system state for an arbitrary state vector. The results can be seen in Table 3. In contrast to the previous experiments, we can see, when observing this, there is almost no difference between the two approaches. This is probably due to the complexity of the algorithm for calculating the diagram value for the state vector, which is given by the number of variables that are in the diagram (number of diagram levels). This number is almost the same in both approaches. Although in the case of a series of diagrams we have to evaluate several diagrams, it seems that this does not play a significant role in the complexity of the calculation.

Table 2. Average number of nodes in a single MDD and in a series of MDDs depending on the number of system components in case of series-parallel systems

n	Single MDD			Series of MDDs		
	3	4	5	3	4	5
500	1,995	5,138	10,756	1,002	1,502	2,002
1,000	4,232	11,436	24,926	2,002	3,002	4,002
1,500	6,562	18,210	40,591	3,002	4,502	6,002
2,000	8,959	25,346	57,424	4,002	6,002	8,002
2,500	11,406	32,749	75,128	5,002	7,502	10,002

Table 3. Average time in milliseconds needed for obtaining state of the system for an arbitrary state vector in a single MDD and in a series of MDDs depending on the number of system components in case of series-parallel systems

n	Single MDD			Series of MDDs		
	3	4	5	3	4	5
500	235	232	244	235	234	246
1,000	467	459	479	467	462	482
1,500	698	687	716	700	690	718
2,000	910	912	950	930	897	951
2,500	1,164	1,137	1,184	1,160	1,139	1,187

5 Conclusion

In this paper, we compared the performance of two approaches for representation of structure function using decision diagrams. One is based on using a single diagram while the other uses a series of diagrams. In the experiments, we monitored the size of the diagrams needed to represent the system and the time needed to calculate system state. The results have shown that the approach based on a series of diagrams is more advantageous in terms of diagram size, and there is no significant difference between the approaches in case of calculating system state. However, it is worth noting that the variable ordering was given by the order in which leaves of the AST used for creation of the decision diagram were visited during the post-order traversal. An interesting question for the future research is whether the approach that uses series of diagrams would still be preferable if the order of variables is chosen differently. In addition to the ordering question, we would also like to identify properties of a system that allow us to represent it efficiently using the series approach. Such analysis could possibly result in generalization of the series approach to other kinds of systems.

Acknowledgement. This work was partially supported by the Slovak Research and Development Agency under the grant "Application of MSS Reliability Analysis for Electrical Low-Voltage Systems" (AMRA, reg. no. SK-PL-21-0003) and by the Ministry of Education, Science, Research and Sport of the Slovak Republic under the grant VEGA 1/0858/21.

References

1. Natvig, B.: Multistate Systems Reliability Theory with Applications. Wiley, Chichester (2011)
2. Rausand, M., Høyland, A.: System Reliability Theory. Wiley, Hoboken (2004)
3. Zaitseva, E.N., Levashenko, V.G.: Importance analysis by logical differential calculus. Autom. Remote. Control. **74**, 171–182 (2013)
4. Zaitseva, E., Levashenko, V.: Multiple-valued logic mathematical approaches for multi-state system reliability analysis. J. Appl. Log. **11**, 350–362 (2013)
5. Yanushkevich, S., Michael Miller, D., Shmerko, V., Stankovic, R.: Decision Diagram Techniques for Micro- and Nanoelectronic Design Handbook. CRC Press, Boca Raton (2005)
6. Bryant, R.E.: Graph-based algorithms for Boolean function manipulation. IEEE Trans. Comput. **C–35**, 677–691 (1986)
7. Kvassay, M., Zaitseva, E., Levashenko, V., Kostolny, J.: Binary decision diagrams in reliability analysis of standard system structures. In: 2016 International Conference on Information and Digital Technologies (IDT), pp. 164–172 (2016)
8. Srinivasan, A., Ham, T., Malik, S., Brayton, R.K.: Algorithms for discrete function manipulation. In: 1990 IEEE International Conference on Computer-Aided Design. Digest of Technical Papers, pp. 92–95. IEEE Computer Society Press (1990)
9. Zaitseva, E., Levashenko, V.: Decision diagrams for reliability analysis of multi-state system. In: 2008 Third International Conference on Dependability of Computer Systems DepCoS-RELCOMEX, pp. 55–62. IEEE (2008)
10. Xing, L., Dai, Y.: A new decision-diagram-based method for efficient analysis on multistate systems. IEEE Trans. Dependable Secure Comput. **6**, 161–174 (2009)
11. Mo, Y., Xing, L., Amari, S.V., Bechta Dugan, J.: Efficient analysis of multi-state k-out-of-n systems. Reliab. Eng. Syst. Saf. **133**, 95–105 (2014)
12. Kvassay, M., Zaitseva, E., Sedlacek, P., Rusnak, P.: Multi-valued decision diagrams in reliability analysis of consecutive k-out-of-(2k-1) systems. In: 2021 IEEE 51st International Symposium on Multiple-Valued Logic (ISMVL), pp. 81–86. IEEE (2021)
13. Kostolny, J., Zaitseva, E., Rusnak, P., Kvassay, M.: Application of multiple-valued logic in importance analysis of k-out-of-n multi-state systems. In: 2018 IEEE 48th International Symposium on Multiple-Valued Logic (ISMVL), pp. 19–24. IEEE (2018)
14. Mrena, M., Kvassay, M.: Comparison of left fold and tree fold strategies in creation of binary decision diagrams. In: 2021 International Conference on Information and Digital Technologies (IDT), pp. 341–352. IEEE (2021)
15. Kolowrocki, K.: Reliability of Large and Complex Systems. Elsevier, London (2014)
16. Mrena, M.: MichalMrena/DecisionDiagrams. https://github.com/MichalMrena/DecisionDiagrams. Accessed 03 Feb 2021

Neural Network Models for the Prediction of Time Series Representing Water Consumption: A Comparative Study

Krzysztof Pałczyński[1(✉)], Tomasz Andrysiak[1], Magda Czyżewska[2], Michał Kierul[3], and Tomasz Kierul[3]

[1] Institute of Telecommunications and Computer Science, Faculty of Telecommunications, Computer Science and Electrical Engineering, Bydgoszcz University of Science and Technology, Al. prof. S. Kaliskiego 7, 85-796 Bydgoszcz, Poland
{krzpal004,andrys}@pbs.edu.pl

[2] Institute of Machine Operation and Transport, Faculty of Mechanical Engineering, Bydgoszcz University of Science and Technology, Al. prof. S. Kaliskiego 7, 85-796 Bydgoszcz, Poland
magczy002@pbs.edu.pl

[3] Research and Development Center SOFTBLUE S.A., Jana Zamoyskiego 2B, 85-063 Bydgoszcz, Poland
{mkierul,tkierul}@softblue.pl

Abstract. In recent years dynamic growth of interests in numerical methods for prediction of water consumption has been observed. The forecasting obtained on the basis of such methods has doubtlessly been becoming very important element of created monitoring systems and management of critical water infrastructure. In herein article the author described the use of computational intelligence methods for water demand forecasting in the context of frequent telemetry readings. The results presented, concentrating on predictions made with the use of three neural network models i.e. linear, recursive and convolutional. The specific structures of neural network models along with their directivity were also considered. The influence of computing sets on the predictions results achieved, in terms of their structure as well as their cardinality was analyzed. It was pointed out that it is advisable to consider computational complexity of the analyzed prediction model in case of implementation into microprocessor architecture of the telemetry model. It was also proved that simple network models are able to manage with problems as effectively as their more complex equivalents in the task of forecasting the water consumption.

Keywords: Neural network models · Time series prediction · Forecasting water consumption · Critical waterworks infrastructure

1 Introduction

Water consumption forecasting is a crucial element of sensible exploitation and supervising of water pipe system infrastructure. It is possible due to performing frequent

W. Zamojski et al. (Eds.): DepCoS-RELCOMEX 2022, LNNS 484, pp. 186–196, 2022.
https://doi.org/10.1007/978-3-031-06746-4_18

measures concerning periodically changeable water abstraction in such a network. Such worked out predictions are mostly used to take up decisions connected with design, development and maintenance of water networks introducing procedures enabling optimization of work of pump houses, water treatment stations or sewage treatment plants [1].

Computer models are more and more often used to perform flow simulations, of type Extended Period Simulation (EPS). It seems to be incorrect – in this context – to use one universal model of water abstraction for all units of a water network. It is advisable to form and use different models, depending on the kind of consumers, the extent of water consumption as well as seasonal water requirement.

Recently, the formation of deterministic and stochastic models of water demand have been carried out by individual consumers. The most frequently used deterministic approach, ascribing model distribution of water consumption to so called reference consumers. Such a consumer then represents a particular, individual model of water abstraction prediction in a relevant period of time.

Prediction we will describe as rational future water consumption within the structure of water pipe system in oncoming time period, where the course of the phenomenon so far and the current state of the whole analyzed system is the basis of such choice. If we demand the prediction to be reliable enough (probability of fulfilling the prediction is close to 1) the prediction horizon ought to be close to the latest observation. Prediction farther than the horizon carries a risk of being overburdened with systematic error. Increasing of prediction horizon with its demanded precision creates basic challenge, when we work on methods of prediction using computational intelligence techniques [2].

Apart from shortening the time in prediction was constructed, its certainty may be increased by widening its tolerance, which means qualitatively, and quantitatively limited field of objective possibilities represented by the prediction.

In the article presented the use of computational intelligence methods for forecasting water demand levels in the context of frequent telemetry readings of water meters located in specific nodes of the water supply network was suggested. The results of short-term forecasts, carried out by means of linear, recursive and convolutional neural network models were presented. The structures and features of exploited models were verified either in terms of their structure or recursive and plexus dependences. It was also proved that simple neural models (that is unidirectional multilayer neural networks) can cope with problems as well as their more complex equivalents (that is recursive and convolutional neural networks) in the tasks of water consumption forecasting.

This paper is organized as follows: after the introduction, in Sect. 2, the related work of neural network prediction of the time series representing water consumption is presented. In Sect. 3 we present a detailed forecast of time series and calculation methodology of the neural networks. The article finishes with experimental results and conclusions.

2 Related Work

Water networks are managed by taking advantage of the instantaneous demand, so the machines depend on the immediate demand for water. There are works [3] in which water resource management focuses on predicting what the water needs will be. As a result, the costs of storage, distribution and processing are reduced. For this purpose, carrier vector linear regression and integrated Autoregressive Integrated Moving Average (ARIMA) were used. Based on the research carried out in Kuwait, ARMIA has Mean Average Percentage Error (MAPE) equal 1.8 and Root Mean Squared Error (RMSE) equal 9.4 in a situation where the linear regression of the support vector has the values equal 0.52 and 2.59, respectively. Therefore, it can be stated that there is a deviation of the forecast water demand in relation to the actual one. Based on the research, it can be concluded that ARIMA performs worse compared to support vector linear regression, which is more accurate and reliable.

In another article [4], ARIMA with and without seasonality, together with artificial neural networks, is used to forecast time series. The ARIMA solution with seasonality was proposed, i.e. SARIMA, neural approaches such as Long-Short-Term Memory (LSTM) or a Multi-Layer Perceptron (MLP) as well as a deterministic model based on the time function. Research was carried out in a service building to forecast the hourly water consumption. Based on the research, the presented model is efficient to improve the accuracy of forecasting water consumption. Consequently, this leads to a minimization of errors by 8.24% in the student error mean and 5.53% in the testing average for all individual or combined models. The deterministic model shows the best performance together with Artificial Neural Network (ANN) and Seasonal Autoregressive Integrated Moving Average (SARIMA).

There are many factors that influence water demand forecasting, which makes it difficult to do with forecasting models. There are articles [5] that propose forecasts based on wavelet noise reduction and ARIMA-LSTM. First, the model removes the interference factors by means of wavelet noise reduction in order to further use the ARIMA model, which aims to obtain the fit residuals and the forecast results. On this basis, the LSTM network is used to train the rest. The ARIMA prediction error is then predicted using the LSTM network to correct the ARIMA prediction result. Based on the research using the average daily water levels in the Chuhe river hydrological station, the presented model in relation to the LSTM network, the ARIMA model is more accurate than others.

Nowadays, the availability of water is declining. Water management and saving is very important also in the agricultural sector, where irrigated agricultural systems are one of the key users of water. Forecasting water consumption by these systems is difficult due to their heterogeneity. In this article [6], the field scale water demand of these systems is forecast for the next week. The Watery Forecaster tool is used, combining artificial intelligence techniques, climate data and satellite remote sensing to automatically build a forecasting model. The aforementioned tool, which was developed in Python, shows the accuracy of models from 17% to 19% and is characterized by a representativeness higher than 80%. The cultivation on the farm is variable, therefore the presented model may respond more accurately. The knowledge gained about the water consumption for the next week contributes to management in the agricultural sectors. This allows you to optimize and reduce water costs.

There are articles [7] that use the Dynamic Gauss Bayesian Network prediction model (DGBN) to forecast water demand in cities. Different DGBN models are compared with regard to their structure and corresponding prediction efficiency. Based on the conducted research, it was noticed that the models that rely on automatic learning of the network structure are not as effective as the models with the structure already designed. The best equivalent of the naive Bayes classifier achieved a 7.6% improvement over the MAPE predictions.

3 The Time Series Prediction

The forecasting process is an important analytical tool in order to determine both future consumption and set the basis for anomaly detection. In this work three different types of neural networks have been examined: fully connected, convolutional and recurrent neural networks. In addition to neural networks, two machine learning algorithms were also assessed i.e., Random Forest [8] and XGBoost [9]. These two algorithms are based on the ensemble of simplified decision trees [10] instead of neurons. Random forest performs regression by each tree soft voting for the result. Whereas XGBoost performs Boosted Trees strategy by training tree to correct the errors of the previous tree. These models were selected as a comparison to neural networks due to their well-known time series forecasting capabilities. In this section the neural networks are described their input data and pre-processing procedure as well.

3.1 Pre-processing of Data Representing Water Consumption

Time series forecasting problem has been resolved in this article by performing the regression on the hourly data gathered from 24-h window. As a result, each model predicted value of water consumption in the next hour based on the water usage from last 24-h. In this particular example desired model can be described by Eq. (1).

$$f : \mathbb{R}^{24} \to \mathbb{R}; \min_{X \in \mathbb{R}^{24}} (|y - f(X)|), \tag{1}$$

where y is the observed value approximated by the regression model. Time series obtained was contaminated by the registering anomalies in water consumption, that had to be corrected in pre-processing. Shapiro-Wilk tests conducted on the distribution of water consumption for each hour in every month suggests that the time series can be treated as normal distribution. As a result, the method of preprocessing the data based on probability from mean of the hour-based distribution was employed. Each value that was higher than 94th percentile of its hourly distribution was replaced by it. Such behavior reduced impact of outliers on training and evaluation of the model. The data prepared in this way represent the analyzed time series. The example of the time series is presented on the Fig. 1. At this point data are cleaned from outliers and divided into 24-h long overlapping time windows, with one hour difference between starts of subsequent time windows. The data is ready to be interpreted by the neural networks and machine learning algorithms.

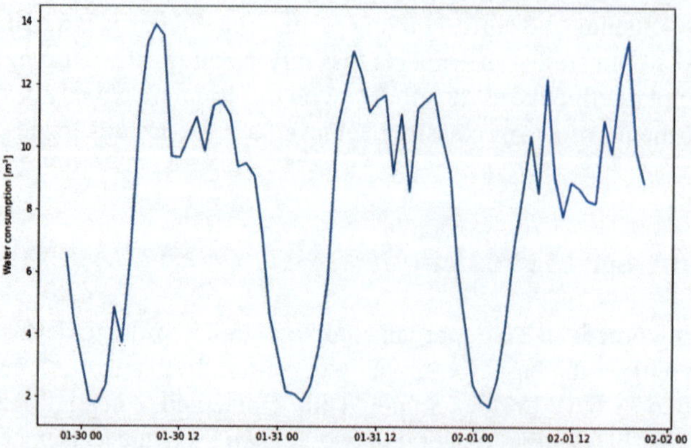

Fig. 1. Example of the cleaned time series.

3.2 Neural Network Models for the Prediction of Water Consumption

In the forecasting processes of the analyzed time series' future values (representing water consumption) more and more often there are used approaches based on different type of artificial neural networks (such as unidirectional multilayer, recurrent or self-organizing networks), which acquire ability to predict due to the course of learning processes. It is known that neural networks provide a possibility to build models mapping complex dependencies between input and output data for phenomena whose structure, rights of action or casual relationships have not been acknowledged to sufficient extent so as to create effective mathematical models. Then, neural analysis and prediction of variables represented in the form of time series requires usage of methods which include changes occurring in time (i.e. character, dynamics, and structure of data). They can also describe regularities connected with them.

In this work three different types of neural networks have been examined: fully connected, convolutional and recurrent neural networks. The next four subsections describe both artificial neural networks and ReLU activation function used for achieving non-linearity.

3.2.1 The ReLU Function

The ReLU function was used in order to achieve non-linearity between subsequent layers of networks. It is described by Eq. (2).

$$ReLU(X) = \max(0, X). \tag{2}$$

The goal of this function is to introduce non-linearity between the subsequent layers. This mechanism is required in order to prevent layers truncation into singular matrix multiplication. There is a multitude of activation functions, however the most commonly used one is ReLU due to its simplicity, computational efficiency and risk reduction of vanishing gradient problem occurrence. Computation of the ReLU function boils down to performing element-wise check of the sign-encoding bit in each value, resulting in computational complexity $O(n)$ and execution of one of the simplest arithmetical operations.

3.2.2 The MLP Neural Network Models with Fully Connected Layers

In this work two fully connected networks have been evaluated: FC-1 with one fully connected layer and FC-2 with two fully connected layers with the addition of the ReLU function between them. The behavior of the layer is presented by the Eq. (3).

$$f(X) = ReLU(XW + B), \tag{3}$$

where W is a matrix of weights and B is a vector of bias. The ith neuron in the layer has weights stored in the ith row of the W matrix and in i^{th} cell of the B vector. Computational complexity of neural network is influenced by chosen matrix multiplication algorithm and number of layers in the network. Naïve matrix multiplication algorithm has computational complexity $O(n^3)$. However, there are more efficient methods available like Strassen algorithm applying divide-and-conquer techniques in order to achieve computational complexity $O(n^{\log_2 7}) \approx O(n^{2.807})$. The amount of layers indicates number of subsequent matrix multiplications resulting in overall computational complexity of MLP network equal to $O(kn^{\log_2 7})$, where k denotes number of layers.

In this work two fully connected networks have been evaluated: FC-1 with one fully connected layer and FC-2 with two fully connected layers with the addition of the ReLU function between them.

3.2.3 The Convolutional Neural Network

The second type of the evaluated networks are the convolutional neural networks [11] (referred as CNN further in the article). Convolutional layers perform operation similar to one described in Eq. 3. The difference is that input tensor is split into the windows and operation described by the Eq. 3 is performed on each window instead on the whole activation map. Convolutional layers contain multiple kernels, each containing its own set of weights and biases and returns one-channel activation map. The layer concatenates results of each kernel into one multi-channel activation map.

Computational complexity of CNN networks is not so easily expressed as the complexity of the MLP network due to its dependance on multitude of factors. Operation of convolution requires summation of the results of element-wise multiplication between values of the signal in computational window and filter, so its computational complexity is equal to $O(km)$, where k is the length of the filter and m is the number of filters. The amount of convolutions performed in one layer is linearly dependent on the length of the input n and amount of filters m resulting in complexity of the network equal to $O(klm^2n)$, where l is the number of layers in the network.

The CNN network evaluated in this work consisted of two convolutional layers and one fully connected layer. The convolutional layers contained 4 and 8 kernels respectively. The multi-dimensional activation map outputted by the last convolutional layer was flattened to the form of the vector before it was inputted to the fully connected layer.

3.2.4 The Recurrent Neural Networks

Increasingly used types of neural networks for forecasting values of time series are currently becoming neural networks in which there is a feedback loop. The loop should undergo the same rules that all the entrances do, i.e. weighting and backward error propagation. Then, the state of individual neurons depends not only on input data but also on the previous state of network *h(t-1)*, which theoretically enables to keep information within the network's structure between consecutive iterations, so in a way it works as some kind of memory. Theoretically, such network should have an ability to react to a set of input data that has already appeared before at the entrance and was preserved in its structure. In practice, it is not that obvious.

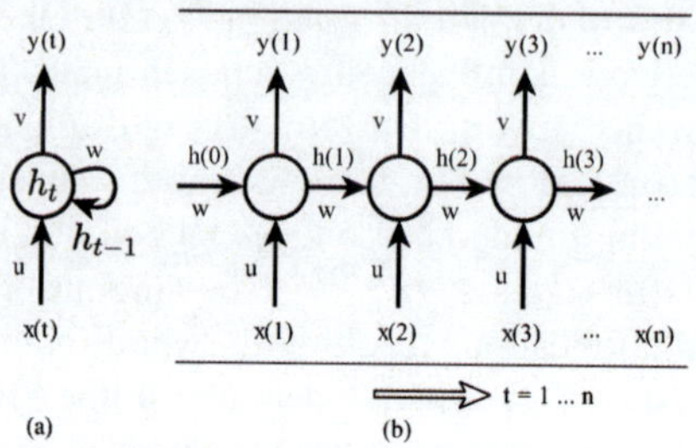

Fig. 2. Scheme presenting a neuron's "developing in time" in a recurrent network.

If we were to imagine an operational scheme of such a network, then in moment *t* every neuron with feedback loop (Fig. 2a) would be as it were an expansion (Fig. 2b) of all its states *h(t-n)*, corrected by weight coefficients *w*. Here is where the main problem lies. Weight coefficients in this chain multiply, and if the chain is long, the result of such operation quickly approaches zero (if $w < 1$) or infinity (if $w > 1$). This phenomenon, known as gradient fading or explosion, in practice, causes that in case of longer sequence the network is not able to learn any valuable data [12]. Solution to this problem was published in work of [13]. It consisted in integration in the structure of each neuron blocks with nonvolatile memory and additional switches controlling data flow. The new network was called Long Short-Term Memory, i.e. network with long memory of short range, which became insensitive to "forgetting" and remembering patterns.

In our article three types of recurrent layers were used i. e. recurrent layers, Long-Short Term Memory layers [14] and Gated-Recurrent Units [15] (GRU). Each recurrent network consists of two recurrent layers and one fully connected. The activation function used between each layer is ReLU.

Recurrent layer is the simplest layer evaluated in this work. It consists of the temporal state cells that process the data and transfer it both to another layer and further in the layer to another cell. This behavior is depicted by the Eqs. 4 and 5:

$$H_t = \tanh(W_h X_t + U_h H_{t-1} + B_h), \tag{4}$$

$$Y_t = \tanh\left(W_y H_t + B_y\right), \tag{5}$$

where W and U denotes weights matrices, B bias vectors and H internal, hidden state in each cell. The intuition behind this layer is that each cell performs the computation on the output of the previous cell and results of each cell is also transformed by another set of weights to convert hidden states into the layer output.

The computation complexity of performing operations to compute hidden state at time t of one RNN cell is equal to $O\left(m^{\log_2 7}\right)$, where m is the number of channels of the input signal. The number of state updates is linearly dependent on the length of the signal n. The amount of hidden states in the network is linearly dependent on the number of cells per layer k and number of layers in the network l resulting in computational complexity of the RNN network equal to $O\left(klnm^{\log_2 7}\right)$.

The LSTM layer is the further evolution of recurrent layer by addition of "forget gates". Due to this modification layer is able to forget information obtained from elements in time series, that are no longer viable in performing regression. It eliminates influence of no longer needed data on the layer and prevent vanishing gradient. The last version of recurrent layer evaluated in this work is layer GRU, which can be summarized as the LSTM layer optimized in order to perform the same task with less parameters.

In this work, the recurrent networks are implemented in two versions: bidirectional [16] and non-bidirectional. The bidirectional network contains bidirectional version of recurrent layers. Bidirectional layer contains two recurrent layers, with difference being that second recurrent layer computes data from the latest sample to the earliest one, instead of from the earliest to latest. Such an approach doubles computational cost of the network in order to increase interpretation efficiency on both earliest and latest input data.

4 Experimental Results

The algorithms have been evaluated using three different metrics: Mean Absolute Percentage Error (MAPE), Mean Squared Error (MSE) and Mean Absolute Error (MAE). In order to reduce the influence of the learning process randomness during best model selection each experiment has been repeated 10 times and the obtained metrics have been averaged. During the repetition of the experiment new training, validation and test sets have been generated, by splitting time data into weeks and selecting randomly selecting the destination of each weekly chunk of data. The training set was formed from 50% of weeks in the dataset, while validation and test sets contained 25% of data each. The results of models evaluation have been presented in the Table 1 and visually depicted on the Fig. 3.

In the Fig. 3 suffix "-B" denotes bidirectional version of the neural network and "RF" is the abbreviation of "Random Forest".

Models, that proved to be the most efficient in this particular task were RNN network and Random Forest machine learning model. The introduction of bidirectional layers increased computational cost without reducing error rate in a significant way. Although, fully connected network with two layers turned out to be the third in terms of highest error rate, the difference in performance between it and other models is negligible due to the much simpler architecture and computational cost. Such a model is desirable for General Purpose Input Outputs (GPIOs) with limited computational power like microcontrollers.

Table 1. Error metrics obtained from evaluation of models.

Network	MAPE	MSE	MAE
RNN	8.92%	0.716	0.622
RNN Bidirectional	9.17%	0.741	0.632
Random Forest	9.17%	0.748	0.631
GRU	9.30%	0.762	0.647
GRU Bidirectional	9.32%	0.759	0.641
XGBoost	9.55%	0.815	0.660
LSTM Bidirectional	9.65%	0.782	0.664
LSTM	9.83%	0.797	0.668
FC-2	10.0%	0.805	0.673
CNN	10.5%	0.880	0.705
FC-1	10.8%	0.902	0.714

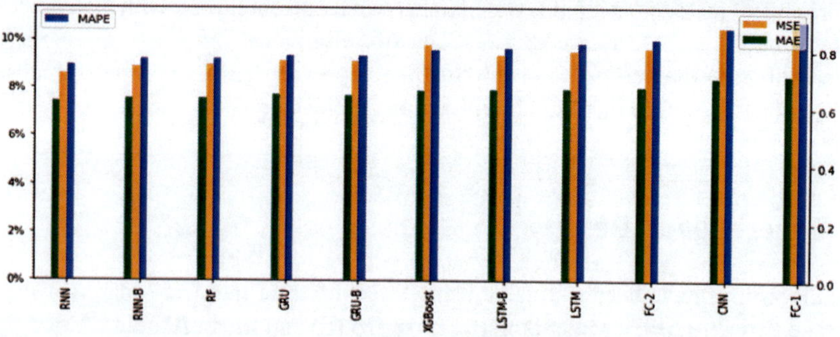

Fig. 3. Visual representation of the models' performance.

The training was performed on the CPU AMD Ryzen 9 3900X with 12 cores and 3.8 MHz clock. The device running the inference of the winner algorithm in production environment is ESP 32 with 2 cores 240 MHz and 4 MB RAM memory storage. This processor is connected to the electronic water meter for gathering measurements to be used for inference. Due to low computational cost of running FC-2 neural network the ESP 32 is able to handle the forecasting and distributed computing is not required. The ESP 32 requires 20 ms to perform forecasting running the FC-2 network.

5 Conclusion

The processes of water consumption forecasting have become the key element of rational strategy of exploitation of water supply systems. From this perspective we can easily notice an outstanding increase of interest in numerical methods of effective prediction of water consumption, which allow us to prepare forecasts within a specified time horizon.

The prepared forecasts enable them to support decisions making, concerning design, extension and maintenance of water supply networks along with introducing procedures enabling optimization of work of pump houses, water treatment stations or sewage treatment plants.

The use of artificial intelligence methods for prediction of particular elements of time series representing specific amounts of water consumption were described in this article. The results obtained focus on predictions made by means of three neural network models (linear, convolutional, recursive) and two machine learning algorithms – that is – Random Forest and XGBoost. In case of neural networks specific connection structures as well as their directivity were taken into consideration. On the other hand, choosing machine learning algorithms, different approaches towards the idea of constructing the set of decision trees for the sake of achieving the best average prediction were taken into account.

The results of experiments confirmed the good quality of the suggested methods' predictions, either the ones based on neural models or the others – using machine learning algorithms. The best results were achieved for RNN with prediction quality index MAPE, which equals 8,9%, whereas the worst prediction result was noted for FC-1 with MAPE that equals 10,8%. It was also stated that simple neural network models were able to cope with problems as efficiently as their significantly more complex equivalents within the task of water consumption forecasting. This conclusion has become outstandingly more empowered when the chosen prediction method needs to be optimized in terms of computational complexity, for the sake of its hardware implementation.

Acknowledgements. This research was supported by the National Centre for Research and Development under the realized fast track - Intelligent Development 2014–2020 (Project POIR.01.01.01-00-0633/21).

References

1. Arregui, F.: Integrated Water Meter Management. IWA (2006)
2. Bilewicz, K.: Smart metering – Inteligentny system pomiarowy. Wydawnictwo Naukowe PWN, Warszawa (2011)
3. Ibrahim, T., Omar, Y., Maghraby, F.A.: Water demand forecasting using machine learning and time series algorithms. In: 2020 International Conference on Emerging Smart Computing and Informatics (ESCI), pp. 325–329 (2020)
4. Ticherahine, A., Boudhaouia, A., Wira, P., Makhlouf, A.: Time series forecasting of hourly water consumption with combinations of deterministic and learning models in the context of a tertiary building. In: 2020 International Conference on Decision Aid Sciences and Application (DASA), pp. 116–121 (2020)
5. Wang, Z., Lou Y.: Hydrological time series forecast model based on wavelet de-noising and ARIMA-LSTM. In: 2019 IEEE 3rd Information Technology, Networking, Electronic and Automation Control Conference (ITNEC), pp. 1697–1701 (2019)
6. Gonzalez Perea, R., Ballesteros, R., Ortega, JF., Moreno, M.A.: Water and energy demand forecasting in large-scale water distribution networks for irrigation using open data and machine learning algorithms. Comput. Electron. Agric. **188**, 106327 (2021)

7. Froelich, W.: Forecasting daily urban water demand using dynamic Gaussian Bayesian network. In: Kozielski, S., Mrozek, D., Kasprowski, P., Małysiak-Mrozek, B., Kostrzewa, D. (eds.) BDAS 2015. CCIS, vol. 521, pp. 333–342. Springer, Cham (2015). https://doi.org/10.1007/978-3-319-18422-7_30
8. Breiman, L.: Random forests. Mach. Learn. **45**, 5–32 (2001)
9. Chen, T., Guestrin, C.: Xgboost: a scalable tree boosting system. In: Proceedings of the 22nd ACM SIGKDD International Conference on Knowledge Discovery and Data Mining, pp. 785–794. ACM (2016)
10. Swain, P.H.: The decision tree classifier: design and potential. IEEE Trans. Geosci. Electron. **15**(3), 142–147 (1977)
11. LeCun, Y., et al.: Backpropagation applied to handwritten zip code recognition. Neural Comput. **1**, 541–551 (1989)
12. Bengio, Y., Simard, P., Frasconi, P.: Learning long-term dependencies with gradi-ent descent is difficult. IEEE Trans. Neural Netw. **5**(2), 157–166 (1994)
13. Hochreiter, S., Schmidhuber, J.: Long short-term memory. Neural Comput. **9**(8), 1735–1780 (1997)
14. Sherstinsky, A.: Fundamentals of recurrent neural network (RNN) and long short-term memory (LSTM) network. Phys. D Nonlinear Phenom. **404**, 132306 (2020)
15. Chung, J., Gulcehre, C., Cho, K., Bengio, Y.: Empirical Evaluation of Gated Recurrent Neural Networks on Sequence Modelling (2014)
16. Schuster, M., Paliwal, K.K.: Bidirectional recurrent neural networks. IEEE Trans. Signal Process. **45**(11), 2673–2681 (1997)

Embedded Systems' Startup Code Optimization

Patryk Pankiewicz[1,2](✉)

[1] Politechnika Śląska, 44-100 Gliwice, Poland
patryk.pankiewicz@gmail.com
[2] GlobalLogic Sp. z o.o. ul., Strzegomska 48, 53-611 Wrocław, Poland

Abstract. One of the priorities of embedded targets is fast boot and wake-up times. After the car is turned on, all the systems should quickly initialize, synchronize and start communication. Automotive AUTOSAR Classic units are the ones with the quickest wake-up time, so they usually also serve as the power masters controlling supply voltage and wake-up signals for the high computation units. There are multiple system requirements creating constraints for the start-up time. The purpose of this paper is to present a memory erase algorithm that was created to reduce the time of bss (block starting symbol) initialization. The research was conducted on two microcontroller architectures – ARM and PowerPC. The algorithm has proven to significantly optimize the start-up time on both architectures. It can be applied to most embedded systems present also in other domains – not only automotive usages.

Keywords: AUTOSAR · Start-up code · Automotive · Assembler · BSS

1 Introduction

In current automotive embedded solutions, there is a high amount of technical requirements, demanding various functionalities and creating difficult constraints for the system architects. On one hand, there are the high computational power functions - multimedia systems, vision processing, high-speed communication, complex data operations, on the other hand, the systems of the car have to be reliable and fulfill safety, security, and timing requirements. Because of this, AUTOSAR Classic and AUTOSAR Adaptive standards were defined. AUTOSAR Classic focuses on hard time requirements, the highest safety requirements - up to ASIL-D (Automotive Safety Integration Level), but relatively low computational power. AUTOSAR Adaptive supports the functions that require very much resources - having the cost of lowering the safety and timing. These two standards combined can create a system that is both very fast, reliable, but also can execute very complicated functions.

The hardware design for such units usually consists of two microcontrollers - one for AUTOSAR Classic-based software and one for AUTOSAR Adaptive. Both are constantly communicating with each other, each realizing its own set of duties.

One of the most important timing requirements is the time from power-up to sending the first communication message or responding to an event. This time is system-specific, but for AUTOSAR Classic solutions, it is in the range of hundreds of milliseconds.

W. Zamojski et al. (Eds.): DepCoS-RELCOMEX 2022, LNNS 484, pp. 197–205, 2022.
https://doi.org/10.1007/978-3-031-06746-4_19

Multiple elements are contributing to the initialization of the unit. AUTOSAR provides a detailed description of the phases, provides a way to configure the order of execution, assign priorities, and with the use of BswM (Base Software Manager) create rules that will control the whole process. However, the element that the author would like to focus on is the "C Init Code". For each embedded unit, the very first element of execution is the startup code for the microcontroller, which is usually written in assembly language.

The content of the startup script usually consists of the following steps:

1. Disabling interrupts and watchdog, to ensure consistent startup code execution.
2. Configuring required microcontroller options concerning memory handling.
3. Erasing BSS (Block Started by Symbol) memory areas.
4. Initializing data in RAM by copying ROM default values.
5. Initializing stack pointer.
6. Configuring exception handlers.
7. Executing microcontroller-specific actions.
8. Jumping to main() function.

Later main() executes EcuM_Init() which then continues with AUTOSAR stack initialization.

The goal of this paper is to inspect the power-up procedure, analyze possible ways to optimize and improve the start-up performance. In Sect. 1 the author introduces the researched area, analyzes the available related scientific papers and justifies the work. Section 2 contains the main problem definition including the industrial measurement setup description. Section 3 presents the main analysis of the issue on two industrial microcontroller architectures – PowerPC and ARM. In Sect. 4, the author describes the main definition of the proposed algorithm, which was implemented and validated – results of the validation are presented in Sect. 5. The last part – discussions and conclusions are described in Sect. 6 which concludes this paper.

1.1 Analysis of Existing Papers

The first step was the analysis of the existing science and industrial papers tackling this area. The microcontroller manufacturers are releasing papers describing software initialization and optimization – [2] focuses on multiple aspects like setting certain microcontroller-specific functions – pipelining, buffering, caching, usage of DMA. [3] specifically handles the start-up code and mostly focuses on the ROM to RAM data copying and proper registers initialization. Science papers are mainly focusing on Android boot time for embedded targets – [4] presents promising results of 65% improvement of start-up time on ARM11 architecture by eliminating unnecessary elements of the initialization procedure, employing special kernel modifications, and modifying the userspace initialization. [5] presents a case of optimizing a digital TV start-up time by reordering resource initialization and I/O interleaving. Finally [6] provides a very interesting analysis of bulk data copying, data caching, DMA usage and proposes a novel architecture that minimized the influence of the biggest bottlenecks in the bulk copying approach.

All of the existing papers, however, are focused on higher abstractions systems, where the time domain is usually expressed in seconds and multiple factors on higher-level abstraction contribute to the initialization time. Even if the analysis touches assembly level, it doesn't focus on low-level embedded targets, especially not AUTOSAR Classic-based systems used for most safety-related systems in cars. Due to the lack of existing papers handling this particular area, the author decided to analyze the topic further.

2 Problem Definition

In the industrial project, the ECU (Electronic Control Unit) was not able to fulfill the requirement of sending the first Ethernet Network Management Frame within 220 ms from startup. Let's analyze the way how AUTOSAR Classic handles this part of the software. EcuM (Electronic Control Unit Manager) [1] is the module that is responsible for the handling of the system states, especially initialization, deinitialization, sleep modes, shutdown, and wake-ups. Proper EcuM configuration greatly influences the startup time of the system. The start-up phase of EcuM is presented in Fig. 1.

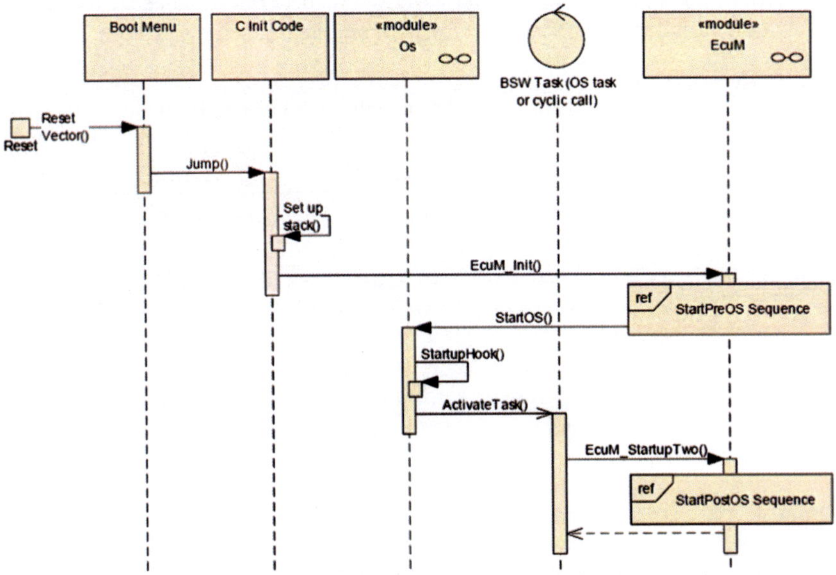

Fig. 1. AUTOSAR system start-up sequence diagram.

The problem was analyzed in two microcontroller architectures - PowerPC and ARM Cortex-M4.

2.1 Measurement Setup

To solve the issue, a timing analysis setup was prepared. Since time measurements were in the range of milliseconds, then the debugger proved to be enough for study.

The measurements hardware setup:

1. PowerPC ECU
2. ARM Cortex-M4 ECU
3. Debugger
4. PC
5. Automotive Ethernet communication device.

The whole start-up sequence present in AUTOSAR systems was analyzed at first, starting from the entry point after POR (Power-on Reset) until the function that sends the Ethernet Frame. After initial analysis and inspection, the measurements showed that a significant contribution to the communication latency was the time of C code initialization- the start-up script. Continuing the inspection, the biggest factor was the BSS section initialization. The results for both architectures are shown in Table 1.

Table 1. Default memory initialization measurements.

	PowerPC	ARM
BSS initialization time [ms]	~30 ms	~300 ms
BSS size [bytes]	512 800 bytes	962 320 bytes
Time of erasing of 1 byte [ns]	~38 ns	~585 ns

With such a high influence on the startup time, this part of code has been identified as the main issue and the candidate for optimization.

3 Analysis

To define a way to clear the memory faster, the default way of operation was inspected. In both solutions, as well as some other projects- the start-up script written in assembler, was using a straightforward approach of clearing the memory in a simple loop, as shown in Fig. 2. The PowerPC solution was using assembler code, the ARM solution employed a C language "for loop" for the same purpose.

To establish a faster way of performing the erasing operation, both architectures had to be inspected.

ARM Cortex-M4

- 13 general-purpose registers,
- 3 special purpose registers.

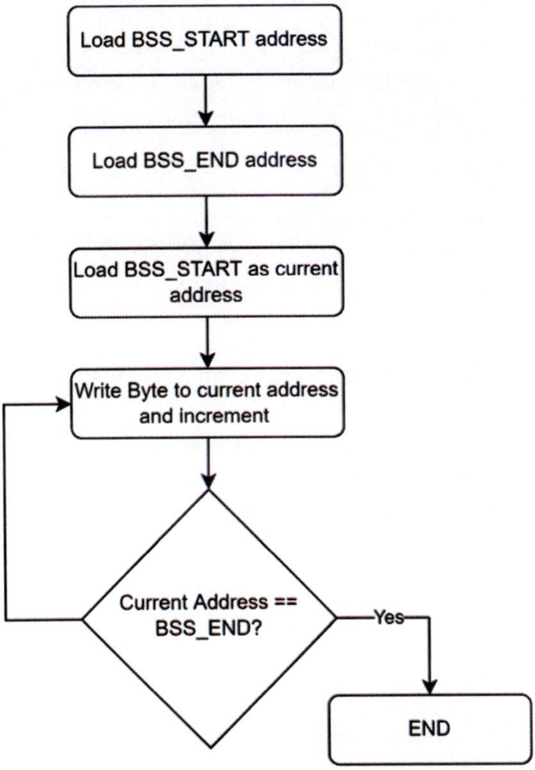

Fig. 2. Default memory initializing procedure.

PowerPC 32-bit architecture.

- 32 general-purpose registers,
- 32 floating point registers,
- special registers for branching, exception handling, and other purposes.

The most straightforward approach to speeding up the process was to increase the quantum of erasing - by default the memory was erased byte by byte on the PowerPC architecture, and word by word on the ARM. After inspecting the available sets of instructions [8, 9], the following orders were identified as potential solutions:

- STM (Store Multiple) for ARM architecture
- STMW (Store Multiple Word) for PowerPC architecture.

These two commands are storing multiple consecutive words from general-purpose registers to a given memory location. With multiple 32 bit registers available, the time of erasing could be greatly decreased - even by 128 times for the PowerPC (32 * 4 bytes registers). However, the author had to analyze the memory access and alignment conditions to create a versatile algorithm. By simply using the store multiple commands, with unknown start and end addresses, the last cells of the memory could have been skipped (if they would be less than the erase quantum). Additional problem is, that in various architectures, unaligned memory access can lead to exceptions.

3.1 Memory Alignment Analysis

Since memory areas that are erased during startup procedure don't have to be perfectly 32-bits aligned, the algorithm has to automatically detect misalignment and handle it accordingly.

PowerPC
For PowerPC STMW instruction there are only the following constraints as per [9]. An exception will be triggered if:

- *The operand of lmw, stmw, lwarx, stwcx., eciwx, or ecowx is not aligned.*
- *The instruction is lmw, stmw, lswi, lswx, stswi, or stswx and the processor is in little-endian mode.*
- *An operand of lmw or stmw crosses a segment or BAT (Block Address Translation) boundary.*

ARM
For ARM STM instruction there are the following constraints as per [10].

- *The word element size aligns the operation.*
- *Non-word operations for multiple commands.*

Additionally, basic sanity checks were implemented before applying the erasing procedure - for example, is the start address smaller than the end address - otherwise, risk of memory length overflow could lead to a critical error of erasing too much memory.

4 The Algorithm Definition

In general, the algorithm was designed to be universal for all microcontrollers' architectures, but of course, due to different available registers and peripherals, it always has to be adjusted for certain usage. The algorithm's flow is shown in Fig. 3.

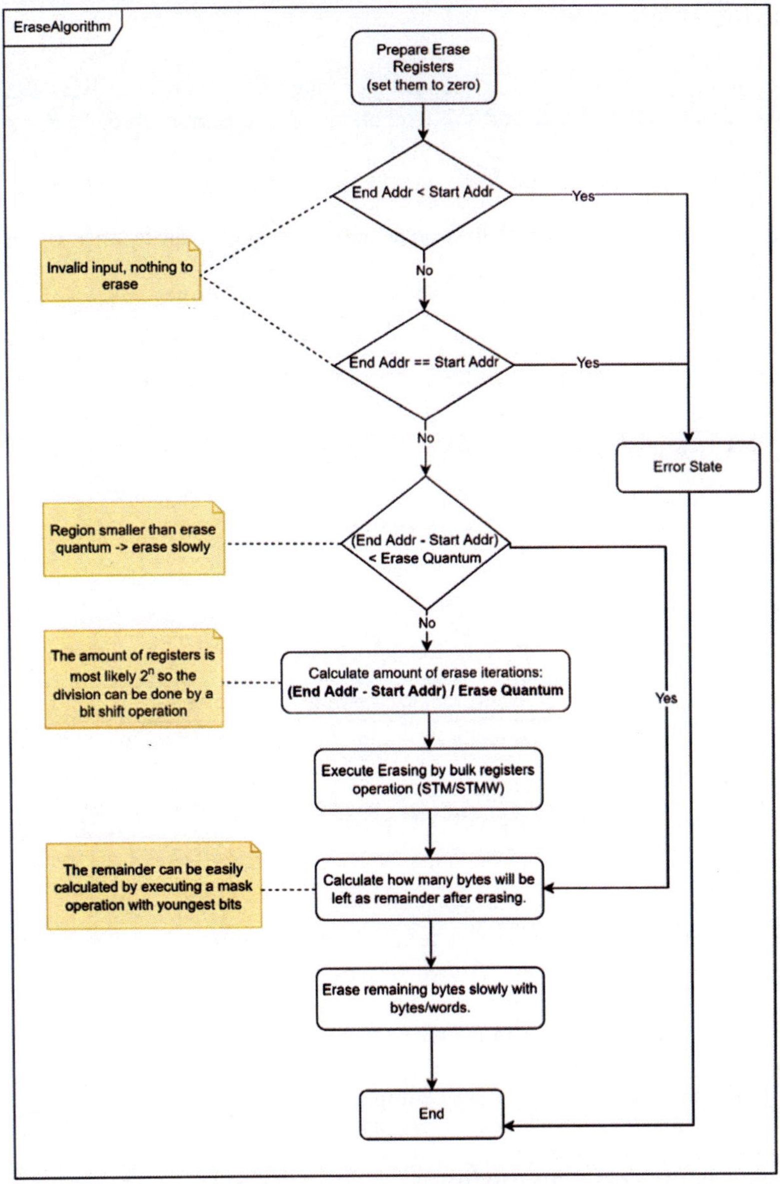

Fig. 3. Erase algorithm flow diagram.

5 Validation

The proposed algorithm was implemented for both architectures – PowerPC & ARM. Both solutions were tested and the execution time was measured. The results are presented in Table 2 and Table 3.

Table 2. Memory initialization measurements after optimization.

	PowerPC		ARM	
	Default	Optimized	Default	Optimized
BSS initialization time [ms]	~19 ms	~2.16 ms	~300 ms	~29.75 ms
BSS size [bytes]	5125800 bytes		962 320 bytes	
Time of erasing of 1 byte [ns]	~38 ns	~4.2 ns	~585 ns	~30.9 ns

The gains were significant:

Table 3. Reduction of memory initialization time.

	PowerPC	ARM
Reduction of erase time [ms]	17.3275 ms	270.25 ms
Reduction of erase time [%]	88.63%	90.08%

The exact results depend on multiple variables – the architecture, the memory region size, the used amount of registers. However, overall the gains are substantial – they provide valuable time for the whole start-up procedure. In the case of ARM architecture, the system requirement of 220 ms until the start of Ethernet communication was impossible to reach, since the memory erase alone took over 300 ms. After applying the modified erase procedure, the requirement can be implemented.

6 Discussion and Conclusions

The presented analysis, measurements, and resulting memory erase algorithm proved to significantly reduce an important part of the microcontroller initialization procedure. The solution was used in two embedded architectures – PowerPC & ARM, which have shown its versatility. The memory-erasing is not specific for automotive AUTOSAR systems – it is present in most embedded targets. The resulting time reduction was in the range of 90%. The author described the measurement and test setup that allows further work on this subject. Possible areas for further work are:

1. Analysis of speed of erasing of aligned vs. non-aligned blocks.
2. Employing bulk erase approach for usage during the execution of the system.
3. Comparing DMA efficiency vs. a bulk copy.

All work was implemented in an industrial solution and proven to work in the field, in a commercial application, in mass-produced cars. It has allowed fulfilling demanding industry requirements concerning very fast initialization and beginning of communication. Additional analysis regions were the error handling and memory alignment topics, which have ensured the required safety of the solution. Appropriate error handling routines were introduced to avoid memory corruption and system exceptions.

The author encourages to employ this solution and apply in it various embedded solutions. The gain of hundreds of milliseconds may be a considerable contribution in cases of high-speed safety-related systems.

References

1. AUTOSAR Homepage, Specification of ECU State Manager. https://www.autosar.org/fileadmin/user_upload/standards/classic/4-3/AUTOSAR_SWS_ECUStateManager.pdf. Accessed 26 Dec 2021
2. NXP Homepage, MPC5676R Software Initialization and Optimization. https://www.nxp.com/docs/en/application-note/AN4324.pdf. Accessed 26 Dec 2021
3. NXP Homepage, MPC5200 Startup Code. https://www.nxp.com/docs/en/application-note/AN2551.pdf. Accessed 26 Dec 2021
4. Singh, G., Bipin, K., Dhawan, R.: Optimizing the boot time of Android on embedded system. In: 2011 IEEE 15th International Symposium on Consumer Electronics (ISCE), pp. 503–508 (2011). https://doi.org/10.1109/ISCE.2011.5973881
5. Jo, H., Kim, H., Jeong, J., Lee, J., Maeng, S.: Optimizing the startup time of embedded systems: a case study of digital TV. IEEE Trans. Consum. Electron. **55**, 2242–2247 (2009). https://doi.org/10.1109/TCE.2009.5373794
6. Jiang, X., Solihin, Y., Zhao, L., Iyer, R.: Architecture support for improving bulk memory copying and initialization performance. In: 2009 18th International Conference on Parallel Architectures and Compilation Techniques, pp. 169–180 (2009). https://doi.org/10.1109/PACT.2009.31
7. ARM Homepage, Cortex-M4 Technical Reference Manual. https://developer.arm.com/documentation/ddi0439/b/. Accessed 26 Dec 2021
8. NXP Homepage, MPCxxx Instruction Set. https://www.nxp.com/docs/en/reference-manual/MPC82XINSET.pdf. Accessed 26 Dec 2021
9. NXP Homepage, Freescale PowerPC Architecture Primer. https://www.nxp.com/docs/en/white-paper/POWRPCARCPRMRM.pdf. Accessed 26 Dec 2021
10. ARM Homepage, ARMv7-M Architecture Reference Manual. https://developer.arm.com/documentation/ddi0403/ed. Accessed 26 Dec 2021

Labeling Quality Problem for Large-Scale Image Recognition

Agnieszka Pilch(✉) and Henryk Maciejewski

Wroclaw University of Science and Technology, Wroclaw, Poland
{agnieszka.pilch,henryk.maciejewski}@pwr.edu.pl

Abstract. Most CNN models trained on the popular ImageNet dataset are created under the assumption that a single label is used per training image. These models realize remarkable performance on the ImageNet benchmark (with top-1 scores over 90%). Despite this, recognition of several categories is not reliable, as models for these categories can be easily attacked by natural adversarial examples. We show that this effect is related to ambiguous, single labels assigned to training and testing data for these categories. The CNN models tend to learn representations based on parts of an image not related to the label/category. We analyze the labeling scheme used to annotate the popular ImageNet benchmark dataset and compare it with two recent annotation schemes - CloudVision and Real labeling schemes, which are both crowd-sourced annotation efforts. We show that these two schemes lead to a very different granularity of annotations; we also argue that new annotations schemes should not rely on the accuracy on current ImageNet benchmarks as the hint for their correctness (at the Real scheme does).

Keywords: CNN · Realibility of deep models · Annotations of ImageNet

1 Introduction

The creation of large data sets has affected the development of machine learning. Researchers were able to use the existing dataset and create a better model based on it. The process of creating large data sets manually and precisely would be time consuming and expensive. Therefore, automatic labeling methods or crowd-sourcing sites are used to create labels for images. These methods are prone to mistakes [11]. Automated process and people can get confused when verifying labels. Therefore, various error corrections are additionally applied to eliminate recognition error or human error [9].

The ImageNet Large Scale Visual Recognition Challenge (ILSVRC) has become a benchmark used in image recognition tasks since 2010 [10]. Recent research has shown that this popular dataset contains questionable, incorrect or ambiguous labels [14]. In [9], the authors demonstrate that in ImageNet and other popular datasets there are such labeling errors: wrong label, multi-label, missing label in class set, ambiguous image. Wrong or ambiguous labels may lead to

W. Zamojski et al. (Eds.): DepCoS-RELCOMEX 2022, LNNS 484, pp. 206–216, 2022.
https://doi.org/10.1007/978-3-031-06746-4_20

spurious correlations learned by CNN models. As shown in [14], CNNs models tend to recognize several ImageNet classes with the model's attention focused on image areas not related to the label. To deal with the ImageNet labelling issues/inconsistencies, new labeling initiatives have been recently developed [1,3].

In this work, we analyze these new labelling schemes, and compare them to the original ImageNet labels. We want to verify which of these schemes may be less susceptible to spurious correlations and thus lead to more robust representations. The main contributions of this work are the following. The most questionable labels, for classes with least robust models, have been selected based on [14]. We took labels with a robustness score below 30%. We showed how new labeling works and compared it with the original labeling process of the ImageNet dataset. We considered two new labeling schemes called "ReaL" and "CloudVision". Then we conducted experiments that show the advantages and disadvantages of relabeling with each of these schemes. Finally, we evaluated the new labels in terms of usability and will suggest ways to improve the labels for better models in the future.

The paper is organized as follows. In Sect. 2, we explain how the robustness score measure works [14], analyze the canonical labeling method [4], and describe two new labeling methods that differ significantly from each other. According to [3], anomalies in ImageNet validation labels are identified and new labels named "ReaL" are created based on this, which eliminate this problem. CloudVision labels were available for the Know Your Data project, which were created by the Cloud Vision API [1] to detect labels in an image. Based on these labels, we later verified their correctness on the original label set. In Sect. 3, we present an analysis of labeling schemes. We examined how the labels are distributed and whether undesired effects occur - repeating the errors of the original labeling. We check the selected labels and report the results. Finally, we discuss labeling methods on how to improve unreliable representations.

2 Methods

2.1 Method Robustness Score

To realize the research, we first used the robustness score method [14], which allows to identify classes whose representations rely on spurious correlations. The score is the percentage of total attention expressed by the saliency map of the model that is in the region of interest. Low values of this measure indicate that there is a mismatch between the object of interest, i.e. the label, and areas on which the model's attention is focused. Figure 1 shows examples of classes with lowest robustness score crs (but high accuracy on testing data). Strong spurious correlations for these examples (e.g., class miniskirt correlates with women's legs, or pickelhaube - with a uniform in sepia), clearly show that labels of these images are either wrong or ambiguous.

In Sect. 3, we will focus on classes/labels with the lowest robustness score measure to validate usability of the new labeling schemes.

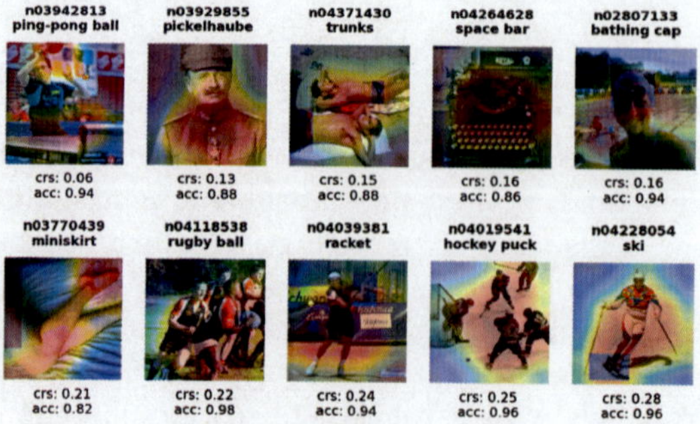

Fig. 1. Selected examples of ImageNet classes with low class robustness score *crs* (image from [14]). Labels for these examples are either wrong (e.g. pickelhaube) or ambiguous (e.g. racket).

2.2 Analysis of Imagenet Labels

ImageNet is a large dataset of photos and labels designed for computer vision tasks. Its task was to contribute to the development of deep neural networks for image recognition [4,5]. There are approximately 14 million images in the collection, approximately 21,000 groups or classes (synsets). ImageNet Large Scale Visual Recognition Challenge (ILSVRC) competitions are organized annually on the basis of a subset of this dataset, which resulted creation of the AlexNet [8], ZFNet [15], GoogLeNet [13], VGG [12] and ResNet [6] networks. ImageNet was created by defining synsets and their trees. Then the photos for each synset were manually collected. After initial deletion of duplicates, approximately 10,000 photos were obtained per synset. Photos were then collected from several online image search engines, based on the synset value, the synset parent value, and the synonym of the sysnet value from the WordNet database (e.g., "Pekingese", "Peke", and "Pekingese dog").

These queries were then translated into other languages e.g. Spanish, Chinese and Italian. After collecting the photos, the dataset was cleaned again from duplicates within the synset. The last step was the verification of labels by users. For this purpose, the Amazon Mechanical Turk (AMT) service was used. The human error related to labeling has been minimized by means of multiple analyses of one photo by many users and the use of a developed algorithm to determine the level of reliability for a given synset. Ultimately, synsets have about 500–1000 images. Images in the subset of ImageNet used as benchmark in ILSVRC-2012 are labeled by one term from the synset.

2.3 New Imagenet Annotation Scheme ReaL

Dataset ReaL [3] was created in response to research on the importance of labels in a dataset. It is based on the ImageNet labels from the ILSVRC-2012 competition. Verification of original labels started with the use of 19 popular CV models and label prediction for all photos. On the basis of this, the photos were grouped into small subsets with identical labels. 256 images were selected and distributed to computer vision experts.

Based on their results, precision and recall (gold standard) were determined. Labels generated by 6 models were selected, which were recalled at approximately 97.1%-symbol for further analysis. Labels from the remaining models have been rejected. After this operation, the average number of labels per photo dropped from 13 to 7. After changing the labels, the photos were again divided into subgroups with identical labels. It is important to note that photos with labels identical to original labels were not subject to human verification. There were 25111 of them. Photos with more than 8 labels were divided according to the WordNet hierarchy into two tasks. Each task was verified by 5 annotators. The animal-related classes got a sixth answer from the expert because the classes that included animals were the most susceptible to ambiguity, according to the dataset developers. Annotators' responses were combined and any errors were eliminated using David and Skene's algorithm. Finally, 57,553 labels were obtained for 46,837 images. A total of 3,163 images were rejected.

2.4 New Imagenet Annotation Scheme CloudVision

The CV labels are taken from the Know Your Data [2] project which uses the Cloud Vision API (to detect labels in an image of different categories). The Cloud Vision API has been trained on large amounts of Google data [7]. The "Cloud Vision" interface mostly returns labels with a computer-generated identifier (MID) corresponding to an entry in the Google Knowledge Graph [1].

3 Analysis of Labeling Schemes

In this section, we compare canonical labeling with two new approaches: CloudVision and ReaL. In the study, we on the original labels with robustness score below 30% for different models [14]. These include e.g.: 'pickelhaube', 'sunglasses', 'miniskirt', 'bathing cap', 'basketball'.

3.1 Treemap Visualization

Treemap is a graphical representation method that allows for hierarchical representation of elements. More frequent labels assigned to original classes are represented by larger rectangles. The color of the bar represents the original (parent) label, colors of rectangles in treemap represent assigned labels. The left part of the Fig. 2 shows CloudVision labels. The clear characteristic of this labeling is the large number of labels per class, and these labels are generally different from the original ImageNet labels. The right part of the Fig. 2 refers to the ReaL labeling. You can see for the most prominent label in each class is the same as the original label (largest rectangle has the same color as the bar). This means that the ReaL labeling repeats the labels from the canonical ImageNet. Note that the same color in Fig. 2 represents the same class (Fig. 3).

Fig. 2. Treemap for Labels, left: Cloud Vision, right: ReaL. Categories and subcategories are marked with specific colors for a rectangle. Cloud Vision has a lot of different labels that don't coincide with ImageNet's canonical labeling. ReaL labeling proposes fewer labels than Cloud Vision, but these labels overlap with ImageNet's canonical labeling.

Fig. 3. Example of a tree map for a selected class pickelhauble, left: Cloud Vision, right: ReaL. CloudVision assigns larger label sets per class, as compared with ReaL. The same holds for other classes, e.g. rugby ball, miniskirt, airship etc.

The labels for Cloud Vision represent general categories such as, for example, for the rugby ball class, the labeling method focuses on the whole image and the most common labels are sports equipment (43), shorts, sports uniform (35), player (31) and ball (26). The minority of labels includes labels such as grass (16), sky (4), motor vehicle (3) and shoe (7). You can clearly see that this labeling method considers a broader context than a small element in the image.

For ReaL labeling example, the labels for the original rugby ball class as follows: rugby ball (47), volleyball (1), sunglass (1), soccer ball (1) and others. These labels are mostly similar to the original labeling. The majority class in ReaL labeling covers the original ImageNet labeling.

3.2 Analysis of ReaL Schema Labeling for Selected Classes

Firstly, we selected labels with robustness score below 30% for analysis. Classes such as miniskirt, diaper, pickelhaube and rugby ball will be presented. In the previous section, we learned that ReaL labeling mostly repeats ImageNet mislabeling. Only a small number of images received different or multiple labels.

Next, we checked how much the new label in ReaL labeling was distributed among the classes from the original labeling. To do this, the proportion for the original labeling was calculated, where it was expressed by the number of occurrences of the new ReaL label with respect to the canonical ImageNet classes. The result is expressed as a percentage, the higher the value, the more often the label occurred in the original labeling. In Fig. 4 we observe recurrence of ReaL

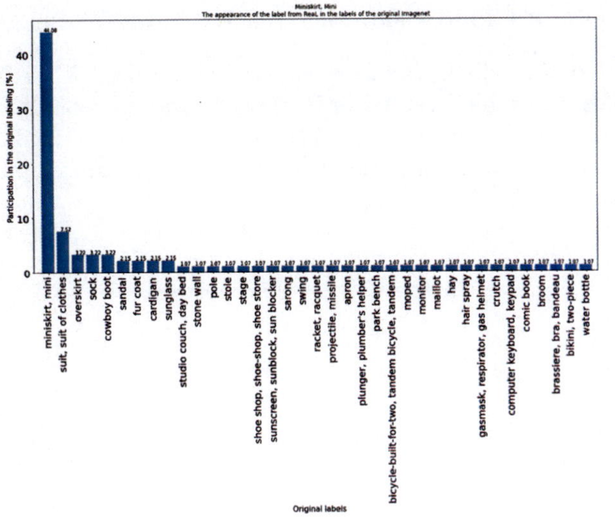

Fig. 4. The appearance of an example ReaL label in the original labels

Table 1. Selected examples of ReaL labeling. **Class** is the canonical ImageNet label. **Soal** is the percentage of the new label in the set of all labels assigned to the class. **Poi** is the percentage on images in the class with the new label. The RL label generally repeats the original label, the other labels provide a some background/context information.

Class	RL	Soal	Poi
Pickelhaube	Pickelhaube	73.13	98
	Military uniform	11.94	16
	Rifle	5.97	8
	Mountain bike	2.99	4
	Picket fence, paling	1.49	2
Miniskirt	Miniskirt, mini	46.07	82
	Sandal	6.74	12
	Cowboy boot	5.62	10
	Jersey, t-shirt, tee shirt	4.49	8
	Sunglasses, dark glasses, shades	3.37	6

labeling with canonical labeling (similar distribution for four example labels). Other labels are few. We can see that for the miniskirt label **(a)** it was noted that in the original labeling it was also found in other classes, e.g. sock, sandal, stole, pole, sunglass, apron and monitor.

The authors of the new ReaL labeling noted that labels from the canonical labeling are error-prone. In addition, they noted that images may contain several other objects in addition to one object. The ReaL labeling is not explicit, but it does repeat errors for labels with low robustness scores [14].

3.3 Analysis of CloudVision Schema Labeling for Selected Classes

In this section, we will introduce CloudVision labeling. CloudVision labeling does not make the same mistake as ReaL and gives completely different labels, unrelated to WordNet labeling. Intersection between classes will be shown with examples of labels detected by Cloud Vision, for example: sky, military uniform, shirt, and knee (Table 2).

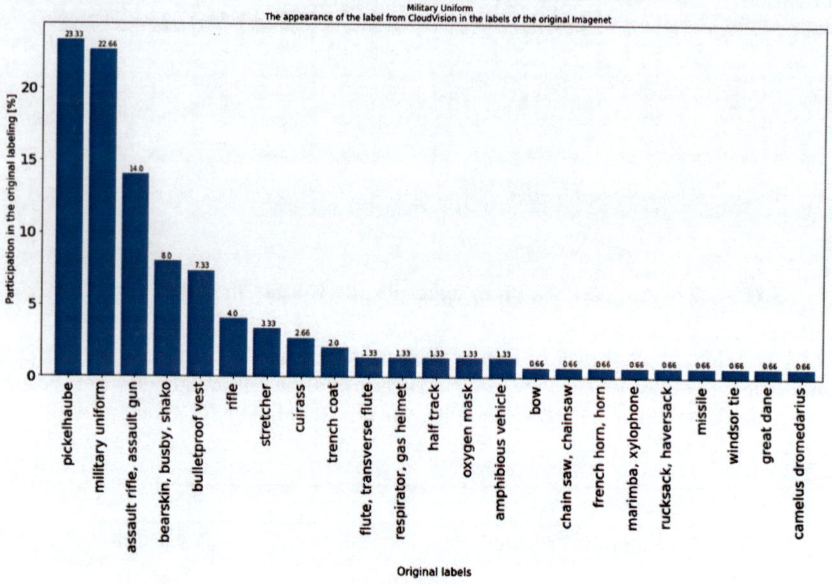

Fig. 5. The appearance of an example CloudVision label in the original labels.

In Fig. 5, we see that labels from the original ImageNet collection such as, for example, pickelhaube, military uniform, assault, bearskin, bulletproof, riffle, and stretcher were classified into the Military Uniform label. This label would partially coincide with the conclusions of the [14], where according to the authors, the pickelhaube label strongly correlates with the uniform in an old-looking monochrome photograph.

Another label we considered is "Sky" because it appeared in many images relative to the old labeling, for example: airship, ski, pole, swimming trunks, sunglasses, maillot, and volleyball. For example, for the original labels volleyball, bow tie, ping-pong ball, shower cap, sunglasses, horizontal bar Cloud Vision labeling gave a shirt label. The new knee label was given to such old labels as miniskirt, dumbbell, maillot, balance beam, diaper, volleyball, rugby ball, and hairspray. This example also shows one of the findings in [14], that the class "miniskirt" strongly correlates with naked female legs (knee here).

Table 2. Selected examples of CloudVision labeling. Columns are defined in Table 1. This labelling scheme focuses on the whole image, not just the ImageNet suggestion. The **Soal** score for new labels is very small for each label, this means that labeling notices even small elements. The **Poi** score is relatively high, meaning that each image has multiple elements, not just the one suggested by ImageNet.

Class	CV	Soal	Poi
Pickelhaube	Military uniform	7.01	70
	Military person	7.01	70
	Soldier	6.41	64
	Headgear	4.81	48
	Non-commissioned officer	4.41	44
Miniskirt	Thigh	6.6	66
	Leg	6.2	62
	Knee	5.4	54
	Waist	5	50
	Sleeve	4.2	42

3.4 Comparison of Labeling Schemes

In this section, we focus on comparing labeling methods. We know that ReaL labeling duplicates ImageNet mistakes. We also know that CloudVision's labeling has its own labels, which is not necessarily similar to ImageNet's original labeling. To compare labeling methods, let's look at the Fig. 6. We see different images, for example in part (a) we have the original "rugby ball", where for Real labeling is the same, and Cloud Vision labeling notes "Sky", "Building", "Skyscraper", "Urban design", "Landmark", "Tower block", "daytime". It may be that this building is shaped like a rugby ball, but in reality in image recognition we would not want our model to classify this image as a "rugby ball". The second image shows a photo where there is a child holding cricket balls, however we can see that in this image we can see many other things besides the bounding box such as for example blouse, pants, child, people and grass. Perhaps Cloud Vision did not include the cricket ball but focused on the background and overall context. In part (b) we have the following original classes: "mouse, computer mouse" and "diaper". Again, ReaL labeling repeated the old labels, while Cloud Vision labeling noted for rugby ball such elements as rugby short, sport uniform, shorts, player, ball game, tournament. For the "mouse" class, CloudVision also focuses on the whole image, which received the following labels: computer, peripheral, cat, input device, personal computer, mouse etc. For the next class "diaper", it can be seen that Cloud Vision does not notice the diaper, but suggests labels such as glasses and knee. For the label "airship" it proposes the following classes: "tire", "smile", "air travel", "wheel", "aircraft", "sunglasses" and "vehicle". This example in part (b) shows that pictures have more elements than the one proposed in the original labeling. The question arises why the selected image has only one label proposed in ReaL and the original labeling, when we can see that there are many other relevant elements in the images. How can our deep network know what to learn from when it relies on such unreliable images? ReaL labeling has a similar labeling approach to the original ImageNet method. Unfortunately, because of this, it makes the same mistakes and is prone to spurious correlations. For CloudVision labeling, there are no analogues in canonical labeling for most labels. It would be necessary to semantically compare these labels and try to match them that way. It is also interesting to consider the question of should the images in these classes duplicate the original label, or should we change it? We also see the dependency of image features appearing everywhere. For example: "sky", "grass" and "cloud".

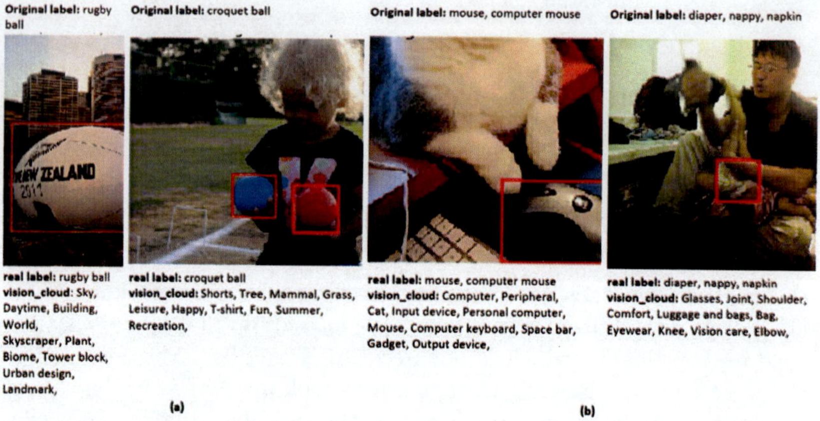

Fig. 6. Representation of the original bounding boxes in images.

4 Conclusion

In this paper, robustness score method was used to select questionable classes. Based on these classes, a verification was later performed on two new labeling methods. It was verified how the new labels work, then a tree map was created that illustrated many of the initial dependencies for these methods. Subsequent levels of the tree map were then checked for questionable classes. It was noticed that CloudVision labeling was wealthy in new labels, previously unknown in ImageNet labeling, and ReaL labeling for questionable labels repeated the old label. For instance, pickelhaube-labeled images, according to ReaL, are still a pickelhaube - this method inherits from the problematic ImageNet class. CloudVision labeling does not inherit from problematic classes, which is clearly an advantage. It would be advisable to focus on semantic category groups as new labels in future work, because using pickelhaube as an example, these annotators have noticed that 70% of the images contain military uniform and military person. Only 50% received the headgear label. In addition, other labels were added to the labels, and the military uniform label alone accounts for only 7% of the total labels for the original labeling.

We also compared ReaL labeling and Cloud Vision labeling in terms of simple statistics: we showed measures such as frequency of a label over all labels assigned to the original category, and percentage of images which received a new label. RL and CV schemes differ in these measures. We also noticed that Cloud Vision, contrary to RL, generally tends to use labels that take a general context, e.g. "sky", "cloud" etc. It can be seen that this labeling focuses on the whole image, so it does not make the same mistakes as canonical ImageNet labeling.

References

1. Cloud vision API. https://cloud.google.com/vision/docs/labels. Accessed 30 Jan 2022
2. Know your data. https://github.com/pair-code/knowyourdata. Accessed 30 Jan 2022
3. Beyer, L., Hénaff, O.J., Kolesnikov, A., Zhai, X., Oord, A.V.D.: Are we done with ImageNet? arXiv preprint arXiv:2006.07159 (2020)
4. Deng, J., Dong, W., Socher, R., Li, L.J., Li, K., Fei-Fei, L.: ImageNet: a large-scale hierarchical image database. In: 2009 IEEE Conference on Computer Vision and Pattern Recognition, pp. 248–255. IEEE (2009)
5. Deng, J., Dong, W., Socher, R., Li, L.J., Li, K., Fei-Fei, L.: ImageNet: a large-scale hierarchical image database. In: 2009 IEEE Conference on Computer Vision and Pattern Recognition, pp. 248–255 (2009). https://doi.org/10.1109/CVPR.2009.5206848
6. He, K., Zhang, X., Ren, S., Sun, J.: Deep residual learning for image recognition. arxiv 2015. arXiv preprint arXiv:1512.03385 (2015)
7. Ilyas, A., Engstrom, L., Athalye, A., Lin, J.: Black-box adversarial attacks with limited queries and information. In: International Conference on Machine Learning, pp. 2137–2146. PMLR (2018)
8. Krizhevsky, A., Sutskever, I., Hinton, G.E.: ImageNet classification with deep convolutional neural networks. In: Advances in Neural Information Processing Systems 25 (2012)
9. Northcutt, C.G., Athalye, A., Mueller, J.: Pervasive label errors in test sets destabilize machine learning benchmarks. arXiv preprint arXiv:2103.14749 (2021)
10. Russakovsky, O., et al.: ImageNet large scale visual recognition challenge. Int. J. Comput. Vision **115**(3), 211–252 (2015)
11. Sambasivan, N., Kapania, S., Highfill, H., Akrong, D., Paritosh, P., Aroyo, L.M.: "Everyone wants to do the model work, not the data work": data cascades in high-stakes AI. In: Proceedings of the 2021 CHI Conference on Human Factors in Computing Systems, pp. 1–15 (2021)
12. Simonyan, K., Zisserman, A.: Very deep convolutional networks for large-scale image recognition. arXiv preprint arXiv:1409.1556 (2014)
13. Szegedy, C., et al.: Going deeper with convolutions. In: Proceedings of the IEEE Conference on Computer Vision and Pattern Recognition, pp. 1–9 (2015)

14. Szyc, K., Walkowiak, T., Maciejewski, H.: Checking robustness of representations learned by deep neural networks. In: Dong, Y., Kourtellis, N., Hammer, B., Lozano, J.A. (eds.) ECML PKDD 2021. LNCS (LNAI), vol. 12979, pp. 399–414. Springer, Cham (2021). https://doi.org/10.1007/978-3-030-86517-7_25
15. Zeiler, M.D., Fergus, R.: Visualizing and understanding convolutional networks. In: Fleet, D., Pajdla, T., Schiele, B., Tuytelaars, T. (eds.) ECCV 2014. LNCS, vol. 8689, pp. 818–833. Springer, Cham (2014). https://doi.org/10.1007/978-3-319-10590-1_53

Identification of Information Needs in the Process of Designing Safe Logistics Systems Using AGV

Honorata Poturaj(✉)

Department of Technical Systems Operation and Maintenance, Faculty of Mechanical Engineering, Wroclaw University of Science and Technology, Wroclaw, Poland
honorata.poturaj@pwr.edu.pl

Abstract. Preparation of the project and implementation of the AGV system is a task that requires appropriate theoretical preparation. The operation of complex anthropotechnical systems depends on the information gathered and the settings made. The necessary data can be extracted from literature, standards or pioneering implementations. In the presented example, the information needs were defined, the adverse events that took place were presented and the paths of events following the occurrence of the information failure (event diagrams) were developed. The analysis developed as part of the research will allow for proper organization and preparation of information in subsequent implementations. The presented approach affects both the stage of implementation and design of the AGV system, as well as the appropriate response to adverse events occurring in the operating system with AGV.

Keywords: AGV · Logistic · Safety

1 Introduction

AGVs are becoming increasingly popular in warehouse logistics and material handling in production systems. It is the result of two factors, which could be observed in modern logistic systems: (1) development of Industry 4.0 and connected to it Logistic 4.0 and (2) increasing difficulties in logistic personnel recruiting and increasing costs of human labor. From the above evidence, in recent years we can observe many implementations of AGVs, which support material delivery system as a part of internal logistics. The implementation of AGVs reduces the participation of employees in the performance of routine logistics operations but does not eliminate the participation of workers in the entire process. AGVs are, mostly, elements of anthropotechnical systems where the machine must cooperate with the human during a given process. For this reason, people who design AGV systems, in addition to the efficiency and reliability of the use of AGVs, must also pay attention to safety in their operation.

The article aims to identify the information needs of the team implementing AGV systems in the logistics service of material flows in the enterprise. The task of such a

W. Zamojski et al. (Eds.): DepCoS-RELCOMEX 2022, LNNS 484, pp. 217–226, 2022.
https://doi.org/10.1007/978-3-031-06746-4_21

team is to create a safe logistic service system. Therefore, the identified information needs should correspond to the risk assessment requirements prepared as part of the implementation. It is worth pointing out that the safety of AGV truck operation in the anthropotechnical system is understood, in this case, as its functioning that does not generate adverse events, which (a) threaten the life or health of participants in the process (operating staff) or (b) cause damage to other elements of the technical infrastructure of this process. The information collected should constitute the basis for the risk analyzes developed concerning the implementation of autonomous solutions in anthropotechnical systems. Information needs were based on ISO standards analyzes, literature review, and case study of research on the implementation of AGV in the logistic service system.

In connection with that, in the second section will be presented literature review and characteristic of chosen ISO standards. In the third section, characteristics of the adopted test method will be described. Then, in Section fourth, the results of the research and conclusions (in discussion form) will be presented. The entire article will be summarized in Sect. 5.

2 AGV in Logistic Systems

The versions of various vehicles and opportunities to adapt the vehicle to the needs of the system caused companies (from different sectors) to decide to implement AGVs in their logistic system.

First, it is important to define what AGV is. AGV is a shortcut from the Automated Guided Vehicle, which 'is an automated guided cart that follows a guided path' [1] or, in different words, 'is a vehicle that runs and navigates by itself without human intervention (…) a driverless vehicle' [2]. AGV also 'refers to a transport vehicle equipped with an automatic guiding device such as electromagnetic or optical, capable of traveling along a prescribed guiding path, with safety protection and various transfer functions, in industrial applications' [3].

Currently, we are observing the faster and faster development of the Logistics 4.0 concept, and its implementation is an important element in the improvement of modern logistics processes. The most frequently used autonomous solutions in logistics are currently unmanned aerial systems and AGV systems [4]. Automated Guided Vehicles are to replace humans in routine operations, enabling increased efficiency of the process. However, the AGV does not eliminate humans from the process, but supports humans in carrying out operations by performing routine activities. This approach allows for the creation of anthropotechnical systems where a human cooperates with a machine. The review of AGV applications in logistics systems operating in production and in warehouses is particularly important for the practical example described in the article.

The AGVs used in the production area are varied in terms of construction but have the same task: deliver the right materials, to the right place, at the right time. Therefore, there is cooperation with the process chain from the receipt of goods to the shipment of the finished product. The correct operation of the entire production system requires constant control, the possibility of adjustment, and convenient adaptation to exceptional situations. Due to the variety of production lines and their transport needs, the following types of AGV are available on the market [2, 5]: (1) underride AGVs, (2) automatic forklifts, (3) unit load AGVs, (4) tugger AGVs, (5) piggyback, (6) towing vehicles, (7) assembly AGVs, (8) heavy load AGVs and (9) special AGVs. The presented division is the most common and is prepared 'by looking at the loads they transport' [5]. The most recognized sector for the use of AGVs is the automotive sector [6], for example Volkswagen [5, 7, 8], BMW [5, 9, 10], Skoda [11] or Mercedes-Benz [12].

AGVs are becoming more attractive in the fields of warehouse and distribution. The benefits of using AGV are (1) increased productivity, (2) smooth transport, (3) lower labor cost, (4) real-time monitoring, (5) reduced number of operations, or (6) better system performance in the same warehouse space [2, 5, 13, 14]. AGVs replace the employee in transport activities, allowing the employee to focus on tasks related to packaging, planning, or warehouse management. Examples of warehouses that use AGV are Amazon [15, 16], Klingspor [17], Freudenberg [18, 19], and Bickford's Group [20]. An interesting solution that uses AGV in the warehouses system is AutoStore [21].

ISO standards are implemented in industries because (1) the company wants to develop their organization, information flow or structure, (2) the company searches for proven information and solutions, or (3) the client requires it. In the design and implementation anthropotechnical systems, enterprises use the guidelines contained in various standards. However, for the purposes of risk assessment related to the implementation of AGV in the logistics process, the following standards should be distinguished. Standards which should be analyzed:

- PN-EN 60300-3-1:2005 and PN-EN 60300-3-12:2011.
- PN-EN ISO 12100-1:2012.
- PN-EN ISO 3691-4:2020.

3 Methodology

Research on the identification of information needs was based on applicable European standards and interviews with representatives of companies that implemented AGV. The research procedure is presented (see Fig. 1).

Analysis of standards both European and German was primary source of information on what is required and how to get the necessary data (from the manufacturer, from information about the system, from the operating manual). Particular attention was paid to the provisions concerning the information necessary to carry out the risk assessment related to the implementation. Then, an structured interviews, carried out on the basis of a questionnaire which was prepared with information from a literature review and standards, was conducted with users of the AGV-based system to check whether the guidelines resulting from the analyzed standards are actually followed and used in the first stages (design and implementation) of the AGV project. On the basis of the collected data, information needs were defined. These information needs should be satisfied to reduce the potential risk that may occur when the AGV system starts to work. The next stage was the testing in the selected system. Defined information needs made it possible to determine what data was not collected and what adverse events occurred. This study will also predict what adverse events may occur when there is a lack of information. The company data and the information contained in the PN-EN ISO 3691-4: 2020 standard allowed one to create event diagrams, which are described in the next part of this article.

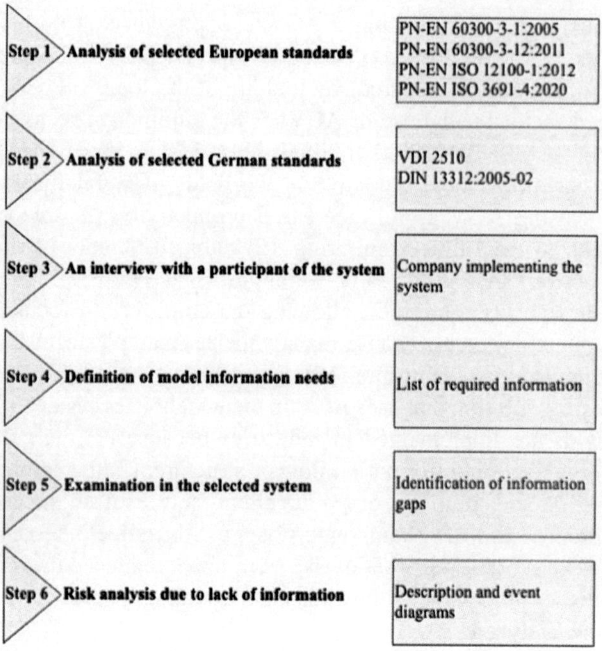

Fig. 1. Stages of research

4 Results and Discussion

AGV was introduced in an enterprise from the automotive industry.to the process of supplying materials. The main task of the vehicle was to deliver materials to the assembly stations and to complete the finished products.

4.1 Information Needs

For the selected company, from the automotive industry, based on the analysis of the system and the analysis of the standards, the following information needs were determined.

- Material flow. To choose the correct type of AGV company must define material flow and describe transport tasks. The vehicle working with the operator was replaced by an automated guided vehicle in the analyzed company. In addition, the AGV was constructed by adding the appropriate module that allows operation without an operator to the vehicle model, which has already been used in the company. Such a transition made it possible to use the information resulting from the operation of the owned tugger, whose task was to transport parts from the warehouse to the assembly lines.
- Handled loads. It is necessary to check its stability to ensure the safe transport of loads. The preparation description of the material is required to define a safe method of

cargo transportation by AGV. This description should include materials, dimensions, weight, the center of gravity, used transport packaging (pallets, KLT boxes, special containers), and unique requirements. In the described example, the information was already possessed and was used again. The adapted trolleys to transport the crates used in the warehouse were attached to the manually operated tugger. Due to the attaching the carriages, there was a possibility to use it with both an automatic and a manually controlled vehicle.

- Environment. The AGV working environment is understood both as climatic conditions (temperature, humidity, pollution) and infrastructure elements (floors, paths, users of the path, elements of equipment). During the implementation of the automotive tugger in the company, the paths had to be widened, the pedestrian paths were refreshed and the entry and exit to the warehouse were revitalized.
- Drive and brake system. This part of the information depends on the manufacturer of AGV, but their settings should be adapted to the security rules in the company.
- Control system. Information need, which is the control system, should be adapted to the existing infrastructure to avoid the need to introduce changes that affect the company's operation. When choosing a vehicle control system, it is important to check which type of navigation will suit them best. It is also important to check the possibilities of positioning and connection with the stationary vehicle management point (remote control station). The automated guided tugger was navigated with a laser positioning system, and the remote control station was placed in the warehouse next to the loading point.
- Modes of operation. The implemented AGV will operate in automatic mode, but it is important to predict and prepare the manual mode's operating and starting conditions. There is also a possibility to apply maintenance mode. All three modes of operation are described in PN-EN ISO 3691-4:2020, and all three were installed in an automated tugger. In this point of preparing the information, the way of starting each of the modes was also described.
- Energy system. Typically, AGVs are equipped with batteries that are charged at stations provided by the manufacturer. In this analysis, information on battery life, the time required for charging, the method of docking in the charging station, or the method of transmitting information about the low battery level by the AGV is necessary. When AGV signals that the battery level is low there are two opportunities (1) battery could be changed, or (2) AGV must go to the charging station. The case analyzed was the first stage of implementing automatic means of transport, so the company implemented only one AGV. Deliveries to assembly lines took place at sufficient time intervals for the tugger to reach the charging station, recharge the battery, and return to scheduled transport tasks.
- Protective devices. The described standards pay special attention to external technical measures that increase the safety of the machine's operation (in the discussed case, AGV). It could be bumpers, personnel and object detection systems, or warning measures. Tugger was equipped with two laser scanners (front and curtain) and dynamic fields interacting with scanners (slowing down and stopping), a light signaling module that works with the vehicle operation status control module, and stop buttons.
- History. Documentation about the history of accidents, hazardous events, or loss of health in workers helps predict adverse events and prepare the system so that they

do not occur. In this information need, the decision-maker may use the AGV history provided by the manufacturer or use information about similar vehicles. In this case, the company owned a history related to the work of a manual tugger and information about residual risks described by the AGV manufacturer.

- User manual. The AGV manufacturer should provide an instruction manual that contains detailed information about the vehicle with the vehicle. The necessary information has been listed and described in PN-EN ISO 3691-4: 2020 and, for example, it is information about: the system installed in the vehicle (software), routine operation, maintenance, applications, required working conditions, modification options, or labeling. This information should be presented and accessible to employees. It is also recommended that employees be trained in safety principles and work in the system using AGV. In the described company, detailed instruction was provided. However, its translation into Polish was not correctly prepared, and some fragments were incomprehensible.

4.2 Adverse Events

The automated tugger worked for 4 months under the supervision of an employee due to three aspects: (1) the application of the vehicle in the company requires adjustment to better match it with the operating system of the assembly lines, (2) every few weeks the AGV underwent an additional service, (3) emergency stop buttons were required for independent operation of the trolley, and the installation of two of them was delayed. The time when AGV worked under supervision allowed one to observe the causes, incident, and effects of adverse events in real time.

In the first days of the AGV's work, an employee supervising the vehicle noticed that the vehicle was automatically shut down due to a software error. The analysis conducted allowed us to conclude that static electricity occurs between the tires and the floor. As a result of the electrification of the vehicle, the module responsible for working in automatic mode stopped working and the vehicle went into sleep mode. The solution to this problem was to install two wire elements on the upper part of the robot, which were responsible for the discharge of charges. This problem did not arise during the operation of the manually controlled vehicle. During preparations for the implementation of the AGV, there was no analysis of the impact of elements of the adjoining infrastructure and vehicle.

Next to one of the assembly stations, the floor was slippery due to the oil waste that spreads from the machine that works there. During the first week of working AGV skidded at this point and, despite automatically applying emergency braking, the vehicle stopped against a wall. Fortunately, the pedestrian path, which is along this wall, was empty then. After this event, the AGV needed repair, during the AGV repair, the company rented a manually controlled tugger from the manufacturer of them. The analysis of the situation resulted in the introduction of the danger zone at the place where there may be oil on the floor. When the vehicle approaches this zone, it automatically reduces speed to drive safely on slippery surfaces, the speed returning to normal when the vehicle leaves the danger zone. An interview with the tug operators was conducted after the described accident and showed that the operators knew about the oil on the floor and slowed down when approaching this place. One can notice here the lack of an appropriate assembly (no interview with the operators of the manually controlled vehicle) and the analysis of the conditions in the assembly hall.

When the automated tugger started working independently, the logistic team noticed an increase in the time needed to complete the path. The system that manages vehicle operation did not show the times that the AGV spent at a given assembly station or on a specific section of the path, but only the total time of the journey. The delays were caused by the downtime of the AGV at the assembly stations, which collected the delivered materials or loaded the assembled products into carriages. AGV stopped at a specific location and waited for confirmation from an employee on the service panel that the materials had been unloaded or loaded. When the vehicle was driven by an operator or worked under supervision, the assembly workers were informed by the operator/supervisor that the vehicle was waiting and when the vehicle worked independently without any worker, it became imperceptible.

4.3 Event Diagrams

Event diagrams were created based on: (1) defined information needs, (2) an interview in a company using AGV to transport materials, (3) information contained in the presented standards, (4) own knowledge and (5) knowledge from articles. Eight events diagrams have been developed; they represent situations resulting from a lack of devolved information needs and were prepared for analyzed company. The list below shows the events that initiated each event diagram and the number of intermediate events that were defined:

- no information about handled loads, 9 intermediate events;
- no information about environment, 31 intermediate events;
- no information about the drive/brake system, 18 intermediate events;
- no information about control system, 22 intermediate events;
- no information about operation modes, 11 intermediate events;
- no information about energy system, 22 intermediate events.

Event diagrams are mainly concerned with situations where one of the vehicle's internal systems is malfunctioning, is damaged, or is being affected by an external factor. The first diagram presents such effects is the information needs related to the protective devices (see Fig. 2). The situation is different in the case of diagrams that show the effects resulting from the influence of the human factor. Errors made by employees may result from negligence, stress, rush, individual employee characteristics or poorly conducted training, and unaccounted knowledge. The two last were defined as user manual informational needs, which is important when implementing a new innovative solution. Meeting the requirements for conducting training and presenting operating instructions is not tantamount to understanding and complying with the new procedures or rules by workers (see Fig. 3).

In the analyzed company, warehouse employees were responsible for loading crates with parts onto trolleys. In this case, the knowledge and skills of the employees have a significant impact on the proper installation of loads for transport (see Fig. 4).

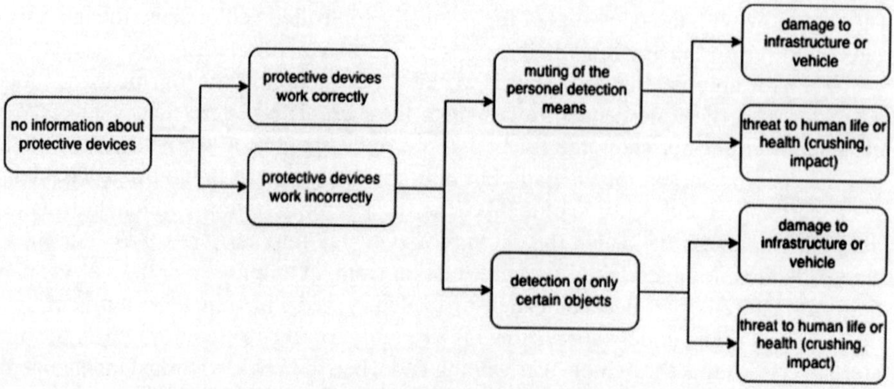

Fig. 2. Event diagram for protective devices information needs

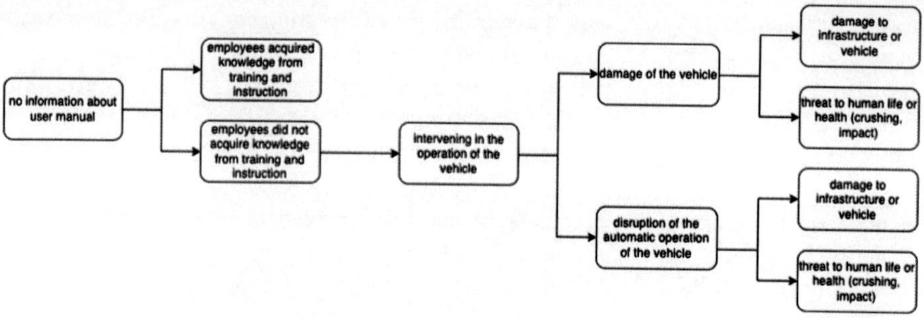

Fig. 3. Even diagram for user manual information needs

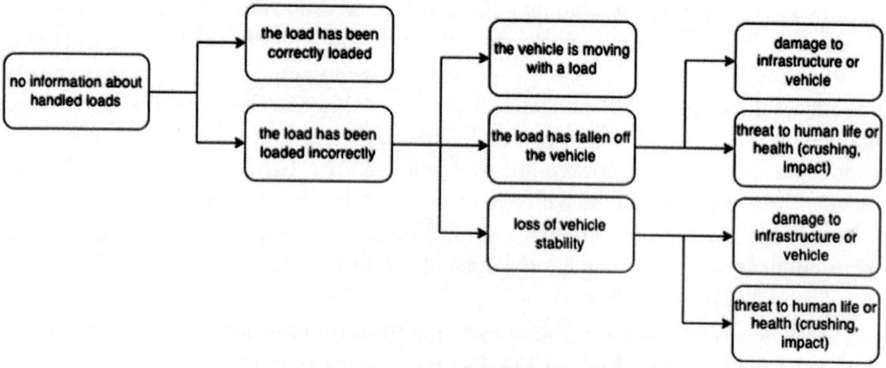

Fig. 4. Event diagram for handled loads information needs

5 Conclusions

The article presents the first stage of research work carried out on the basis of one AGV implementation as part of cooperation with the company. In the examined company, the lack of processing and analyzing information was noticed with respect to the working environment of the automatic vehicle. The people working on the AGV implementation project assumed that if the vehicle with the operator has been working in the hall, it has been enough to adapt the hall infrastructure (path width, elements of the vehicle navigation system for the vehicle) to the automatic solution. The errors resulting from this assumption were visible in the first weeks of using the automatic tugger. In addition, the research, and in particular the developed event diagrams, showed what adverse events may occur, which makes it possible to implement corrective actions in the system.

In further research, an analysis of other implementation projects is planned, which will allow us to determine the frequency of occurrence of events resulting from information deficiencies. Analysis of other implementations will also allow event diagrams to be supplemented with the probability of the occurrence of individual intermediate events, which will enable not only qualitative but also quantitative analyzes. Additionally, it will allow for a better characterization of the possible outcomes of these events.

References

1. Fazlollahtabar, H., Saidi-Mehrabad, M.: Autonomous Guided Vehicles Methods and Models for Optimal Path Planning. Springer, Cham (2015). https://doi.org/10.1007/978-3-319-14747-5s
2. KMC SRLS: What the Heck is an Automated Guided Vehicle. (2019)
3. Wang, C., Mao, J.: Summary of AGV path planning. In: 2019 IEEE 3rd International Conference on Electronic Information Technology and Computer Engineering, pp. 332–335. Institute of Electrical and Electronics Engineers Inc. (2019)
4. Tubis, A.A., Ryczyński, J., Żurek, A.: Risk assessment for the use of drones in warehouse operations in the first phase of introducing the service to the market. Sensors **21** (2021)
5. Ullrich, G.: Automated Guided Vehicle Systems. Springer, Berlin (2015). https://doi.org/10.1007/978-3-662-44814-4
6. Tubis, A.A., Poturaj, H.: Challenges in the implementation of autonomous robots in the process of feeding materials on the production line as part of logistics 4.0. Logforum **17**, 411–423 (2021)
7. Piątek, Z.: Volkswagen wdraża systemy AGV i automatyzuje logistykę wewnętrzną, https://przemysl-40.pl/index.php/2018/09/06/volkswagen-wdraza-systemy-agv-i-automatyzuje-logistyke-wewnetrzna/. Accessd 27 Jan 2022
8. NDC Solutions: Truck Factory - Hangzhou and Volkswagen Logistics - AGV (2021)
9. Kocher: Projects. https://www.kocher.biz/en/projects.html. Accessed 25 Jan 2022
10. Kinexon: With Kinexon Brain, AGVs Become Smart Transport Robots. https://kinexon.com/bmw/. Accessed 28 Jan 2022
11. Asseco: Latest state-of-the-art logistics robots from a Slovak manufacturer deployed in Skoda Auto. https://asseco.com/en/news. Accessed 25 Jan 2022
12. KUKA: Daimler setzt auf flexibles fahrerloses Transportsystem. https://www.kuka.com/de-de/branchen/loesungsdatenbank/2020/01/agv-solution_biw-trucks. Accessed 22 Jan 2022
13. Tang, H., Cheng, X., Jiang, W., Chen, S.: Research on equipment configuration optimization of AGV unmanned warehouse. IEEE Access. **9**, 47946–47959 (2021)

14. Silva, T., et al.: Simulation and economic analysis of an AGV system as a mean of transport of warehouse waste in an automotive OEM. In: IEEE Conference on Intelligent Transportation Systems, Proceedings, ITSC. pp. 241–246. Institute of Electrical and Electronics Engineers Inc. (2016)
15. Roser, C.: The Amazon Robotics Family: Kiva, Pegasus, Xanthus, and more. https://www.allaboutlean.com/amazon-robotics-family/. Accessed 28 Jan 2022
16. Kim, E.: Amazon's $775 million deal for robotics company Kiva is starting to look really smart. https://www.businessinsider.com/kiva-robots-save-money-for-amazon-2016-6. Accessed 24 Jan 2022
17. Michel, R.: Mobile automation takes on expanding requirements. https://www.mmh.com/article/mobile_automation_takes_on_expanding_requirements. Accessed 2 Jan 2022
18. Lots of Bots: Success Story: Corteco relies on Swisslogs CarryPick (2020)
19. Swisslog: Freudenberg Italy: On-time delivery for spare parts. https://www.swisslog.com/en-us/case-studies-and-resources/case-studies/2020/05/freudenberg_italy. Accessed 27 Jan 2022
20. Bickford's Group. https://www.dematic.com/en/downloads-and-resources/case-studies/featured-case-studies/bickfords-group/. Accessed 20 Jan 2022
21. Swisslog: AutoStore: Space saving storage and order picking system for small parts. https://www.swisslog.com/en-us/products-systems-solutions/asrs-automated-storage-retrieval-systems/boxes-cartons-small-parts-items/autostore-integrator. Accessed 26 Jan 2022

Optimizing the Maintenance Strategy for Offshore Wind Turbines Blades Using Opportunistic Preventive Maintenance

Panagiotis M. Psomas(✉), Agapios N. Platis, and Vasilis P. Koutras

Department of Financial and Management Engineering, University of the Aegean, 82100 Chios, GR, Greece

{ppsomas,platis,v.koutras}@aegean.gr

Abstract. In this paper, an opportunistic preventive maintenance policy for the rotor blades of an offshore wind turbine system subjected to multiple types of internal and external damages is proposed. Internal damages (such as fatigue, wear) are generally caused by system deterioration, whereas the external damages (such as wind, icing, waves and lightnings) are ought to harsh weather conditions. The system of the wind turbine can experience various levels of deterioration of power output due to the different damages in the wind turbine rotor blades. Additionally, at any time, the system can experience a sudden failure due to extreme weather conditions. Thus, the system can enter a state of lower or higher wind energy production due to wind speed variation or can enter a degraded state due to damage in the wind turbine blades, or even the failure state due to a sudden failure. When a failure occurs, a repair procedure is carried out and the system is restored to its initial fully operational state. The proposed model incorporates minor, major, and opportunistic maintenance policies. The model refers to the blades of a wind turbine and the endmost aim is to determine the optimal minor and major maintenance rates such that the asymptotic availability is maximized. A numerical example based on empirical data is used to illustrate the proposed model as well as the effectiveness of the maintenance strategy.

Keywords: Maintenance strategy · Markov modelling · Wind turbine availability

1 Introduction

The wind energy industry has experienced an extensive growth during the past decades due to the evolution of the technology. Its contribution in energy market keeps increasing during the last 20 years. The target of the European Wind Energy Association is wind energy being half of Europe's electricity by 2050, with wind energy capacity rising from 220 GW today to up to 1300 GW [1]. However, due to the lack of space on land and the better quality of the wind at the sea, the installation of wind turbines is changing from onshore to offshore locations [2]. As offshore wind turbines are located at remote sites with limited accessibility, they are exposed to severe environmental conditions that affect the integrity of the system and the demand for high reliability [3]. Operation and

W. Zamojski et al. (Eds.): DepCoS-RELCOMEX 2022, LNNS 484, pp. 227–236, 2022.
https://doi.org/10.1007/978-3-031-06746-4_22

maintenance costs of offshore wind turbines contribute about 25–30% of the total energy generation costs, which can be estimated to be 50% higher than that of the onshore farm [4].

The complex engineering systems reveal multiple possible states from perfectly operational to totally failed, during their functioning time [5]. Wind turbines are complex electromechanical systems usually having a design lifetime of 20–30 years. The main parts of a wind turbine are rotor and hub, several bearings, gearbox, generator, brakes, control system and a part that balances the electricity. Analyses of field failure data collected from various databases[1,2,3] show that the rotor blades are among the most critical components in offshore wind turbines. The damages of rotor blades can be categorized into two types: internal and external [6]. Internal damages are caused by system degradation due to mechanical stress. External damages can be a sudden variation in wind speed or a sudden failure due to high sea waves, lightning strikes or meteorological icing that stops the wind turbine and requires system replacement. In order to improve the output performance of the wind turbine system which decreases with deterioration, the system should be maintained whenever it is considered necessary. Maintenance optimization is an essential issue for industries that utilize physical assets due to its impact on costs, and performance [7]. Maintenance actions can be carried out to keep or restore a turbine system into a normal functioning condition or even to extend its lifetime. At offshore areas with higher wind speeds, weather conditions can affect the reliability of wind turbines substantially and reduce the life of turbines' key components. Tavner et al. [8] show that there is a strong relationship between the number of failures and wind speeds.

Opportunistic maintenance (OM) is also widely used in mechanical engineering systems [9], to reduce the overall maintenance cost. Recent research on opportunistic maintenance (OM) could provide significant benefits for wind turbine maintenance practice. OM refers to a particular type of preventive maintenance which is performed when an opportunity for lower maintenance cost occurs. OM is usually considered for the operating components of a system when corrective or preventive maintenance activity occurs on other components [10, 11]. This type of opportunity is often referred to as internal, unlike to external opportunities which are caused by different weather events, which can stop the operation of a system and create a chance for maintenance. Bhaba R. Sarker and Tasnim Ibn Faiz [12] described an opportunistic preventive maintenance strategy for offshore wind turbines. In their strategy, maintenance activities are set off by the failure of the components (rotor, bearing, gearbox, generator) and the technicians can take the opportunity to preventively replace or perform maintenance on functioning components. Kurt and Kharoufeh [13] investigated the maintenance optimization problem under a controlled Markovian environment. The deterioration of the system depends fully on the environment process, while in contrast the environmental changes did not depend on the current level of deterioration. Besnard et al. [14] proposed an opportunistic maintenance strategy for offshore wind turbine systems based on failures, corrective maintenance activities and wind forecasts. They presented an optimization model with

[1] http://www.offshore-wmep.de.

[2] http://www.windustry.org/resources/wind-stats-newsletter.

[3] http://www.elforsk.se.

a series of constraints aiming at minimizing the cost when preventive maintenance is performed.

Wind turbine operation is obviously affected by wind speed [15]. Due to the nature of wind intensity the proposed model in this paper is based on the categorization of wind energy production in five categories and on the transitions between them. Speeds near or below the cut-in speed can significantly decrease or stop the production and hence create low-cost maintenance windows. To better illustrate the maintenance actions and its effects, an opportunistic maintenance model considering external opportunities to maintenance activities of the wind turbine is developed. The external opportunities considered in the model are time-windows (discretized as days) in which wind speed is too low for the operation of the wind turbine. The paper presents an approach that maximizes the availability, through the optimization of the minor maintenance rate and major maintenance rate at the different categories of wind energy production.

The remainder of the paper is organized as follows: In Sect. 2 the wind turbine system with opportunistic maintenance is described in detail. In Sect. 3 the asymptotic availability for the turbine system and the optimization problem are defined. An illustrative case is presented in Sect. 4 and conclusions are presented in Sect. 5.

2 Wind Turbine System with Minimal, Major, and Opportunistic Maintenance

2.1 Model Description and Assumptions

A wind turbine system is considered, and it is assumed that the condition of the blades can be classified in one of the following categories: "perfect operational state", "minor degradation", "advanced degradation", "major degradation", and "failure". At the failure state ("failure"), a repair procedure is carried out and the wind turbine returns to an as good as new state ("perfect operational state"). Apart for the "perfect operational state" and the "failure" state, the wind turbine system can in one of the abovementioned degraded conditions due to deterioration. By assuming that the sojourn time in any system state follows an exponential distribution, the system's evolution in time can be modeled by a continuous time Markov process $\{Z(t), t \geq 0\}$ [16]. Five states are associated with the proposed deterioration classification for rotor blades, O representing a perfect operational state, D_1 representing a minor degradation state, D_2 representing an advanced degradation state, D_3 representing a major degradation state and F representing the failure state.

The size of blades has generally been increasing rapidly during the last decades and for this reason they are subjected to high stress. The blades are subject to a number of defects such as the separation of laminate plies and the loss of connection between the face and core in a sandwich panel [17, 18].

Due to the complexity and the high dependence of wind energy systems on climatic and environmental factors, there is the need to perform maintenance actions to ensure the efficient operation of wind turbines and effectively utilizing the wind energy such that maximum power can be generated. The proposed maintenance strategy is based on the following assumptions. We assume that the states of the model that the system can

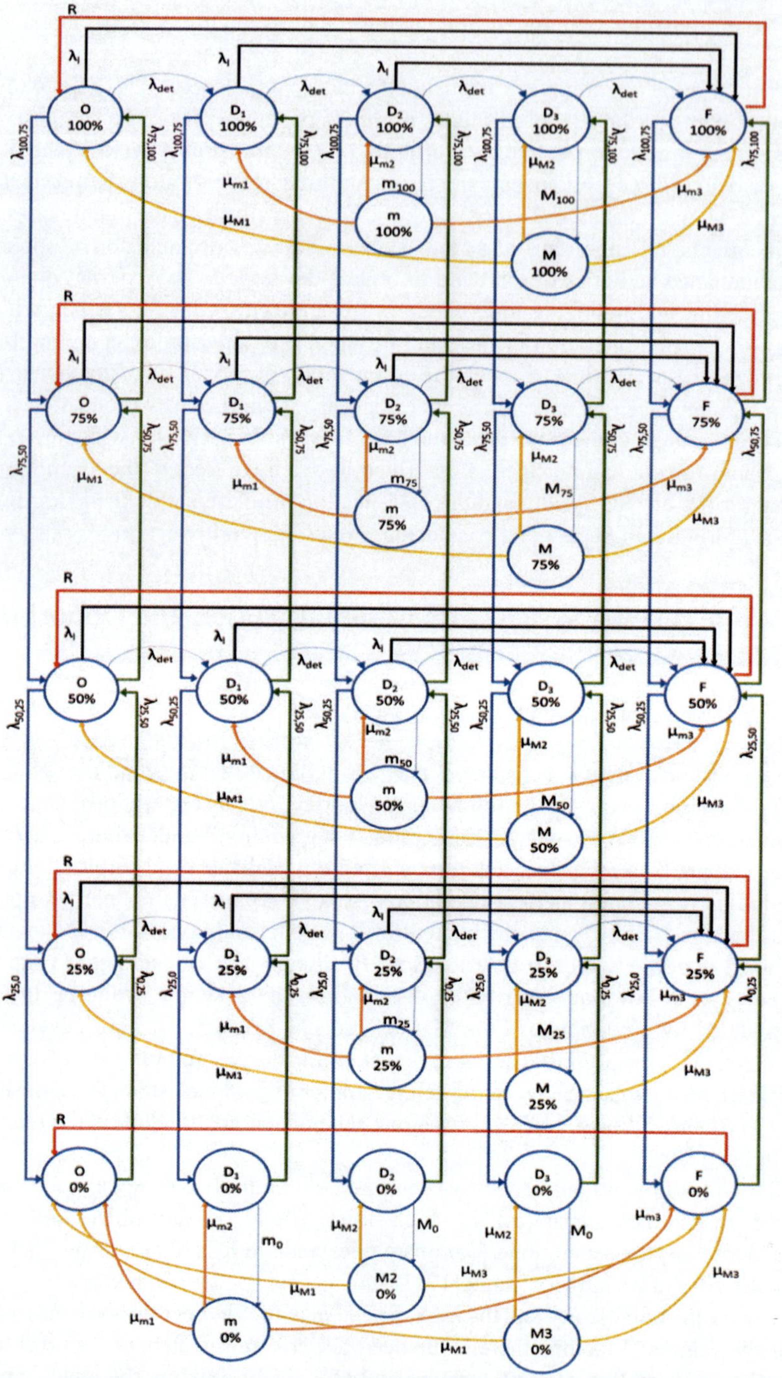

Fig. 1. State transition diagram for the wind turbine degradation system under maintenance

operate in different wind intensities can be represented by a finite set of discrete states, $S = \{O, D_i, F\}$, $i = 1, 2, 3$, as it can be seen in Fig. 1. The first category describes a state in which there is excessive wind, and the wind turbine is capable to produce the 100% of its rated capacity. The second category describes a state in which the wind is very strong, and the turbine is capable to produce 75% of its rated capacity. The third category describes a state where the wind turbine is capable to produce 50% of its rated capacity. Then, in fourth category prevails light wind, and the wind turbine is capable to produce 25% of generator's rated capacity. Finally, the fifth category describes the case in which the wind speed is very low and as a result the wind turbine cannot actually produce electricity.

In this approach, we consider a system that starts to operate in its perfect functioning condition, denoted by state (O, 100%), where the wind turbine is capable to produce 100% of turbine's rated capacity (see Fig. 1). The wind turbine's blades level of degradation increases with system's operational time. As time evolves, the system can enter a state of lower or higher wind energy production due to the wind speed variation or it may enter a degraded state due to a deterioration of the wind turbine. As the degradation progresses to the next level, we assume that the system from states (D_i, 100%), (D_i, 75%), (D_i, 50%), (D_i, 25%), $i = 1, 2, 3$ may enter the adjacent deterioration states (D_{i+1}, 100%), (D_{i+1}, 75%), (D_{i+1}, 50%), (D_{i+1}, 25%), $i = 1, 2$ respectively. Additionally, it is also assumed that the wind turbine system may experience a sudden failure which can be caused due to external factors such as lightning strikes. In this case, the system from any state (j, 100%), (j, 75%), (j, 50%), (j, 25%), $j = O, D_1, D_2, F$, enters the total failure state (F, 100%), (F, 75%), (F, 50%), (F, 25%) respectively. However, in case the wind turbine is in the last deterioration states (D_3, 100%), (D_3, 75%), (D_3, 50%), (D_3, 25%), it is assumed that the system enters the adjacent total failure states (F, 100%), (F, 75%), (F, 50%), (F, 25%).

Two different types of preventive maintenance actions are proposed to be performed to prevent the turbine system from a total failure and extend its lifetime. Either minor (perfect) maintenance is performed, when the system is in states (D_2, 100%), (D_2, 75%), (D_2, 50%), (D_2, 25%), and the system is restored to the previous degradation stages (D_1, 100%), (D_1, 75%), (D_1, 50%), (D_1, 25%) respectively, or major (perfect) maintenance is carried out when the system is in states (D_3, 100%), (D_3, 75%), (D_3, 50%), (D_3, 25%) and the system is restored to an as good as new state (O, 100%), (O, 75%), (O, 50%), (O, 25%) respectively. The type of maintenance actions that should be implemented depends on the level of system degradation. Note that while the rotor blades are perfectly functioning, no maintenance action takes place. Additionally, we assume, that when the wind turbine is in state (D_1, 100%), (D_1, 75%), (D_1, 50%), (D_1, 25%), it can operate at an acceptable level and no maintenance action is initiated.

As can be obtained in Fig. 1, we introduce two more concepts. Firstly, as far as minor maintenance is concerned, it is assumed that this action may not manage to be properly completed and thus be imperfect. In this case, an imperfect minor maintenance fails to restore the turbine back to its previous degradation state and instead, the system returns to its current deterioration state. Thus, an imperfect minor maintenance action causes a transition from states (m, 100%), (m, 75%), (m, 50%), (m, 25%) to states (D_2, 100%), (D_2, 75%), (D_2, 50%), (D_2, 25%) respectively. Besides imperfect minimal maintenance, the case where minor maintenance is badly performed mainly due to human factor, is also considered. In this case the turbine experiences a total failure. Therefore, the minor maintenance action is characterized as failed. Failed minor maintenance is modeled in Fig. 1 by the transitions from states (*m*, 100%), (*m*, 75%), (*m*, 50%) and (*m*, 25%) to the total failure states (*F*, 100%), (*F*, 75%), (*F*, 50%) and (*F*, 25%) respectively. Correspondingly, major maintenance action can also turn out to be imperfect. In this case the turbine cannot be restored to the perfect functioning state, but instead it returns to its current deterioration states (D_3, 100%), (D_3, 75%), (D_3, 50%) and (D_3, 25%). Similarly, with minor maintenance, it is assumed that major maintenance can fail as well. In this case, the wind turbine enters to the total failure state and a transition from states (*M*, 100%), (*M*, 75%), (*M*, 50%), (*M*, 25%) to states (*F*, 100%), (*F*, 75%), (*F*, 50%), (*F*, 25%) is carried out. Finally, in the case that the wind turbine fails, a repair process is initiated after which the system is restored to the perfectly functioning states (*O*, 100%), (*O*, 75%), (*O*, 50%), (*O*, 25%) and (*O*, 0%) for all the categories of energy production.

2.2 The Proposed Opportunistic Maintenance Strategy

The proposed opportunistic maintenance strategy consists in performing preventive maintenance only at the last category of wind energy production (0%), because the wind speed is very low and as a result the wind turbine cannot produce energy. In particular, when the deterioration level reaches state (D_1, 0%) a minor maintenance action takes place and restores the turbine back to its previous state which is (*O*, 0%). If the minor maintenance action is imperfect the system returns back to the deterioration state (D_1, 0%). Moreover, if the minor maintenance is badly performed (failed minor maintenance), the system enters the total failure state (*F*, 0%). Additionally, when the degradation level reaches state (D_2, 0%) a major maintenance action is implemented and restores the system to state (*O*, 0%). Similarly with the minor maintenance, it is assumed that major maintenance can be imperfect; in this case the system returns to current deterioration state (D_2, 0%). Finally, when the deterioration level reaches the last degraded state (D_3, 0%), a major maintenance is initiated and the system is restored to the initial state (*O*, 0%). We assume again that when the major maintenance is imperfect the system returns back to the deterioration state where the maintenance was initiated, while when the major maintenance is badly performed, either in (D_2, 0%) or in (D_3, 0%), the system enters the total failure state (*F*, 0%). Consequently, according to the proposed strategy, the case when the wind speed is between 0-3*m/s* can be considered as an opportunity to preventively maintain the rotor blades.

3 Asymptotic Availability and Optimization

The aim of this work is to determine a maintenance policy for the proposed model that maximizes the wind turbine system availability. Generally, the availability of a system can be defined as the probability that the system is operational at time *t*: $Av(t) = Pr(\text{system is functioning at instant } t)$ [19]. For the proposed model, let *E* be the state space of the Markov process $\{Z(t), t \geq 0\}$ that describes the evolution of the wind turbine system in time, and it can be partitioned into two subsets: the subset U which contains the operational states [(O, 100%), (D_1, 100%), (D_2, 100%), (D_3, 100%), (O, 75%), (D_1, 75%) (D_2, 75%), (D_3, 75%), (O, 50%), (D_1, 50%), (D_2, 50%) (D_3, 50%), (O, 25%), (D_1, 25%), (D_2, 25%), (D_3, 25%)] and the subset D which contains, the rest of the states which are non-operational states, with $E = U \cup D$, $U \cap D = \varnothing$, $U \neq \varnothing$, $D \neq \varnothing$. Note that subset D contains the maintenance states, the failure states of the system and states where the wind energy production is 0%. Therefore, the availability of the system at time t can be further defined by the following equation:

$$Av(t) = \Pr(Z(t) \in U) = \sum\nolimits_{i \in U} p_i(t) \tag{1}$$

where $p_i(t)$ is the probability that the system is in state i at time t.

When referring to continuously running systems, the quantity of interest is the asymptotic availability. The asymptotic or steady state availability for an ergodic Markov chain can be computed as follows:

$$Av = \lim_{t \to \infty} Av(t) = \lim_{t \to \infty} \sum\nolimits_{i \in U} p_i(t) = \sum\nolimits_{i \in U} \lim_{t \to \infty} p_i(t) = \sum\nolimits_{i \in U} \pi_i \tag{2}$$

where π_i, is the steady state probability of state i.

Under the assumption that the major maintenance rate is lower than the minor maintenance rate, that is M < m, the aim is to maximize the wind turbine system asymptotic availability with respect to variables m and M.

Thus, the optimization problem to be solved for the asymptotic availability is provided in Eq. (3):

$$\begin{aligned} &\max_{m,M} Av(m, M) = \max_{m,M} \sum\nolimits_{i \in U} \pi_i(m, M) \\ &s.t. \quad 0 < M < m \end{aligned} \tag{3}$$

As it can be obtained in Eq. (3), the steady state probabilities of the operational states depend on the maintenance policies m and M. Solving the optimization problem (Eq. (3)) provides the optimal maintenance policy (m^*, M^*) which maximizes the availability measure.

4 Case Study: Experimental Results and Analysis

The data used consider the wind variation through a year and have been obtained from the HNMS (Hellenic National Meteorological Service) for 2020 and for the specific Limnos Island which are presented in Table 1. Also, the table shows the number of days that the wind speed was at each of the wind speed categories and the corresponding probability that the wind speed belongs to each of these categories. This probability is expressed by the ratio of the number of days for each category to the number of days in the year. By assuming that the time spent in each of the wind speed categories is exponentially distributed, the constant transition rates among different wind intensities can be estimated using maximum likelihood estimators by the data from HNMS. Table 2 presents the non-zero estimated wind speed transition rates.

Table 1. Wind speeds categories for Limnos Island

Days	Wind speed [m/s] (Production Percentage)	Probability of wind speed remaining at the corresponding number of days
71	12–15 (100%)	0.194
75	9–12 (75%)	0.205
104	6–9 (50%)	0.284
50	3–6 (25%)	0.136
65	0–3 (0%)	0.178

Table 2. The transition rates below indicate how many times the wind speed changes in a year

$\lambda_{0,25} = 11\ y^{-1}$	$\lambda_{25,50} = 20\ y^{-1}$	$\lambda_{50,75} = 25\ y^{-1}$	$\lambda_{75,100} = 22\ y^{-1}$
$\lambda_{25,0} = 10\ y^{-1}$	$\lambda_{50,25} = 18\ y^{-1}$	$\lambda_{75,50} = 27\ y^{-1}$	$\lambda_{100,75} = 22\ y^{-1}$

Wind turbine's rated output power is assumed to be $P_R = 2$ MW. The deterioration rate of the wind turbine is assumed constant and equal with $\lambda_{det} = 0.52\ y^{-1}$. A sudden failure due to lightning strikes is considered and for this reason lightning strikes are modeled by a direct transition to the failure state as shown in Fig. 1. It is assumed that the lightning strike rate is $\lambda_l = 0.01\ y^{-1}$. Table 3 summarizes the values of the model parameters. The repair rate (R) is equal with 146.19 y^{-1}. The values of the minor maintenance rates ($m_{100}, m_{75}, m_{50}, m_{25}, m_0$) for each of the corresponding wind energy production categories (100%, 75%, 50%, 25%, 0%) and these of the major maintenance rates ($M_{100}, M_{75}, M_{50}, M_{25}, M_0$) for the corresponding wind energy production categories are also presented in Table 3. Finally, the perfect, imperfect, and failed minor maintenance rates ($\mu_{m1}, \mu_{m2}, \mu_{m3}$), as well as the perfect, imperfect, and failed major maintenance rates ($\mu_{M1}, \mu_{M2}, \mu_{M3}$) are given in Table 3. The minor and major maintenance duration and the repair time are exponentially distributed and the time to a failed maintenance is modelled by exponential distributions too.

Our strategy is to implement most maintenance policies during the inactivity of the wind turbine i.e., when the wind category is 0%. Our goal is to evaluate the minor maintenance rate m_0 and the major maintenance rate M_0 where wind energy production is 0% and in this case the wind turbine cannot produce energy. In order to optimize the wind turbine's asymptotic availability, the optimization problem of Eq. (3) is solved. The maximum asymptotic availability under the assumption that $M < m$ is 0.8205 and is achieved for $m_0 = 5.34$ and $M_0 = 1.95$. In other words, in order to achieve the maximum availability for the system, regarding the opportunistic maintenance strategy, the minor maintenance at wind energy production 0% should be performed 5.34 times a year and the major maintenance 1.95 times a year. All experiments were implemented using Matlab R2020a on a PC equipped with a 2.30 GHz Intel Core i5-and 8.00 GB of RAM.

Table 3. Summary of the model parameters [y^{-1}]

$\lambda_{det} = 0.52$	$M_{50} = 3.12$	$m_{50} = 8.52$	$\mu_{m3} = 4.38$
$\lambda_l = 0.01$	$M_{25} = 1.49$	$m_{25} = 4.08$	$\mu_{M1} = 277.77$
$R = 146.19$	$M_0 = 1.95$	$m_0 = 5.34$	$\mu_{M2} = 8.77$
$M_{100} = 2.13$	$m_{100} = 5.82$	$\mu_{m1} = 859.64$	$\mu_{M3} = 5.84$
$M_{75} = 2.25$	$m_{75} = 6.15$	$\mu_{m2} = 13.15$	

5 Conclusions

In this paper a wind turbine system with major, minor, and opportunistic maintenance is studied. We also consider minor and major maintenance actions that beyond perfect can also be imperfect or failed. We are interested in determining the optimal minor and major maintenance rates that maximize the asymptotic availability.

For the proposed model, we provided the appropriate theoretical framework to calculate and optimize the availability. The decision variables of the optimization consist of the minor and major maintenance rates for the wind turbine system. Thus, the corresponding solutions can be used for planning a maintenance strategy that would improve the turbine's availability. The innovation of this work consists in providing an integrated model for degrading wind turbine systems with major, minor, and opportunistic maintenance and sudden failures along with the different wind speed conditions and the wind energy production for each state.

The proposed maintenance policies can be generalized in a future work by considering other kinds of connection such as among internal and external damages, components and failure states and material properties and crack size of the rotor blades. Additionally, omitting the repair rate as well as the minor and major maintenance rates in the case when the turbine's energy production is 100%, 75% can be also considered in a future work, since in these cases technical staff safety issues arise. Furthermore, for the proposed model the minimum days where the wind speed is between 0 and 3 m/s can be calculated in order to establish that the availability of an opportunistic maintenance model will be higher than the availability of the current model performing minor and major maintenance. In addition, more performance measures can be evaluated such as the expected energy not supplied (EENS).

References

1. EWEA: European Wind Energy Association (EWEA), Offshore Development Wind Energy, P. 212. The facts Routledge. Taylor and Francis, London (2009)

2. Esteban, M.D., Diez, J.J., Lopez, J.S., Negro, V.: Why offshore wind energy? Renew. Energy **36**, 444–450 (2011)
3. Kovacs, G., Erdos, Z., Viharos, J., Monostori, L.: A system for the detailed scheduling of wind farm maintenance. CIRP Ann. Manuf. Technol. **60**, 497–501 (2011)
4. Zhu, W., Fouladirad, M., Berenguer, C.: A predictive maintenance policy based on the blade of offshore wind turbine. In: 2013 Proceedings Annual Reliability and Maintainability Symposium (RAMS) (2013)
5. Platis, N.A., Koutras, P.V., Malefaki, S.: Achieving high availability levels of a deteriorating system by optimizing condition based maintenance policies. In: Steenbergen S., van Gelder, P., Vrouwenvelder, A. (eds.) Safety, Reliability and Risk Analysis: Beyond the Horizon, pp. 829–837 Taylor & Francis Group, London (2014)
6. International Standard IEC61400-1: Wind turbines – Part1: design requirements (2005). https://webstore.iec.ch/. (2005)
7. Andrawus, J.A., Watson, J., Kishk, M.: Wind turbine maintenance optimization: principles of quantitative maintenance optimization. Wind Eng. **31**(2), 101–110 (2007)
8. Tavner, P.J., Edwards, C., Brinkman, A., Spinato, F.: Influence of wind speed on wind turbine reliability. Wind Eng. **30**, 55–72 (2006)
9. Kumar, G., Maiti, J.: Modelling risk based maintenance using fuzzy analytic network process. Exp. Syst. Appl. **39**(11), 9946–9954 (2012)
10. Lirong, C., Haijun, L.: Opportunistic maintenance for multi-component shock model. Math. Methods Oper. Res. **63**(3), 493–511 (2006)
11. Zhao, H., Yan, S., Zhang, X.: Deterministic opportunistic replacement maintenance strategy for wind turbine. Acta Energiae Solaris Sin. **35**(4), 568–574 (2014)
12. Sarker, B.R., Faiz, T.I.: Minimizing maintenance cost for offshore wind turbines following multi-level opportunistic preventive strategy. Renew. Energy **85**, 104–113 (2016)
13. Kurt, M., Kharoufeh, J.P.: Monotone optimal replacement policies for a Markovian deteriorating system in a controllable environment. Oper. Res. Lett. **38**(4), 273–279 (2010)
14. Besnard, F., Patriksson, M., Stromberg, A., Wojciechowski, A., Bertling, L.: An optimization framework for opportunistic maintenance of offshore wind power system. In: Proceedings of the 2009 PowerTech International Conference. Bucharest, Romania (2009)
15. Eggen, O., Rommetveit, O., Reitlo, A., Midtbø, E.O.: Handbook on condition monitoring of wind turbines. In: Proceedings of the *European Wind Energy Conference*, Marseille, France, 16–19 March, 2009
16. Ross, S.M.: Stochastic Processes, 2nd edn. Wiley, Hoboken (1996)
17. Wedel-Heinen, J., Hayman, B., Brønsted, P.: Materials challenges in present and future wind energy. Mater. Res. Soc. Bull. **33**(4), 343–354 (2008)
18. Lekou, D.J., Vionis, P.: Report on Repair Techniques for Composite Parts of Wind Turbine Blades Knowledge Centre WMC, Tech. Rep. ENK6-CT2001-00552 (2002)
19. Kishor, S.: Trivedi, Andrea Bobbio: Reliability and Availability Engineering. Cambridge University Press, London (2017)

Estimation of Ethereum Mining Past Energy Consumption for Particular Addresses

Przemysław Rodwald(✉)

Department of Computer Science, Polish Naval Academy,
Śmidowicza 69, 81-127 Gdynia, Poland
p.rodwald@amw.gdynia.pl

Abstract. There is a demand for cryptoassets analysis tools among law enforcement. Most solutions are focused on tracking and tracing money flows. The popularity of energy-hungry blockchains, where illegal mining activities are growing rapidly, shows a need to estimate past energy consumption. This paper presents an online system that helps with this estimation based on publicly available data for Ethereum. It could become a tool for forensics investigators dealing with ETH based cases and starting point for further development of other proof-of-work based cryptocurrencies.

Keywords: Ethereum mining · PoW · Energy consumption

1 Introduction

Cryptocurrencies are gaining both recognition and acceptance. Except for many advantages of blockchain-based currencies, including decentralization, privacy, speed of transactions, there is one crucial factor indicated by opponents - high energy consumption. For proof-of-work (PoW) based cryptoassets, security stems from a robust incentive system. Participants, called miners, are required to provide computationally expensive proofs of work, and they are rewarded according to their efforts. The estimation of overall electricity used for cryptocurrency mining is enormous [10,13]. Even though cryptocurrency mining is not an illegal activity, criminals find new ways to use it for illicit behaviours. Energy stealing or illegal hardware usage are the most frequent use-cases. For years, criminals have been siphoning power from neighbours or straight from the grid to provide the lighting and heating for growing marijuana. The exact process of stealing power could be beneficial to a cryptocurrency miner because it would remove a significant proportion of the cost of mining [9]. Such cases are well known, both worldwide [1–3,6,12] and nationwide [4,19].

Our vision was to propose an easy to use online system based on publicly available data, which helps criminal investigators estimate the amount of energy used for past cryptomining.

W. Zamojski et al. (Eds.): DepCoS-RELCOMEX 2022, LNNS 484, pp. 237–244, 2022.
https://doi.org/10.1007/978-3-031-06746-4_23

In the following, in Sect. 2, we first briefly introduce the cryptocurrency mining phenomenon. In Sect. 3, we provide some background information explaining possible scenarios during an investigation for illegal cryptocurrency mining, next our approach to energy consumption estimation is explained. Then, in Sect. 4, the technical design and dashboard are presented. Finally, in Sect. 5, known challenges, limitations and further development direction are mentioned.

2 Cryptomining

The blockchain concept, introduced by Nakamoto [15] and originally designed to be the backbone of the Bitcoin protocol, has since been applied to a dozen hundreds of altcoins (alternative coins). Most of them use a proof-of-work algorithm based on cryptographic hashing, proposed earlier by Douceur [5]. The miners compete against each other to solve the problem of finding a random value (nonce) which hashed with the information about all transactions within the block and hash of the previous block, fulfils the network's current difficulty requirements, i.e. is less than a specified target value. The network's difficulty is dynamically adjusted so that the time to mine a block varies between certain pre-established time bounds. The incentive to take part in these competitions is the award - miners are compensated for their efforts with both newly minted coins and transactions fees.

One of the key distinctions between different cryptocurrencies is the hash function used to solve the puzzle and the average time of generating new blocks. Bitcoin uses SHA-256 algorithm and needs approximately 10 min, while Ethereum uses Ethash and needs on average 15 s. The rationale behind choosing memory-intensive hash is an attempt to disadvantage miners with dedicated ASICs.

2.1 Mining Pools

The ratio of difficulty to the overall network mining power consequences that solo miners are expected a high variance of payouts which depends on their share of the total mining power. Therefore, miners form so-called mining pools, where all participants mine concurrently and share their revenue whenever one of them creates a block. Mining pools play a crucial role in todays' public based systems [8]. They are typically implemented as a pool manager and a cohort of miners. The pool manager joins the particular cryptocurrency system as a single miner and outsources the work to the miners. The pool manager estimates each miner's to evaluate the miners' efforts. When a miner successfully mines a new block, it sends it to the pool manager, which publishes this to the cryptocurrency system. The pool manager thus receives the entire revenue of the block and distributes it fairly according to its members' power. Experience with various cryptocurrencies shows that many pools are open, allowing anyone to join using a public Internet interface. Since the appearance of the first mining pools for most cryptocurrencies, the fraction of blocks mined by solo miners has significantly declined and is negligible today.

2.2 Mining Rewards

In Ethereum, there are three types of block rewards: static block reward, uncle block reward, and nephew block reward. The first one, static reward, is the constant value equal to 3 Ethers, sent to its miner as an economic incentive. To explain uncle and nephew rewards, we introduce the concepts of regular and stale blocks. A block is called regular if it is included in the system main chain and is called stale block otherwise [16]. The uncle and nephew rewards are unique in Ethereum, do not exist in Bitcoin. An uncle block is a stale block that could be described as a direct child of the system main chain, which means that the parent of an uncle block is always a regular block. An uncle block receives a certain reward if it is referenced by some future regular block, called a nephew block, through the use of reference links. The values of uncle and nephew rewards depend on the DISTANCE between the uncle and nephew blocks. For uncle block, the reward decreases starting from $\frac{7}{8}$ of the static block reward ($\text{DISTANCE} = 1$), through $\frac{6}{8}$ ($\text{DISTANCE} = 2$) and ending with 0 ($\text{DISTANCE} > 6$). The nephew reward is always $\frac{1}{32}$ of the block reward. In addition to block rewards, miners can also receive gas costs as all transaction fees.

The miner could estimate his expected profit based on the current state of the ethereum network (hashrate, block reward, average block time). They are web calculators, such as whattomine.com[1], coinwarz.com[2], where users could easy calculate suspected profits. Network hash rate varies over time, so such calculators provide just an estimation based on current values. But in this paper we are interested in past profits, that's calculations should be based on archive data. This set of data, provided by Ethereum explorers, consists of following: $hashrate$ - the historical processing power of the Ethereum network; $blocktime$ - the historical average time taken in seconds for a block to be included in the Ethereum blockchain; $blockcount$ - the historical number of blocks produced daily on the Ethereum network; $blockreward$ - total daily Ether supplied to the Ethereum network; $transfee$ - total historical number of Ether paid as a transaction fee for the Ethereum network. Having those historical data and assuming that: $hashpower$ - indicates average miners processing power and T - mining time in seconds, the $miningreward$ - reward paid to miner for his effort, could be calculated by the following formula:

$$miningreward = \frac{hashpower}{hashrate} \times \frac{(blockreward + transfee)}{blockcount} \times \frac{T}{blocktime}$$

3 Investigating Scenarios

With the growing popularity of cryptoassets and accompanying increase in their value, the interest in mining continues to grow. It is clearly visible in the charts presenting hashrate for particular cryptocurrencies (e.g. BTC[3], ETH[4]). This means that more and more computing power is required to generate the unit of a cryptocurrency. The power consumption, energy cost, and wear and

[1] https://whattomine.com.
[2] https://www.coinwarz.com.
[3] https://blockchain.com/charts/hash-rate.
[4] https://etherscan.io/chart/hashrate.

tear on the hardware are increasing. Digging with own computers and legal energy becomes less and less profitable. Resourceful cryptominers soon began looking for new ways to mine cryptocurrencies more profitably. One approach is to find extra hardware, and the other is to find a free energy source. Eskandari et al.[7] refer to cryptojacking - that is malware specially designed for mining cryptocurrencies and installed on a computer or mobile device and then uses its resources to mine cryptocurrency. In 2018 cryptojacking was even recognized as the biggest cyber security threat [11]. Even though in March 2019, Coinhive, the service that enables websites worldwide to use browser CPUs to mine Monero (XMR), has been shut down, approaches in which users' webpages are attacked through malicious web mining programs involve a wide range of attacks. In addition to the web mining services provided by Coinhive, many other service providers and even mining scripts written by attackers are still available [18].

Based on the author's experience as a court expert in actual criminal cases, there is an increase in miners using electricity at public institutions. Miners who are consuming the resources, both hardware and energy, of state-owned enterprises, government agencies, and universities or research institutes. After the seizure of suspected mining rigs in such institutions, investigators are facing two possible scenarios: with and without logs provided by mining software.

3.1 Scenario with Mining Software Logs

In the first scenario, when on the seized hard-drive, the computer investigator identified a mining software with enabled logging, the estimation of consumed energy could be done directly from data log. In many cases, those log files provide information not only of mining duration but also of energy consumed by GPUs. Such a script written in PHP script language has been designed by the author and shared[5] with the community. Forensics investigators should keep in mind that even if some log files were found, other log files might have been deliberately and successfully deleted.

3.2 Scenario Without Mining Software Logs

In the second scenario, where cryptominer tries to hide evidence of illicit mining by disabling logging in mining software or permanently deleting log files, an investigator possesses only information about ETH address. In the most popular ETH mining software like T-RexMiner[6], PhoenixMiner[7] or older Claymore[8] such data could be find in the .bat file (e.g. *-wal 0xdba4c80e8a1298228 d31d822dae069fd624d7b16*). The same file provides information about mining pool (e.g. *-pool eu1.ethermine.org:4444*). Having identified ETH address and mining pool, an investigator could possess information about payouts from a particular mining pool website. Pool websites provide this data in structured .csv files, the starting point for further energy consumption estimation.

---

5 https://rodwald.pl/cmepce/EN/LOG/.

6 https://github.com/trexminer/T-Rex/.

7 https://phoenixminer.org/.

8 https://github.com/Claymore-Dual/Claymore-Dual-Miner.

4 System Design

Cryptocurrency Mining Energy Past Consumption Estimator (CMEPCE) is designed and implemented as a modular platform, where adding a new cryptocurrency or mining pool should be relatively straightforward. The overall data pipeline could be divided into blockchain data sources, uploaded payouts file analyser and main calculating. The data sources module collects relevant data from a particular blockchain explorer. As a source of archive data from the Ethereum network, publicly available blockchain explorer etherscan.io has been chosen. Payouts analytical module parses .csv files entered by the user from mining pools. Finally, calculating module aggregate data from both sources and, based on the daily interval for ETH according to the formula presented in Sect. 2.2, provide output data. Directly from the formula average miner's hashpower is calculated, and then under the assumption that every 50 MH/s need 200 W of energy, consumed energy is calculated. The system is implemented in PHP language and is available on the dedicated website[9].

Much emphasis was also placed on the ease of use of the system because its users do not have to be IT-oriented (police officers, criminal investigators, prosecutors). A screenshot of the CMEPCE dashboard is presented in Fig. 1.

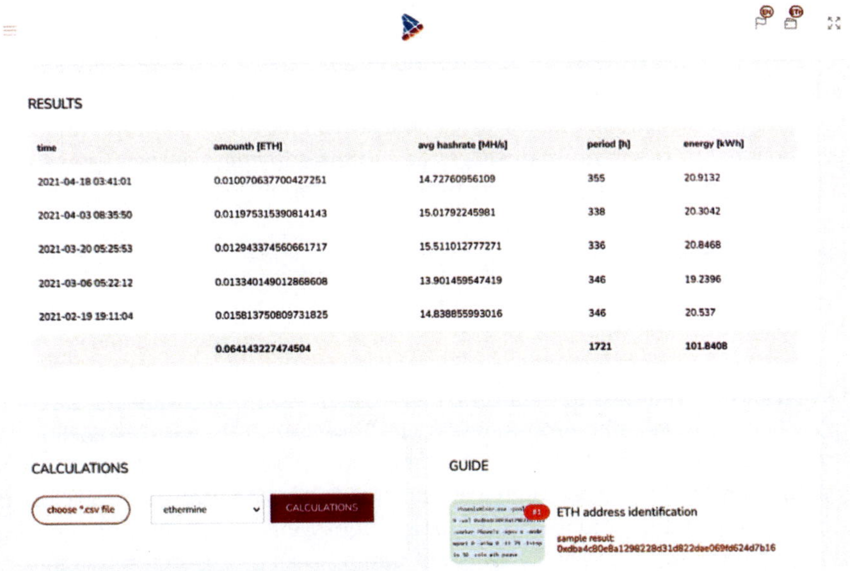

Fig. 1. Screenshot of the Cryptocurrency Mining Energy Past Consumption Estimator (CMEPCE) dashboard.

[9] https://rodwald.pl/cmepce.

5 Discussion

5.1 Limitations

There are several limitations to the proposed calculation. Firstly, the assumption is based on constant energy consumption to reach the speed of 50 MH/s for the miner. Energy consumption strongly depends on the GPU family (Nvidia vs Radeon) and the particular model of GPU. The number of GPUs per mining rig influences the calculations as well. This simplification will be replaced by more precise estimates, where a user would have an option to select a particular GPU model, which is well known for mentioned earlier mining profitability calculators. Secondly, data from mining polls do not present information about starting time of mining activities. The first record is when the first payout has been made. We assumed in our calculation that the time between starting and the first payment is the same as the time between the first and second payments. Thirdly, we assumed constant mining pool fee value (1%) and constant software mining fee value (1%). Both variables are pool or software dependant. Typically, pools may charge between 1% and 3% as pool fees, and each mining software mines to the developer's wallet every hour for a short period. Mining software fee ranges from 0% up to 5%[10]. Finally, we are not considering the reward distribution system implemented by the mining pool. There are multiple ways to design such a system in pooled mining [17]. Some popular schemes include: Pay-per-share - where the expected reward per share is paid, Pay-per-last-N-shares - the last N submitted shares are considered for payment when a block is found. While there are differences among these schemes (and their variations), all of them aim to distribute the reward such that the payoff of an individual miner is proportional to the number of shares he submitted, which in turn is proportional to his computational power contributed to the pool [14].

5.2 Directions

Since its launch, Ethereum was planned to migrate to the proof-of-stake model. Once the protocol has fully migrated, there won't be any revenue to be made from ethereum mining. But there are a lot of other cryptocurrencies that support GPU-based mining, so miners probably switch "their" software to start mining other cryptoassets. The proposed CMEPCE system is designed in such a way to easily adapt to the new circumstances and add analysis of other cryptocurrencies.

Looking for the suggestions coming from the potential users of CMEPCE, the system could be updated with further improvements. A more comprehensive range of cryptoassets, a wider range of mining pools, or more precise hardware-dependent energy cost calculations belong to the sample proposals.

6 Conclusions

With the growing popularity of cryptoassets, crypto-oriented illegal activities are also on the rising trend. CMEPCE is a pilot response to the increasing need to

[10] https://2cryptocalc.com/mining-software.

support cryptocurrency investigations. Investigations where illegal mining is an issue. With the identified demand for new functionalities, cryptoassets and mining pools, the author will continue developing CMEPCE as part of his research and court expert activities.

References

1. Alexandre, A.: Germany: suspects arrested for stealing electricity in crypto mining operation (2019). https://cointelegraph.com/news/germany-suspects-arrested-for-stealing-electricity-in-crypto-mining-operation. Accessed 09 Jan 2022
2. Bloomberg: Hundreds of banned crypto miners were siphoning power at China's state firms (2021). https://www.bloomberg.com/news/articles/2021-10-15/china-finds-banned-cryptominers-siphoning-power-at-state-firm. Accessed 09 Jan 2022
3. Crawley, J.: UK jails crypto miner for 13 months for stealing electricity: report (2021). https://www.coindesk.com/policy/2021/10/14/uk-jails-crypto-miner-for-13-months-for-stealing-electricity-report/. Accessed 09 Jan 2022
4. De Saro, M.: Polish police discover bitcoin-mining operation inside own department (2021). https://beincrypto.com/police-discover-mining-operation-in-headquarters/. Accessed 09 Jan 2022
5. Douceur, J.R.: The Sybil attack. In: Druschel, P., Kaashoek, F., Rowstron, A. (eds.) IPTPS 2002. LNCS, vol. 2429, pp. 251–260. Springer, Heidelberg (2002). https://doi.org/10.1007/3-540-45748-8_24
6. Erazo, F.: Malaysian crypto miners were caught stealing electricity from the state (2020). https://cointelegraph.com/news/malaysian-crypto-miners-were-caught-stealing-electricity-from-the-state. Accessed 09 Jan 2022
7. Eskandari, S., Leoutsarakos, A., Mursch, T., Clark, J.: A first look at browser-based cryptojacking. In: 2018 IEEE European Symposium on Security and Privacy Workshops (EuroS&PW), pp. 58–66. IEEE (2018)
8. Eyal, I.: The miner's dilemma. In: 2015 IEEE Symposium on Security and Privacy, pp. 89–103. IEEE (2015)
9. Furneaux, N.: Investigating Cryptocurrencies: Understanding, Extracting, and Analyzing Blockchain Evidence. Wiley, Hoboken (2018)
10. Gallersdörfer, U., Klaaßen, L., Stoll, C.: Energy consumption of cryptocurrencies beyond bitcoin. Joule **4**(9), 1843–1846 (2020)
11. Gorman, B.: Cryptojacking: a modern cash cow (2018). https://www.symantec.com/blogs/threat-intelligence/cryptojacking-modern-cash-cow. Accessed 09 Jan 2022
12. Heal, J.: Indonesian student arrested in South Korea for illicit crypto mining on college computers (2019). https://coinrivet.com/indonesian-student-arrested-in-south-korea-for-illicit-crypto-mining-on-college-computers/. Accessed 09 Jan 2022
13. Li, J., Li, N., Peng, J., Cui, H., Wu, Z.: Energy consumption of cryptocurrency mining: a study of electricity consumption in mining cryptocurrencies. Energy **168**, 160–168 (2019)
14. Luu, L., Saha, R., Parameshwaran, I., Saxena, P., Hobor, A.: On power splitting games in distributed computation: the case of bitcoin pooled mining. In: 2015 IEEE 28th Computer Security Foundations Symposium, pp. 397–411. IEEE (2015)
15. Nakamoto, S.: Bitcoin: a peer-to-peer electronic cash system. Decentralized Bus. Rev. 21260 (2008)

16. Niu, J., Feng, C.: Selfish mining in ethereum. arXiv preprint arXiv:1901.04620 (2019)
17. Rosenfeld, M.: Analysis of bitcoin pooled mining reward systems. arXiv preprint arXiv:1112.4980 (2011)
18. Shih, D.H., Wu, T.W., Hsu, T.H., Shih, P.Y., Yen, D.C.: Verification of cryptocurrency mining using ethereum. IEEE Access **8**, 120,351–120,360 (2020)
19. Zak, K.: Illegal cryptocurrency mine at the police school in legionowo (2021). https://www.rmf24.pl/fakty/polska/news-,nId,5545873. Accessed 09 Jan 2022

Reducing Development Time of Embedded Processors by Using FSM-Single and ASMD-FSMD Techniques

Valery Salauyou(✉)

Bialystok University of Technology, Wiejska 45A, 15-351 Bialystok, Poland
v.salauyou@pb.edu.pl

Abstract. Currently, there is a reduction in the lifetime of embedded systems. This requires a shorter time to market, that is, a shorter time to design system-on-chip (SoC) whose core component is the embedded processor. This paper discusses three techniques for designing embedded processors on the field programmable gate array (FPGA): the traditional approach, the FSM-single technique and the ASMD-FSMD technique using the example of designing the PIC-processor. It has been shown that although the FSM-single and ASMD-FSMD techniques are slightly inferior to the traditional approach in terms of implementation cost, they allow significantly to reduce (5–8 times) the development time. In addition, the ASMD-FSMD technique allows to increase the performance of processors, in some cases by 40%.

Keywords: Development time · Embedded processor · Design technique · Finite state machine (FSM) · Finite state machine with datapath (FSMD) · Algorithm state machine with datapath (ASMD) · Field programmable gate array (FPGA) · Verilog language

1 Introduction

Modern field programmable gate arrays (FPGAs) for executing programs and algorithms often include processors called embedded processors. The need to design a processor on the FPGA may arise when

- the processor implements the application-specific instruction-set;
- the processor is required to execute one particular program;
- the original processor with its instruction set is implemented.

Developing a new processor requires a lot of time and effort from the developer. The question arises: is it possible to reduce the development time of a new processor? This paper is intended to answer this question.

Usually, when developing a new processor, a single-cycle processor project is first built. The prototype single-cycle processor tests the main ideas (concepts) that form

W. Zamojski et al. (Eds.): DepCoS-RELCOMEX 2022, LNNS 484, pp. 245–254, 2022.
https://doi.org/10.1007/978-3-031-06746-4_24

the basis of the processor being developed, examines the effectiveness of the processor architecture, clarifies the processor instruction set, etc. Then, on the basis of the single-cycle processor, multi-cycle processors are built, which are the basis for creating pipeline processors.

In this paper we will consider three techniques of embedded processor design: the traditional approach, the FSM-single technique and the ASMD-FSMD technique, using the PIC-processor as an example. For each technique, the design capabilities of one-, two-, three-, and four-cycle PIC-processors are explored.

The design of embedded processors is an important task in the development of FPGA-based embedded systems. A large number of papers are devoted to this topic. Here we will consider only some of them. In [1], the LISA language is proposed to describe application-specific instruction set processor (ASIP) this allows you to automatically create the software and hardware parts of the embedded system on the FPGA. The FPGA design methodology for the RISC pipeline processor core of embedded systems is discussed in [2]. In [3], the ASP is created by extending the instruction set of the base processor. An overview of methodologies and algorithms for designing multiprocessor system-on-chip (MPSoC) is given in [4]. In [5], the architecture and the design methodology for FPGA for real-time embedded systems are described. A technique for designing a fault-tolerant processor that uses the reconfiguration property of modern FPGAs is discussed in [6]. The technique of designing a service-oriented system on a chip, which combines embedded processors and hardware accelerators, is discussed in [7]. In [8], the RISC-V processor for vector systems on the FPGA is described. The techniques for designing the RISC-V processors based on high-level synthesis using the C language and an universal verification methodology (UVM) are offered in [9]. The reconfigurable architecture of the embedded processor on the FPGA to reduce power consumption is discussed in [10]. The same problem is solved by reconfiguring at the thread and memory level in [11].

However, the problem of reducing the development time of embedded processors is not considered in any of the reviewed techniques. This paper is devoted to solving this problem.

The paper is organized as follows. Section 2 discusses the PIC-processor. Section 3 describes the traditional approach to the processor design and Sect. 4 describes the FSM-single technique. Section 5 deals with the ASMD-FSMD technique. Section 6 presents the results of experimental studies.

2 PIC - Processor

The instruction set architecture of the popular microcontroller PIC16F84A [12] is taken as the basis of the implemented PIC-processor. In this architecture one of the operands always is in an accumulator W, and the other operand arrives from a register file. The location of the result is determined by the value of the bit d, which is in the instruction code. When d = 0, the result is placed in the accumulator W, and when d = 1, the result is written back to the register file.

The PIC-processor [13], unlike the microcontroller PIC16F84A, uses the traditional data memory of the type RAM. To interact the processor with data memory, two additional instructions are entered into the PIC-processor instruction set: lw to load the value to the accumulator W from the data memory and sw to store the value of the accumulator W into the data memory. To be able to branch programs, two new instructions were added to the PIC-processor instruction set: gotoz to go to the address of the instruction memory if the result of the previous operation is zero, and gotonz to go to the address if the result of the previous operation is not zero. The PIC-processor instruction set is shown in [13].

3 Traditional Approach to Embedded Processor Design

The technique for designing processors on the FPGA is described in [14]. This technique will be called the *traditional approach*, with which we will compare the proposed techniques.

The main feature of the single-cycle processor is that all instructions are executed during one clock cycle. The design of a single-cycle processor begins with the definition of processor memory elements. Such elements are the program counter (PC), the instruction memory (IM), the data memory (DM), the register file (RF), and the accumulator W.

The design of a single-cycle processor consists in the development of the circuit for determining the address of the next instruction (PC-logic), the operating unit (datapath), the arithmetic logic unit (ALU) and the control unit (controller). All these devices are developed in parallel. In this case, separate groups of the PIC-processor instructions are sequentially considered and the necessary components are added to the structure of each unit to implement this group of instructions. The detailed description of the design process for the single-cycle PIC-processor using the traditional approach and the Verilog-code of the design are presented in [13].

The datapath components of the single-cycle and multi-cycle processors largely are the same. The main differences between a single-clock and a multi-clock processor are in their control units. The multi-cycle processor control unit in case of the traditional approach is presented as a finite state machine (FSM) of the Moore type.

Designing of the multi-cycle processor consists in sequential determination of the FSM states: fetching the instruction, decoding the instruction, as well as implementing the certain groups of the instructions. In the multi-cycle processor, many the datapath components are the same as those of the single-cycle processor components. The differences are the addition to the datapath of the registers, which necessary to build the multi-cycle processor. The detailed description of the design process for the multi-cycle PIC-processor using the traditional approach and the Verilog-code of the design are presented in [13].

4 The FSM-Single Design Technique for Multi-cycle Processors

Developers spend a lot of time designing the control unit, since many components of the datapath are standard functional blocks, but the control unit for each project needs to be designed again every time. The question arises: is it impossible to use the control unit of the single-cycle processor in the multi-cycle processor project? This is possible, and such a technique is called *FSM-single* (combining a FSM and the controller of one-clock processor).

The structure of the four-cycle PIC-processor, which was created using the FSM-single technique, is shown in Fig. 1a. In this structure, the controller is completely the same as the controller of the single-cycle PIC-processor, and the datapath is the same as the datapath of the four-cycle PIC-processor from Sect. 3. In the structure of the PIC-processor in Fig. 1, the FSM is added, which generates signals F, D, E and M, corresponding to the stages Fetch, Decode, Execute, and Memory of the four-cycle processor.

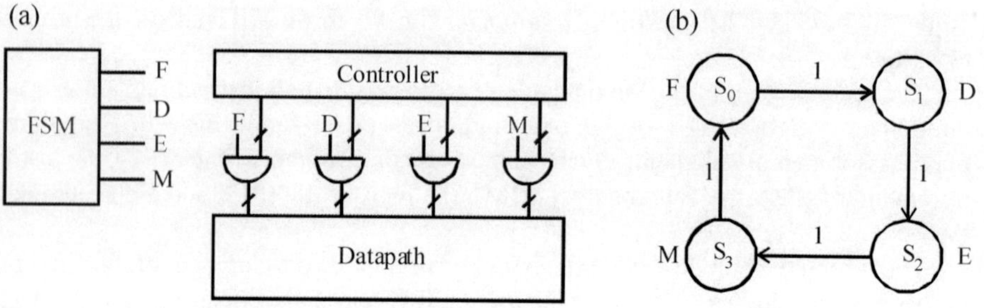

Fig. 1. The four-cycle PIC-processor built according to the FSM-single technique: a – the structure, b – the FSM graph

Details of building the multi-cycle PIC-processor using the FSM-single technique are given in [13].

5 ASMD-FSMD Design Technique for Embedded Processors

The development of digital devices by means of the ASMD-FSMD technique consists in creation a chart of an algorithmic state machine with datapath (ASMD) and in the description of the Verilog-code in the form of the finite state machine with datapath (FSMD) [15].

The ASMD chart consists of ASMD blocks. Each ASMD block describes the behavior of the FSMD in one state during one clock cycle. The ASMD block includes one state box (rectangle) and can have several decision boxes (rhombs) and conditional output boxes (ovals).

For a Moore-type machine, the operations, which are performed in this FSMD state, are written inside the state box (rectangle). For the Mealy-type machine, the operations, which are performed on FSMD transitions, are written inside the conditional output boxes (ovals). Logical expressions are written inside the decision boxes (rhombs). The outputs of the decision box are denoted by the digits 0 and 1, which correspond to transitions in the case of a false and true value of the logical expression. The rectangles or the ovals can contain any register operations, and the rhombs can contain any logical expressions that are allowed in the Verilog language. The ASMD-FSMD technique can be represented as the following algorithm.

Algorithm. The ASMD-FSMD technique for designing of digital devices.

1. The FSMD states are determined.
2. The ASMD block is constructed for each FSMD state.

   a. The logical expressions are written in the decision boxes.
   b. For the Moore FSMD, the register operations are written in the state box.
   c. For the Mealy FSMD, the register operations are written in the conditional output boxes.

3. The ASMD blocks are connected to each other by means of arcs in accordance with the algorithm of the device operation. Each output of the ASMD block can be connected to only one input of this or other ASMD block.
4. If necessary, the ASMD is modified to increase the performance or reduce the area of the designed device. For example, the algorithm loops are analyzed and the ASMD is changed in such a way as to minimize the number of states in the loop.
5. Directly based on the ASMD chart, the FSMD code is built in the Verilog language. In Verilog code, the variables correspond to the device registers. The logical expressions in the *if* statements correspond to the logical expressions checked in the ASMD decision boxes. The actions performed in ASMD blocks are described as procedural blocks *begin...end*. The operations performed in the rectangles (for Moore FSMD) are described first in the block *begin...end*, and the operations performed in the rhombs (for Mealy FSMD) are described in the corresponding places of the statements *if* (possibly using the operator brackets *begin...end*).
6. The FSMD is implemented using the appropriate design tool.
7. End.

Figure 2a shows a general view of the ASMD chart for implementing the instruction set of the single-cycle PIC-processor. The ASMD chart in Fig. 2a consists of one ASMD block for the Mealy FSMD. The instruction codes (i_1,…,i_k) are written in the rhombs, and the operations (ex_1,…,ex_k) required to implement the particular instruction (in the form of assignment statements) are written in the ovals.

The ASMD chart of the four-cycle PIC-processor consists of four ASMD blocks, which correspond to the stages of the four-cycle processor (Fig. 2b). The ASMD block Execute in Fig. 2b repeats the ASMD chart for a single-cycle PIC-processor, except for the instructions lw and sw that are implemented in the block Memory.

The two-cycle PIC-processor is built by combining the blocks Fetch and Decode, and also the blocks Execute and Memory in Fig. 2b. The three-cycle PIC-processor can be built either by combining the blocks Fetch and Decode, or by combining the blocks Execute and Memory in Fig. 2b.

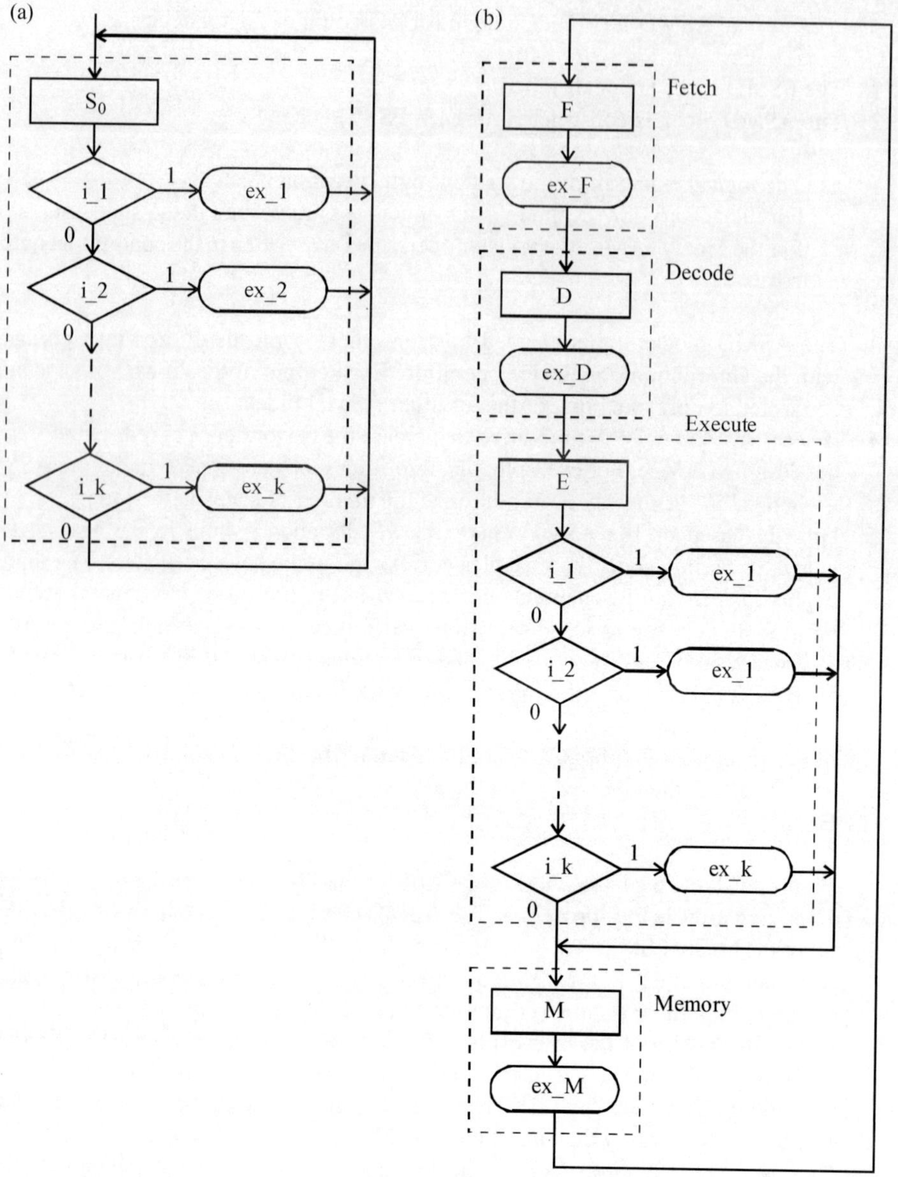

Fig. 2. The ASMD chart: a - for the single-cycle PIC-processor; b - for the four-cycle PIC-processor

6 Experimental Research

The following PIC-processor designs have been developed to test the effectiveness of the embedded processor design techniques reviewed:

- PIC_1_cycle_Tr, PIC_2_cycle_Tr, PIC_3_cycle_Tr, PIC_4_cycle_Tr – single-, two-, three-, and four-cycle PIC-processor designs built using the traditional approach;
- PIC_2_cycle_FSM, PIC_3_cycle_FSM, PIC_4_cycle_FSM – two-, three-, and four-cycle PIC-processor designs built using the FSM-single technique;
- PIC_1_cycle_ASMD, PIC_2_cycle_ASMD, PIC_3_cycle_ASMD, PIC_4_cycle_ASMD – single-, two-, three-, and four-cycle PIC-processor designs built using the ASMD-FSMD technique.

Studies of the effectiveness of the techniques considered were carried out when implementing the PIC-processors on the family FPGA Cyclone IV E using the system Quartus Prime Standard version 21.1. The PIC-processor designs were investigated with data width N equal to 4, 8, 16, 32, 64 and 128 bits.

The results of the studies are given in Table 1, where L_T, L_F, L_A are the number of FPGA logic elements used (an implementation cost) in the case of the traditional approach, as well as using the techniques FSM-single and ASMD-FSMD, respectively; F_T, F_F, F_A are the maximum synchronization frequency of the PIC-processor designs using various techniques; L_F/L_T, L_A/L_T, F_F/F_T and F_A/F_T are ratios of corresponding parameters. In Table 1, the processor designs are denoted as follows: PIC_k_cycle_N, where k is the number of processor cycles, N is the width of the data bus.

Table 2 shows the arithmetic mean values of the respective parameter ratios for the convenience of comparing the various techniques with each other.

Analysis of the Table 2 shows that the traditional approach is slightly (by 3–4%) superior to the FSM-single technique in terms of implementation cost. This result is predictable because the FSM is added to the PIC-processor structure. The traditional approach also surpasses the FSM-single technique in speed (by 19–21%). This is due to the fact that the signal delay of the processors built using the FSM-single technique increases by the value of the signal delay in the FSM (Fig. 1).

Comparing the traditional approach with the ASMD-FSMD technique shows that the traditional approach exceeds the ASMD-FSMD technique in terms of implementation cost (by 9–15%), but is inferior in speed, with the exception of four-cycle processors. In some cases, the superiority of the ASMD-FSMD technique in speed over the traditional approach reaches 40% (the example PIC_2_cycle_128 in Table 1).

However, the main advantage of the FSM-single and ASMD-FSMD techniques over the traditional approach is a significant reduction in the design development time. To evaluate the development time under the traditional approach and with using FSM-single and ASMD-FSMD techniques, all projects were created by one developer. The time in minutes spent developing each design is shown in Table 3, where DT_T is the development time in the case of a traditional approach; DT_F is development time in case of using the FSM-single technique; DT_A is development time in case of ASMD-FSMD technique; DT_T/DT_F and DT_T/DT_A are ratios of corresponding parameters; mid is the average value.

Table 1. The research results of the PIC-processor design techniques

Processor	L_T	L_F	L_A	F_T	F_F	F_A	L_F/L_T	L_A/L_T	F_F/F_T	F_A/F_T
PIC_1_cycle_4	1002	-	1197	70.67	-	71.89	-	1.19	-	1.02
PIC_1_cycle_8	1653	-	1991	66.80	-	69.25	-	1.20	-	1.04
PIC_1_cycle_16	2978	-	3494	67.85	-	66.97	-	1.27	-	0.99
PIC_1_cycle_32	5561	-	6498	53.58	-	57.33	-	1.17	-	1.07
PIC_1_cycle_64	10804	-	12456	52.59	-	51.83	-	1.15	-	0.99
PIC_1_cycle_128	21290	-	24359	39.08	-	39.54	-	1.14	-	1.01
PIC_2_cycle_4	914	1011	1344	84.40	73.85	79.47	1.11	1.47	0.88	0.94
PIC_2_cycle_8	1571	1656	2143	79.79	65.52	73.03	1.05	1.36	0.82	0.92
PIC_2_cycle_16	2909	2940	3497	67.28	68.85	75.34	1.01	1.20	1.02	1.12
PIC_2_cycle_32	5503	5593	6104	60.02	50.21	69.59	1.02	1.11	0.84	1.16
PIC_2_cycle_64	10706	10787	11300	56.11	53.19	68.61	1.01	1.06	0.95	1.22
PIC_2_cycle_128	21202	21262	21834	42.69	40.95	59.96	1.00	1.03	0.96	**1.40**
PIC_3_cycle_4	920	1015	1359	83.17	70.46	76.09	1.10	1.48	0.85	0.91
PIC_3_cycle_8	1571	1656	2160	76.90	65.79	74.43	1.05	1.37	0.86	0.97
PIC_3_cycle_16	2870	2962	3502	74.58	68.03	71.41	1.03	1.22	0.91	0.96
PIC_3_cycle_32	5509	5602	6157	60.33	56.14	69.31	1.02	1.12	0.93	1.15
PIC_3_cycle_64	10718	10801	11304	58.15	52.24	68.29	1.01	1.05	0.90	1.17
PIC_3_cycle_128	21190	21274	21849	43.04	39.78	59.41	1.00	1.03	092	1.38
PIC_4_cycle_4	898	1002	1325	109.15	88.87	70.49	1.12	1.48	0.81	0.65
PIC_4_cycle_8	1577	16832	2198	97.16	91.89	69.52	1.07	1.39	0.95	0.72
PIC_4_cycle_16	2897	2990	3477	97.42	81.56	67.82	1.03	1.20	0.84	0.70
PIC_4_cycle_32	5506	5601	6111	92.20	79.79	67.34	1.02	1.11	0.87	0.73
PIC_4_cycle_64	10739	10831	11260	89.63	60.87	63.71	1.01	1.05	0.68	0.71
PIC_4_cycle_128	21188	21295	21550	50.87	48.61	55.63	1.01	1.02	0.96	1.09

Table 2. The arithmetic averages ratios of the implementation cost and the performance for PIC-processors

Processor	Mid (L_F/L_T)	Mid (L_A/L_T)	Mid (F_F/F_T)	Mid (F_A/F_T)
PIC_1_cycle	-	1.19	-	1.02
PIC_2_cycle	1.03	1.20	0.91	1.13
PIC_3_cycle	1.04	1.21	0.90	1.09
PIC_4_cycle	1.04	1.21	0.85	0.77

Table 3 shows that using the FSM-single technique, the development time of the PIC-processors is reduced by 7 to 8 times. This large reduction in the development time is due to the fact that for the design of multi-cycle processors using the FSM-single

Table 3. The development time of the PIC-processors (in minutes)

Processor	DTT	DT_F	DT_A	DT_T/DT_F	DT_T/DT_A
PIC_1_cycle	10560	-	1440	-	7.33
PIC_2_cycle	840	103	120	8.16	7.00
PIC_3_cycle	6240	837	960	7.46	6.50
PIC_4_cycle	980	135	150	7.26	6.53
Mid				7.66	6.84

technique, the controller and most datapath components of a single-cycle processor are used, and only the simple FSM is being developed.

Using the ASMD-FSMD technique allows you to reduce the development time of the PIC-processors by 6–7 times. The significant reduction in the development time when using the ASMD-FSMD technique is due to the fact that there is no need to develop and test all components of the datapath, the controller, as well as the top-level module. The design of the processor, built according to the ASMD-FSMD technique, consists of one module in the Verilog language, plus the modules of the memory elements.

7 Conclusions

The paper presents two embedded processor design technologies, FSM-single and ASMD-FSMD, which can significantly reduce development time. The FSM-single technology is very simple and is designed for designing multi-cycle processors when a single-cycle processor design is already available. The ASMD-FSMD technology is an universal design technology of digital devices and is effective when the device can be represented as the datapath and the FSM. It is shown that both technologies allow to reduce significantly (by 6–8 times) development time of embedded processors on FPGA.

The results obtained allow to draw the following conclusions:

- if the main optimization criterion is the implementation cost, then the traditional approach is out of competition;
- if the main optimization criterion is the performance, then you should choose between the traditional approach and the ASMD-FSMD technique;
- if the development time is important and there is already a single-cycle processor design built using the traditional approach, then multi-cycle processors can be quickly developed using the FSM-single technique;
- if the main criterion is the development time and there is no a single-cycle processor design, then use the ASMD-FSMD technique.

Acknowledgements. The present study was supported by a grant WZ/WI-IIT/4/2020 from Bialystok University of Technology and founded from the resources for research by Ministry of Science and Higher Education.

References

1. Hoffmann, A., et al.: A novel methodology for the design of application-specific instruction-set processors (ASIPs) using a machine description language. IEEE Trans. Comput. Aided Des. Integr. Circ. Syst. **20**(11), 1338–1354 (2001)
2. Gschwind, M., Salapura, V., Maurer D.: FPGA prototyping of a RISC processor core for embedded applications. IEEE Trans. Very Large Scale Integr. (VLSI) Syst. **9**(2), 241–250 (2001)
3. Sun, F., et al.: A scalable synthesis methodology for application-specific processors. IEEE Trans. Very Large Scale Integr. (VLSI) Syst. **14**(11), 1175–1188 (2006)
4. Wolf, W., Jerraya, A.A., Martin, G.: Multiprocessor system-on-chip (MPSoC) technology. IEEE Trans. Comput. Aided Des. Integr. Circ. Syst. **27**(10), 1701–1713 (2008)
5. Oliveira, A.S.R., Almeida, L., de Brito Ferrari, A.: The ARPA-MT embedded SMT processor and its RTOS hardware accelerator. IEEE Trans. Ind. Electr. **58**(3), 890–904 (2009)
6. Psarakis, M., Apostolakis, A.: Fault tolerant FPGA processor based on runtime reconfigurable modules. In.: 17th IEEE European Test Symposium (ETS), pp. 1–6. IEEE, Annecy, France (2012)
7. Wang, C., et al.: Service-oriented architecture on FPGA-based MPSoC. IEEE Trans. Parallel Distrib. Syst. **28**(10), 2993–3006 (2017)
8. Johns, M., Kazmierski, T.J.: A Minimal RISC-V vector processor for embedded systems. In.: Forum for Specification and Design Languages (FDL), pp. 1–4. IEEE, Kiel (2020)
9. Chen, J., et al.: Design and verification of RISC-V CPU based on HLS and UVM. In.: IEEE International Conference on Computer Science, Electronic Information Engineering and Intelligent Control Technology (CEI), pp. 659–664, IEEE, Fuzhou (2021)
10. Tamimi, S., et al.: An efficient SRAM-based reconfigurable architecture for embedded processors. IEEE Trans. Comput. Aided Des. Integr. Circuits Syst. **38**(3), 466–479 (2018)
11. Jain, S., Lin, L., Alioto, M.: Processor energy-performance range extension beyond voltage scaling via drop-in methodologies. IEEE J. Solid-State Circuits **55**(10), 2670–2679 (2020)
12. Wilmshurst, T.: Designing Embedded Systems With Pic Microcontrollers: Principles and Applications. Elsevier, Oxford (2006)
13. Salauyou, V.V.: Functional Block Design of Embedded Systems on FPGA. Hotline– Telecom, Moscow (2020) (in Russian)
14. Harris, S.L., Harris, D.: Digital Design And Computer Architecture, ARM Morgan Kaufmann, San Francisco (2013)
15. Solov'ev, V.V.: ASMD–FSMD technique in designing signal processing devices on field programmable gate arrays. J. Commun. Technol. Electr. **66**(12), 1336–1345 (2021)

Influence of Accelerometer Placement on Biometric Gait Identification

A. Sawicki(✉)

Faculty of Computer Science, Bialystok University of Technology, Bialystok, Poland
a.sawicki@pb.edu.pl

Abstract. This paper examines the impact of acceleration sensor localization on the performance of gait-based person identification. The novelty of the presented approach is the extension of the study to different types of ground substrates. A publicly available data corpus used in this study included gait cycles acquired using six sensor locations: trunk, right/left thigh, right/left shin, and right wrist. The results of two experiments conducted using the Support-vector machine (SVM) classifier are presented in the study. In the first one, classifiers were trained and validated using gait samples acquired on surfaces such as concrete, grass, cobblestone, slope, and stairs. In the second experiment, training was performed with pavement gait and validated with other substrates samples. For both presented scenarios, a pair of sensors located on the right thigh and right shank achieved the highest average identification rates.

Keywords: Biometric · Gait · Accelerometer

1 Introduction

The aim of biometric gait analysis systems is to identify persons based on the acquired motion signals. Regardless of the data acquisition method (RGB/depth camera, accelerometer, etc.), the gait cycle is influenced by environmental factors such as clothing, footwear, walking surfaces, terrain slope, and obstacles [1]. In the case of wearable sensors like Inertial Measurement Unit (IMU), an additional factor affecting measurement is the sensor mounting location [2]. The literature demonstrates studies of wearable sensor selection intended to maximize the metrics of gait biometric systems. Such approaches try to solve the question of how many sensors there should be and where they should be placed in order to build the most efficient biometric system. These considerations, however, are quite limited in nature; they apply to one type of data acquisition surface by default. A gap in the literature concerns the design and development of systems intended to operate in real-world environments where the user walks on grass, pavement or concrete.

The aim of this study is to investigate the effect of IMU sensor mounting on the performance of behavioral biometric systems tested on various surface types. The paper provides an objective comparison of the effectiveness of systems based on a pair of accelerometers placed at different locations (e.g., left thigh and right wrist). The novelty

W. Zamojski et al. (Eds.): DepCoS-RELCOMEX 2022, LNNS 484, pp. 255–264, 2022.
https://doi.org/10.1007/978-3-031-06746-4_25

of the presented work is that the analysis is performed in terms of different types of tested surfaces. The conducted experiments will allow building high accuracy biometric systems, which can operate in real-life scenarios with not always ideal even surface.

Two main experiments were conducted in the study. The first allowed to observe how the biometric system will behave under favorable conditions, where samples are available for any substrate. In the second case, a learning set was constructed from gait cycles for the pavement surface and tested on the other surface types. Which allowed verifying how the system would behave under an unfavorable environment.

The article is structured as follows: Sect. 2 presents the literature review, and Sect. 3 details the methodology. The experiment results are presented in Sect. 4. Section 5 is the conclusion of the paper.

2 Literature Review

Sensor Placement

Reviewing the literature in the field of biometric systems and gait analysis we can encounter review publications summarizing the current state of the art in the form of meta-analysis. Existing research focuses on two main groups of solutions, an experimental approach in which subjects wear a group of sensors (often in the form of dedicated suits [3, 4]) and real-life scenarios applications based on smartphone or smartwatch devices [5, 6].

A review publication [7] gives a general overview of the placement of accelerometers mentioning: shoes, ankles, hands, waist, wrists, but also about trouser pockets or breast pockets. However, the advantages of the different mounting approaches are debatable, as the specific listed methods use different datasets, often relatively small. On the other hand, comprehensive studies [3, 4] are developed in which experiments are performed with a single multi-sensor data collection. In [3], the effect of sensor placement and a number of sensors on person identification rates was investigated using a proprietary data corpus. In which data acquisition was performed with sensors located on the wrist right, upper arm, right side of the pelvis, left thigh, and right ankle. The authors [3] indicated that in the case of single-sensor solutions, the best identification results were achieved for sensors located on the right side of the pelvis. With increasing the number of sensors and using a pair of sensors, the best identification rates were obtained for the right side of the pelvis and the ankle joint.

On the other hand, in work [4] the authors did not conduct the problems of person identification, but the detection of heel strike and toe-off moments in the gait cycle. In this study, sensors located on the feet, shin, and trunk were used. This work indicated that the sensors located on the feet and shin allowed the best metrics to be achieved, thus providing the most relevant information. Works such as [3, 4] contributed a great deal to the development of the discipline, while the experiments performed in them were conducted for a fixed type of substrate. A natural continuation of the research is to verify whether a change in the substrate, e.g. grass, concrete, affects the sensor location's optimal selection.

Different Types of Substrate
To date, many approaches to gait analysis have been published, few of which address multiple ground types. One of the most important works in this area is [8, 9]. The authors [8] implemented biometric systems for surfaces such as stone plates, gravel, ground, and grass. It should be emphasized, however, that the research did not describe the final identification metrics, but the possibility of separating individual gait cycles. For hard surface types, the data representing individuals' gait cycles were separated and, therefore should be distinguishable. However, the experiments were carried out only for a group of 5 people, and the work used one sensor located in the pelvic area.

On the other hand, the paper [5] compares identification systems for surfaces such as flat ground, gravel, and grass walk. In this case, the authors developed a typical identification statement along with the Equal Error Rate (EER) metric. The best quality of identification was observed for gravel, flat surface, grass, and the worst for ramp (named as inclined walk). In this case, also a single sensor was used in the form of a cell phone placed at the hip area.

The works [5, 8] address the interesting aspect of the dependence of substrate type on gait-based person identification systems. For both works, however, only a single acceleration sensor was used. It is a natural continuation of the current study to expand the research by analyzing how the variate substrate will affect more complex multi-sensor systems.

3 Methodology

The objective of the conducted work was to investigate the association of the location of the accelerometer on the body and the type of substrate on the accuracy of gait-based biometric systems. Two experiments were conducted as part of the current study. In the first one, the identification efficiency with two sensors and eight types of the substrate was examined. Training and test set included recordings on each type of substrate. The second experiment was more restrictive, learning was carried out using gait cycles recorded on a flat surface and validation using other types of surfaces.

3.1 Data Corpus Description

In the study a publicly available human gait body was used [9]. We utilized this database in previous experiments [10], in which, however, we developed a single-sensor biometric system. The human gait was acquired using the MTw Awinda device, which consisted of 6 IMU units attached to the body with velcro. The locations of the individual sensors are: right and left shin, right and left thigh, right wrist, and back of the torso. Figure 1 shows the installation of all sensors with selected measurement axes marked.

The acquisition was carried out with the participation of 30 subjects who walked on surfaces such as pavement, concrete, grass, cobblestone, slope/ramp, stairs.

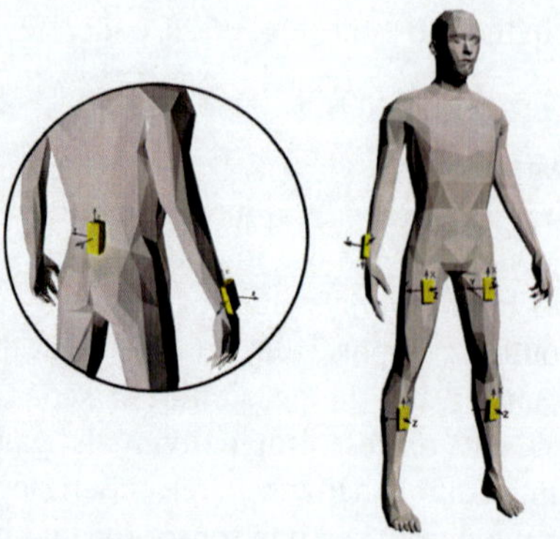

Fig. 1. Location of six three-axis accelerometers on the subject body

3.2 Segmentation and Signal Preprocessing

Motion tracking sessions were represented as block recordings containing multiple gait cycles as well as initial/final periods of immobility. The segmentation process adapted the methodology described in [3]. The detection of gait cycles was performed using manual inspection of the magnitude of the right shin accelerometer signals. Details of the segmentation are made available in the public repository [https://github.com/aleksandersawicki/Influence-of-accelerometer-placement-on-biometric-gait-identification].

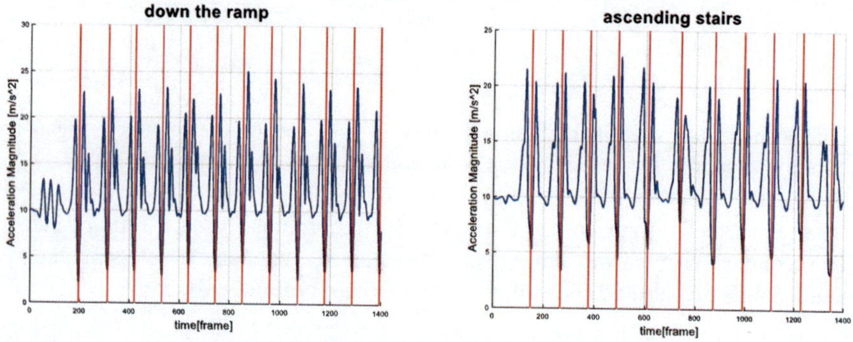

Fig. 2. Block recording segmentation, vertical lines indicate beginnings of gait cycle

Figure 2 presents the accelerometer signals recorded for the same participant while walking on different types of substrate. The graphs show the time series of magnitudes of the accelerometer signals, the vertical lines indicate the detected beginnings of gait cycles. It can be observed that despite the dissimilarity of movements such as moving down the ramp or climbing stairs, the signal has characteristic local extrema. The applied approach works sufficiently effectively in this case. Then, due to the influence of the sensor mounting method on its measurements [11], additional filtering of the signal with the orientation invariant gait matching algorithm [12] was applied.

3.3 Feature Extraction and Classifier Selection

The gait cycles analyzed within the present biometric studies were collected in diverse conditions i.e. walking on sidewalk, grass, ascending/descending stairs. A somewhat similar issue is the analysis and recognition of Human Activity Recognition (HAR), in which activities such as walking/lying standing are detected. Seeing the similarities, it was decided to use the features of signals taken from the HAR domain [13]. This is motivated by the assumption that if these features carried useful information about the type of activity, they may be promising in the field of person identification as well. For each of the three accelerometer axes, feature vectors were used in the form of the metrics: Mean, Standard Deviation (STD), Median, Root Mean square (RMS), averaged derivatives, Skewness, Kurtosis, Energy, Mean Crossing Rate (MCR). In the classification process, an SVM classifier was deployed. It should be noted, however, that it was allowed to use models with different configurations for each validation trial. The learning process used the *GridSearch* implementation of the model parameter search feature. The search process for radial basis function iteratively assumed C$\in${300,500,700,1000}, and gamma$\in${0.001,0.0050,0.01,0.02}.

4 Results

Experiment I, Training and Validation for Each Surface Type
Figure 3 presents a comparison of the accuracy of biometric systems depending on the location of the motion sensors and the type of substrate. The Y-axis shows the combinations of the mountings resulting from the selection of 2 of the 6 locations (right/left calf, right/left thigh, trunk, wrist). The X-axis is related to the used substrate type (change from the most regular to the most fluctuating). The numerical values in each cell represent the average and standard deviation F1-score determined from 5 cross-validations.

On the basis of the presented graphic, we can notice two essential dependencies. First, identification is quite regular for the first six types of ground and significantly worse for walking on stairs. Moreover, among the resulting combinations of sensor pairs, the system based on trunk and wrist (Tk,Wt) and right thigh and wrist (Tr,Wt) sensors give worse results than the other sensor configurations. For these two cases, there is also the largest standard deviation of results for all studied cases (surface:"*Up strairs*"; std:0.13,0.11). Furthermore, a typical standard deviation of the F1-score for the "*Pavement*", "*Flat Even*", "*Cobblestone*", "*Grass*", "*Slope Up*" and "*Slope Down*" cases is at least 0.02. For the stair moving cases, it is typically at least 0.05.

To answer the question "Where on the body should two accelerometers be mounted to get the best biometric identification results?" we used an additional classifier ranking adapted from [14]. An indicator from 1 to 15 (number of combinations) was assigned for each row in each column, depending on the classification score. The lower the ranking value, the better the biometric system did compare to the others. The higher the value, the worse the ranking compared to the entire compilation. Table 1 shows the results of the proposed approach, with the columns to the right showing the average results for the entire dataset and excluding stair motion.

	Pavement	Flat Even	Cobblestone	Grass	Slope Up	Slope Down	Up stairs	Down stairs
Tk,Tr	0.96±0.04	0.98±0.04	0.92±0.04	0.98±0.02	0.96±0.04	0.98±0.02	0.87±0.04	0.78±0.05
Tk,Tl	0.98±0.04	0.96±0.05	0.96±0.03	0.95±0.03	0.97±0.04	0.98±0.04	0.83±0.07	0.73±0.11
Tk,Sr	0.99±0.02	0.97±0.02	0.93±0.05	0.97±0.04	0.98±0.04	0.96±0.03	0.80±0.03	0.82±0.06
Tk,Sl	0.98±0.02	0.97±0.03	0.95±0.05	0.99±0.02	0.96±0.05	0.98±0.02	0.81±0.07	0.85±0.06
Tk,Wt	0.93±0.03	0.87±0.09	0.88±0.05	0.90±0.05	0.88±0.08	0.88±0.07	0.76±0.13	0.69±0.08
Tr,Tl	0.98±0.04	0.96±0.04	0.94±0.02	0.95±0.03	0.95±0.03	0.96±0.03	0.80±0.04	0.67±0.11
Tr,Sr	0.99±0.02	0.98±0.02	0.96±0.02	0.99±0.02	0.97±0.02	0.97±0.02	0.80±0.05	0.80±0.09
Tr,Sl	0.98±0.02	0.98±0.04	0.92±0.03	0.99±0.02	0.96±0.02	0.99±0.02	0.84±0.06	0.78±0.05
Tr,Wt	0.95±0.05	0.93±0.04	0.89±0.04	0.90±0.03	0.94±0.04	0.94±0.02	0.78±0.11	0.65±0.05
Tl,Sr	0.99±0.02	0.97±0.02	0.92±0.06	0.97±0.04	0.96±0.04	0.99±0.02	0.77±0.07	0.78±0.09
Tl,Sl	0.99±0.02	0.98±0.02	0.90±0.05	0.99±0.02	0.97±0.02	0.99±0.02	0.80±0.06	0.84±0.08
Tl,Wt	0.97±0.04	0.97±0.05	0.92±0.04	0.88±0.07	0.94±0.04	0.96±0.04	0.82±0.07	0.72±0.05
Sr,Sl,	0.99±0.02	0.99±0.02	0.92±0.04	0.98±0.02	0.95±0.05	0.97±0.05	0.81±0.06	0.74±0.05
Sr,Wt	1.00±0.00	0.96±0.05	0.88±0.02	0.95±0.03	0.96±0.03	0.96±0.02	0.72±0.05	0.73±0.07
Sl,Wt	0.98±0.02	0.97±0.04	0.89±0.04	0.97±0.02	0.93±0.04	0.95±0.05	0.66±0.09	0.80±0.06

Fig. 3. F1-score classification measure, for 15 sensor pairs and 8 surface types. Remark: Tk – Trunk; TR – Tight left; RL – Thigh right; SR – Shin right; SL – Shin Left, Wt – Wrist right.

Based on Table 1, we can observe that for the pairs Tr,Sr (Tight right, Shin right), Tl,Sl (Tight left, Shin left), and Tk,Sl (Trunk, Shin left) the highest identification rates were achieved. This indicates that the Thigh and Shin sensor combinations provide the most information about gait. Furthermore, it can be observed that the sensor pair Tr,Sr gives higher results to Tr,Sl. At the same time, the pair Tl,Sl provide higher results than Tl,Sr. This indicates that to achieve higher identification efficiency, sensors should be placed on the same limb.

From Table 1, it is possible to try to approximate the classification results for applications based on smartphones and smartwatches. For this purpose, it should be assumed that locations Tr (Tight Righ), Tl (Tight Left) with some accuracy may reflect trouser pockets (where a smartphone is placed), and location Wt (Wrist Right) may imitate signals coming from a smartwatch device. Identification with these sensor pairs performed average compared to the others. However, it should be noted that in all cases except for the stairs, identification performance above 0.8 F1-score was achieved. When comparing the Tr,Wt and Tl,Wt pairs directly, it can be seen that the second sensor location guarantees better identification quality. Tl,Wt achieves higher or equivalent identification rates for 7 out of 8 substrate types.

Experiment II, Pavement Surface Training, Validation for Other Substrate Types
A natural continuation of the research is to consider the case in which the learning process

Table 1. Ranking of biometric systems [14], training for each substrate

	Pavement	Flat even	Cobble stone	Grass	Slope up	Slope down	Up stairs	Down stairs	Mean rank	M. rank w/o stairs
Tk,Tr	3.5	5.5	13.0	6.5	5.0	8.5	1.0	7.0	6.3	7.0
Tk,Tl	13.0	11.0	9.0	3.0	5.0	1.5	3.0	11.0	7.1	7.1
Tk,Sr	8.0	7.5	4.0	1.0	11.5	5.0	10.0	3.0	6.3	6.2
Tk,Sl	6.0	2.5	9.0	6.5	5.0	3.0	5.0	1.0	**4.8**	**5.3**
Tk,Wt	15.0	13.0	15.0	15.0	15.0	15.0	13.0	13.0	14.3	14.7
Tr,Tl	11.5	11.0	9.0	11.0	9.5	4.0	8.0	14.0	9.8	9.3
Tr,Sr	3.5	2.5	4.0	3.0	7.5	1.5	7.0	4.0	**4.1**	**3.7**
Tr,Sl	3.5	2.5	9.0	6.5	2.0	6.5	2.0	8.0	5.0	5.0
Tr,Wt	14.0	14.0	14.0	12.5	14.0	12.0	11.0	15.0	13.3	13.4
Tl,Sr	8.0	9.0	4.0	6.5	2.0	6.5	12.0	6.0	6.8	6.0
Tl,Sl	3.5	2.5	4.0	3.0	2.0	11.0	9.0	2.0	**4.6**	**4.3**
Tl,Wt	10.0	15.0	12.0	12.5	11.5	8.5	4.0	12.0	10.7	11.6
Sr,Sl	1.0	5.5	4.0	10.0	7.5	10.0	6.0	9.0	6.6	6.3
Sr,Wt	11.5	11.0	1.0	9.0	9.5	14.0	14.0	10.0	10.0	9.3
Sl,Wt	8.0	7.5	9.0	14.0	13.0	13.0	15.0	5.0	10.6	10.8

Remark: Tk – Trunk; TR – Tight left; RL – Thigh right;
SR – Shin right; SL – Shin Left, Wt – Wrist right.

of classifiers is carried out for one type of substrate, and the validation is performed on the other types of surfaces. For this purpose, experiments were conducted in which a learning set contains gait cycles recorded on the Pavement surface. Figure 3 presents the results of the simple validation, per sensor pair and gait surface type.

Based on Fig. 3, we can observe that the classifiers trained on the pavement surface performed well in identifying samples flat even, cobblestone and grass surfaces. Worse for the inclined ramp and in a disqualifying way for moving up and downstairs. The motion patterns collected on the pavement surface are significantly different from the signals collected for moving on stairs. In order to accurately compare the classifications, the ranking method [14] was used once again.

From Table 2, it can observe that excluding the staircase movement, the identification was most successful for the pairs Tr,Sr (Thigh right, Shin right), Tl,Sr (Thigh left, Shin right), and Tl,Sl (Thight left, shin left). Again, the shin and thigh sensor pairs carried the most information about gait. In addition, it can be noted that Tr,Sr sensor pairs give better results than Tr,Sl. In contrast, the configuration of Tl,Sr gives better results than Tl,Sl. It is not evident in this case whether it is preferable to use sensors placed on one limb or on the opposite limb.

	Flat Even	Cobblestone	Grass	Slope Up	Slope Down	Up stairs	Down stairs
Tk,Tr	0.779	0.736	0.723	0.559	0.488	0.098	0.122
Tk,Tl	0.824	0.778	0.723	0.569	0.539	0.083	0.087
Tk,Sr	0.778	0.764	0.750	0.743	0.580	0.111	0.066
Tk,Sl	0.784	0.733	0.703	0.666	0.583	0.096	0.088
Tk,Wt	0.751	0.697	0.691	0.647	0.389	0.162	0.067
Tr,Tl	0.818	0.810	0.686	0.463	0.545	0.061	0.098
Tr,Sr	0.817	0.827	0.784	0.688	0.643	0.107	0.100
Tr,Sl	0.810	0.759	0.731	0.603	0.606	0.089	0.071
Tr,Wt	0.782	0.792	0.660	0.587	0.521	0.128	0.121
Tl,Sr	0.800	0.818	0.785	0.623	0.742	0.100	0.095
Tl,Sl	0.816	0.778	0.747	0.616	0.715	0.094	0.072
Tl,Wt	0.780	0.809	0.717	0.579	0.586	0.107	0.111
Sr,Sl,	0.772	0.690	0.665	0.654	0.635	0.049	0.051
Sr,Wt	0.764	0.796	0.737	0.719	0.677	0.140	0.105
Sl,Wt	0.797	0.736	0.681	0.672	0.600	0.101	0.065

Fig. 4. F1-score classification measure, for 15 sensor pairs and 7 surface types. Remark: Tk – Trunk; TR – Tight left; RL – Thigh right; SR – Shin right; SL – Shin Left, Wt – Wrist right

Table 2. Ranking of biometric systems [14], training for Pavement substrate

	Flat even	Cobble stone	Grass	Slope up	Slope down	M. rank w/o stairs
Tk,Tr	11.0	11.5	7.5	14.0	14.0	11.6
Tk,Tl	1.0	7.5	7.5	13.0	12.0	8.2
Tk,Sr	12.0	9.0	3.0	1.0	10.0	7.0
Tk,Sl	8.0	13.0	10.0	5.0	9.0	9.0
Tk,Wt	15.0	14.0	11.0	7.0	15.0	12.4
Tr,Tl	2.0	3.0	12.0	15.0	11.0	8.6
Tr,Sr	3.0	1.0	2.0	3.0	4.0	**2.6**
Tr,Sl	5.0	10.0	6.0	10.0	6.0	7.4
Tr,Wt	9.0	6.0	15.0	11.0	13.0	10.8
Tl,Sr	6.0	2.0	1.0	8.0	1.0	**3.6**
Tl,Sl	4.0	7.5	4.0	9.0	2.0	**5.3**
Tl,Wt	10.0	4.0	9.0	12.0	8.0	8.6
Sr,Sl	13.0	15.0	14.0	6.0	5.0	10.6
Sr,Wt	14.0	5.0	5.0	2.0	3.0	5.8
Sl,Wt	7.0	11.5	13.0	4.0	7.0	8.5

Remark: Tk – Trunk; TR – Tight left; RL – Thigh right; SR – Shin right; SL – Shin Left, Wt – Wrist right

When comparing the two pairs Tr,Wt and Tl,Wt, which can imitate smartphone sensors (placed in the trouser pocket) and smartwatch sensors (placed on the right wrist), it can be seen that placing the sensors in the left pocket and right wrist (Tl,Wt) provides higher identification results than placing the sensors on the same side of the body (Tr,Wt).

5 Conclusions

This paper presents an analysis of the effect of acceleration sensor localization on the performance of biometric identification systems. The novelty of the presented approach compared to [3] is the extension of the study to different types of substrates. A publicly available data corpus used in this study included gait cycles collected from six sensor locations: Tk - Trunk; TR - Tight left; RL - Thigh right; SR - Shin right; SL - Shin Left, Wt - Wrist right, and eight surface types. A study was conducted in which pairs of sensors were selected from six montage possibilities, resulting in a total of 15 input data combinations. In the first experiment, gait cycles collected on each substrate type were used for training and validation (Fig. 3 and Table 1). In the second case, the gait collected on the pavement surface was used to train the decision module, and validation was done using the walks collected on the other surface types (Fig. 4 and Table 2).

It can be concluded that for both experiments the placement of the Tr,Sr (Thigh right, Shin right) sensors allowed the highest average identification rates (3.7 rank Table 1, 2.6 rank Table 2). In the literature [3], the highest classification accuracy was obtained for the combination of sensors located in the right side of pelvis and right ankle, with lower rates for the left thigh and right ankle case. The database used in our experiments did not include right ankle signals. However, both our study and the publication [3] revealed that the location of sensors on one side of the body allows achieving the highest identification rates.

It should be emphasized that systems based on sensors placed on shin or trunk are theoretically experimental. It is difficult to imagine wearing sensors in these locations on a daily routine. The pairs Tr,Wt (Thigh right, Wrist right) and Tl,Wt (Thight Left, Wrist right), which may imitate smartphone sensors (placed in the trouser pocket) and smartwatches (placed on the right wrist), can be used in real life scenerios. For both performed experiments, sensor placement on opposite sides of the body (Tl,Wt) achieved better identification rates than placement on the same side of the body (Tr,Wt). However, It was not possible to fully compare the sensor montage by absence of left wrist sensor in used datasets. Additionally, publications [3] determined identification rates for only the left thigh and right wrist with no sensor on the right thigh and left wrist. This makes it difficult to come to a fully objective conclusion.

Highlighting the application potential, attention can also be drawn to the use of behavioral biometrics in virtual reality applications [15]. Created approaches aimed at user authentication not based on the password, which can be stolen but on information about the movement. Identification in this case is based on the tilt signals of the VR glasses, which are determined using the accelerometer readings.

Finally, from Fig. 4 it can be observed that the biometric systems trained using gait on a flat surface perform quite well on an uneven surface such as grass, cobblestone but less efficiently on an inclined ramp. The results are in full agreement with the literature[5]. For ascending and descending stairs in both experiments (Fig. 3 and Fig. 4), identification rates are worse than for the other ground types. Therefore, it is not recommended to develop biometric systems using this type of surface.

Acknowledgment. This research was funded in whole or in part by National Science Centre, Poland 2021/41/N/ST6/02505. For the purpose of Open Access, the author has applied a CC-BY public copyright licence to any Author Accepted Manuscript (AAM) version arising from this submission.

References

1. Sprager, S., Juric, M.B.: Inertial sensor-based gait recognition: a review. Sensors (Basel). **15**(9), 22089–220127 (2015). https://doi.org/10.3390/s150922089
2. Atallah, L., et al.: Sensor placement for activity detection using wearable accelerometers. In: International Conference on Body Sensor Networks, pp. 24–29 (2010). https://doi.org/10.1109/BSN.2010.23
3. Zhang, Y., et al.: Accelerometer-based gait recognition by sparse representation of signature points with clusters. IEEE Trans. Cybern. **45**(9), 1864–1875 (2015)
4. Panebianco, G.P., et al.: Analysis of the performance of 17 algorithms from a systematic review: influence of sensor position, analysed variable and computational approach in gait timing estimation from IMU measurements. Gait Posture. **66**, 76–82 (2018)
5. Muaaz, M., Nickel, C.: Influence of different walking speeds and surfaces on accelerometer-based biometric gait recognition. In: 2012 35th International Conference on Telecommunications and Signal Processing, pp. 508–512 (2012)
6. Zhao, Y., Zhou, S.: Wearable device-based gait recognition using angle embedded gait dynamic images and a convolutional neural network. Sensors (Basel) **17**(3), 478 (2017)
7. Wan, C., Wang, L., Phoha, V.V.: A survey on gait recognition. ACM Comput. Surv. **51**, 1–35 (2018)
8. Sprager, S., Zazula, D.: Impact of different walking surfaces on gait identification based on higher-order statistics of accelerometer data. In: 2011 IEEE International Conference on Signal and Image Processing Applications, pp. 360–365 (2011)
9. Luo, Y., et al.: A database of human gait performance on irregular and uneven surfaces collected by wearable sensors. Sci. Data **7**, 1–9 (2019)
10. Sawicki, A., Saeed, K.: Application of LSTM networks for human gait-based identification, theory and engineering of dependable computer systems and networks. Adv. Intell. Syst. Comput. 1389 (2021). https://doi.org/10.1007/978-3-030-76773-0_39
11. Subramanian, R., Sarkar, S.: Evaluation of Algorithms for Orientation Invariant Inertial Gait Matching. IEEE Transactions on Information Forensics and Security. p. 1. https://doi.org/10.1109/TIFS.2018.2850032 (2018)
12. Gadaleta, M., Rossi, M.: IDNet: smartphone-based gait recognition with convolutional neural networks. Pattern Recogn. **74**, 1–10 (2018)
13. Hamdi M. M., Awad M. I., Abdelhameed M. M., Tolbah F. A.: Lower limb gait activity recognition using Inertial Measurement Units for rehabilitation robotics. In: International Conference on Advanced Robotics (ICAR), pp. 316–322 (2015)
14. Demšar, J.: Statistical comparisons of classifiers over multiple data sets. J. Mach. Learn. Res. **7**, 1–30 (2006)
15. Olade, I., Fleming, C., Liang, H.N.: BioMove: biometric user identification from human kinesiological movements for virtual reality systems. Sensors **20**, 2944 (2020). https://doi.org/10.3390/s20102944(2020)

Comparison of Orientation Invariant Inertial Gait Matching Algorithms on Different Substrate Types

A. Sawicki([✉]) and K. Saeed

Faculty of Computer Science, Bialystok University of Technology, Bialystok, Poland
{a.sawicki,k.saeed}@pb.edu.pl

Abstract. This paper concerns biometric wearable identification systems based on the Inertial Measurement Unit (IMU). Applications of this type are affected by the negative phenomenon of the accelerometer mounting method influencing sensor readings. The study includes an objective comparison of invariant inertial gait matching algorithms reported in the literature. The novelty of the paper is the comparison of the methods for gait cycles captured on different types of ground. The research provides identification results for a publicly available data corpus that includes recordings of gait on pavement, grass, cobblestone, etc. Validation of identification score was performed using the classical SVM algorithm as well as recurrent LSTM neural networks The experiments revealed that for both tested classifiers, Gadaleta's method allows achieving the best identification rates - independently of the type of substrate on which the subject walks on.

Keywords: Accelerometer · Biometric · LSTM · SVM · IMU

1 Introduction

The purpose of biometric systems is to identify people on the basis of selected behavioral or physiological features or a combination of them. Scientific research on the development and implementation of the application attracts the attention of both academic and industrial environments [1]. Recently, there has been a rapid development of applications related to human gait-based identification using wearable sensors. The intensification of works in this area is rationalized by: the possibility of applications with the use of devices such as smartphones or smartwatches [2], lack of active interaction of the participant with the system (as in the case of fingerprinting) [3], difficulties in intentionally gait imitating [4]. Despite the numerous advantages of biometric systems based on accelerometers, the use of these sensors involves a certain inconvenience - the received signals have a "footprint" associated with the device montage. The literature describes many algorithms to eliminate this disadvantage, although it is complicated to measure their effectiveness or impact on biometric systems.

The aim of the paper is to investigate the use of orientation invariant gait matching algorithms on the performance of IMU-based biometric systems. The paper includes

W. Zamojski et al. (Eds.): DepCoS-RELCOMEX 2022, LNNS 484, pp. 265–275, 2022.
https://doi.org/10.1007/978-3-031-06746-4_26

an objective comparison of methods reported in the literature [5–8] using a data corpus containing gait cycles collected on different substrate types. To the best of the authors' knowledge, studies on the association of gait surface and types of signal preprocessing methods have not been published. Conducted experiments allowed to build high accuracy biometric systems that can operate in real scenarios and answer a question "Whether the orientation invariant gait matching algorithms should be selected individually to the substrate type?".

2 Literature Review

The readings of the triaxial accelerometers depend on the orientation of the sensor at the time of the measurement. From the theoretical point of view, sensors of this type are capable of measuring acceleration in a local reference system [12] which can be modeled by orientation in the form of Euler angles, quaternions, or rotation matrices. In biometric systems based on Inertial Measurement Units (IMU), this dependence presents a certain inconvenience, the received signals have a specific "footprint" associated with the installation of the device. In real scenarios, it is unrealistic to have an identically positioned sensor multiple times. Both for experiments involving inertial motion capture systems (the goal of our work) or more practical ones, such as a smartphone placed in a pocket, it is not possible to even place the sensor perfectly twice. Therefore there are developed methods to eliminate the unfavorable phenomenon, known as "orientation invariant inertial gait matching algorithms".

One of the simplest algorithms belonging to the group of these methods is the modification of the three-axis accelerometer measurements reading into a magnitude time series [5]. This solution eliminates the influence of the sensor orientation on its measurement, at the cost of losing some information. The dimensionality reduction ultimately affects person identification metrics.

A second, rather standard approach, is to convert the accelerometer measurements from the local sensor coordinate system to the global North West Up (NWU) Earth coordinate system [6, 7]. However, the presented approach has a significant disadvantage - the processed signals have a "footprint" related to the place of data acquisition. Depending on whether the subject is facing North, South, East, or West, the global N and W axes will show different accelerations even with perfectly repeated walk. A transformation of this type was proposed in [6], with the integration of an additional calibration quaternion associated with the acquisition location. This allowed eliminating the previously mentioned footprint and was equivalent to artificially rotating individual subjects in the same Earth direction. A disadvantage of the method [6] is the requirement to know the direction towards the North for a given acquisition location. However, when data is collected within a single fixed location, this factor can be ignored.

On the other hand, it is proposed in [7] to consider only the transformed vertical signal and the norm of the signals in the vertical plane for identification. The undoubted advantage of approach [7], in comparison to [6], is the complete elimination of information about the place of data acquisition. The original signals from three-dimensional space are transformed to two-dimensional space. Consequently, some data are irretrievably lost. For example, the individual participant's preference to turn right or left during gait performance cannot be detected. Information that could potentially positively influence the outcome of the identification process is reduced.

On the other hand, in [8] the authors proposed another very interesting approach to develop orientation invariant inertial gait matching algorithms. They decided to manually estimate the local reference system, based only on information about accelerometer measurement data during a single gait cycle. The first axis of the estimated coordinate system was related to the gravitational acceleration and indicated roughly the vertical direction, the second axis was related to the variance analysis and indicated the direction of motion (the direction of gait has the highest variance), and the last axis was calculated from the vector product of the previous two.

This proposed approach satisfied the assumption of orthonormality and the fixed right/left convention of the new coordinate system. As in the previous methods [6, 7], the accelerometer signals were finally transformed from a local system to some type of global system. This is undoubtedly an interesting solution, whose advantages include the lack of a requirement for accurate sensor orientation (the coordinate system is created using raw accelerometer signals). On the other hand, it is not entirely clear whether the creation of a coordinate system based only on a small part of the data, i.e. a single gait cycle (1–1.5 s) is an optimal solution, or whether better results could not be obtained by using several gait cycles for this purpose.

The literature also contains descriptions of new rather unique signal transformation methods. For example, in [9] new algorithms based on the Kabsha transform are proposed. This method differs significantly from the previously discussed examples [5–8] and requires each participant should have a fixed reference gait pattern. The Kabsha algorithm is used to create a matrix of dimension 3×3 such that the test signal approximates the reference signal as closely as possible. The matrix is determined between each reference gait cycle and each test gait cycle. Each gait test cycle has as many versions as there are reference gait patterns. Which of these versions is considered useful depends on the distance measure. Solutions of this type are characterized by high computational resources and the absence of the possibility of using an independent classifier. As indicated by other research articles [10], the results are comparable to other methods.

The proposed research focuses on comparing the performance of biometric systems with different gait matching algorithms for gait signals collected on different surfaces. The results of presented work may be useful for building high accuracy biometric systems that can operate in real scenarios. To the best of our authors' knowledge, no study associating the performance of gait matching algorithms with surface types such as grass, cobblestone, and concrete has been published to date. The most extensive study of a similar issue is presented in [10]. However, this work only focuses on comparing algorithms for a fixed substrate type. In the presented research, methods such as Magnitude [5], Hoang [7], Global transformation [6] and Gadaleta [8] were implemented. Which is associated with the use of methods mainly omitted in [10]. The developed study can be considered in the context of the continuation of the research started in [10].

3 Methodology

3.1 Dataset Description

A publicly available corpus of human gait dedicated to the evaluation of biometric systems, with the unique feature of varying ground types was used in the study. The dataset contains gait cycles collected with 30 participants on 8 types of ground: pavement, concrete, cobblestones, grass, slope (walking up and down), and stairs (ascending and descending) [11]. During a single tracking session, the participant performed 6 walking trials on a specified surface, where each trial consisted of several gait cycles. Data acquisition was performed using a Xsens MTw Awinda inertial motion capture system, which consisted of six Inertial Measurement Unit (IMU). Each sensor provided synchronized measurement of acceleration, angular velocity, magnetic field strength, and orientation signals at a frequency of 100 Hz. The six sensors were located respectively on the torso, right wrist, right/left thigh, and right/left shin. Single time series was described by general Eq. (1):

$$\boldsymbol{x} = (x_1, x_2 \ldots x_n)^T \tag{1}$$

For further signal processing, the measured values of the triaxial accelerometer $\boldsymbol{A}$ and the sensor orientation time series $\boldsymbol{Q}$ were used. Featured data were represented as a combination of multiple time series (2):

$$\begin{aligned} \boldsymbol{A} &= \left[\boldsymbol{a_x}, \boldsymbol{a_y}, \boldsymbol{a_z}\right]^T \\ \boldsymbol{Q} &= \left[\boldsymbol{q_w}, \boldsymbol{q_x}, \boldsymbol{q_y}, \boldsymbol{q_z}\right]^T \end{aligned} \tag{2}$$

where:

a_x, a_y, a_z – x, y, z axis measurement values of the accelerometer;

q_w, q_x, q_y, q_z– orientation timeseries in quaternion form.

It should be noted that the dataset in the raw form included angular velocity information. Measurements of this type are not the subject of any examined orientation invariant inertial gait matching algorithms. Therefore, these type of signals were completely omitted from further processing.

3.2 Orientation Invariant Inertial Gait Matching Algorithms

In this paper, four popular preprocessing algorithms are implemented and examined. From the diagram presented in Fig. 1, it can be seen that the Magnitude Transform algorithm [5] performs dimensionality reduction. The original $3 \times n$ signal is replaced by one dimensional vector of length n. The Hoang et al. [7] and Global Transform [6] use an additional information about sensor orientation to converted local acceleration measurements to a global reference frame. However, in the Hoang et al. [7] algorithm additional dimension reduction step is performed. The last implemented Gadaleta et al. [8] method involves creating a new coordinate system based on variance analysis.

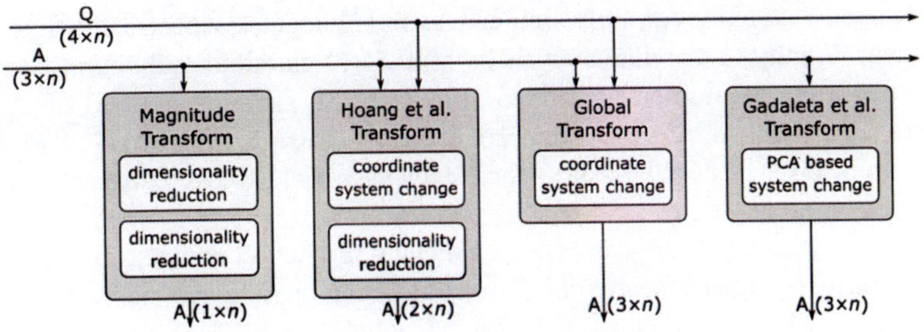

Fig. 1. Four examined algorithms dataflow diagrams

Figure 1 presents a data flow diagram that demonstrates the dimension of the input and output signals for each algorithm. Methods such as Hoang et al. [7] or Global transform [6] required additional information about the sensor orientation, although for each investigated method the output signal is a modification of the acceleration measurements. Due to publication restrictions, it is not possible to provide implementation details of each method within the scope of this paper. Therefore, the critical aspects of each algorithm are discussed in Sect. 2, while the reference to the original papers [5–8] is encouraged.

3.3 Gait Cycle Detection

The dataset [11] in its original representation contained motion signals in a form of block recordings. Individual cycles having different durations occurred in an unsegmented form. We have developed a dedicated segmentation algorithm based on expert knowledge in our previous work [13]. Segmentation required a well-described gait pattern of the accelerometer's OZ axis measurement. It should be noted that in previous studies, gait analysis was performed only on a flat surface for which literature knowledge could be obtained. In the current research due to the varying gait characteristics including moving upstairs, walking on grass, it was not possible to obtain external knowledge of expected gait patterns.

Due to the variance of gait cycles with respect to the surface area, a robust segmentation method adapted from [14] was applied. The gait cycle was extracted by manual local mimes detection in right shin accelerometer signal. This process involved using the magnitude signal determined according to Eq. (3).

$$a_{mag}(i) = \left(a_x(i)^2 + a_y(i)^2 + a_z(i)^2\right)^{1/2}, i\epsilon\langle 1, n\rangle \tag{3}$$

The use of signal magnitude similarly as in [7] negates aspects of different sensor placement. Finally to provide analysis by both SVM classifier and neural network, the gait cycles were resampled to a fixed length of 128 samples. A total of 17351 gait samples (Pavement - 2119; Flat even- 1658; Grass - 1807; Cobblestone- 2046; Slope up -3213; Slope down -3366; Upstairs -1562; Downstairs -1580) were extracted during the segmentation process.

Since the gait detection was done in a manual inspection manner, in order to ensure a repeatability by other researchers, frame indexes indicating gait starts are included in the repository (https://github.com/aleksandersawicki/Comparison-of-orientation-invariant-inertial-gait-matching-algorithms-on-different-substrate-types).

3.4 Classification

In the presented research, both types of approaches were developed: a classical SVM classifier and a Deep Learning method. The first presented approach used feature vector consisting of fundamental metrics such as: mean, standard deviation, energy, sequential absolute differences, kurtosis, skewness, median, minimum, maximum. Furthermore, feature selection process was guided by a literature example [15].

In the case of the deep learning classifier, as in our previous work [13], a recurrent LSTM network was applied. The classifier's input data were in the form of $S \times 128$ accelerometer time series processed by one of the inertial gait matching algorithms (Fig. 1), where $S \in \{1,2,3\}$. Neural network developed in this study consisted of two LSTM type cells with 32 hidden neurons. Model was trained at a learning rate equal to 0.0025 for 300 epochs using the ADAM optimizer. Similar to the SVM classifier was examined using 5-fold cross-validation tests.

4 Results

The first of the conducted experiments concerned the biometric identification system based on the measurement data of the sensor located on the right thigh. Table 1 presents a comparison of the influence of the tested algorithms on the identification performance for two types of the examined classifier. For each type of surface, the best result is presented in bold, while the second achieved result is underlined. In the case of the SVM classifier, the best results were achieved for the Gadaleta et al. method, while the second-best result was obtained for the Global Transform algorithm.

For the deep learning approach, the Gadaleta et al. algorithm was again the most successful, while the second-best result was observed for the Hoang et al. methods. It can be seen quite distinct differences between the results for the classical and the deep approach.

In the next step, the identification results of biometric systems based on SVM classifiers will be discussed. The presented metrics will be additionally related to different locations of the measurement sensor. Figure 2 shows the identification results in the F1-score metric for various ground types. Each subplot include a comparison of the impact of one of the four gait matching algorithms (Sec. 3.2) on identification performance. Bar

Table 1. F1-score metric value for identification with right thigh sensor

		Pavement	Flat even	Grass	Cobble stone	Slope up	Slope down	Upstairs	Downstairs
SVM	Mag.	0.83 ± 0.04	0.76 ± 0.04	0.61 ± 0.05	0.61 ± 0.04	0.82 ± 0.02	0.64 ± 0.04	0.48 ± 0.03	0.41 ± 0.02
	Hoang	0.88 ± 0.03	0.85 ± 0.01	0.74 ± 0.06	0.72 ± 0.03	0.85 ± 0.06	0.80 ± 0.03	0.58 ± 0.03	0.55 ± 0.03
	Global	0.90 ± 0.04	0.88 ± 0.01	0.83 ± 0.04	0.79 ± 0.04	0.87 ± 0.05	0.86 ± 0.06	0.58 ± 0.05	0.69 ± 0.02
	Gadaleta	**0.92 ± 0.02**	**0.93 ± 0.01**	**0.88 ± 0.04**	**0.85 ± 0.04**	**0.91 ± 0.01**	**0.89 ± 0.02**	**0.74 ± 0.03**	**0.65 ± 0.03**
LSTM	Mag.	0.83 ± 0.04	0.81 ± 0.03	0.76 ± 0.03	0.69 ± 0.05	0.82 ± 0.02	0.78 ± 0.03	0.52 ± 0.03	0.46 ± 0.02
	Hoang	0.88 ± 0.03	0.87 ± 0.02	0.80 ± 0.04	0.79 ± 0.04	0.85 ± 0.06	0.85 ± 0.02	0.61 ± 0.02	0.52 ± 0.03
	Global	0.90 ± 0.04	0.81 ± 0.03	0.74 ± 0.03	0.73 ± 0.04	0.87 ± 0.05	0.86 ± 0.05	0.49 ± 0.04	0.61 ± 0.03
	Gadaleta	**0.92 ± 0.02**	**0.88 ± 0.05**	**0.86 ± 0.02**	**0.84 ± 0.03**	**0.91 ± 0.01**	**0.90 ± 0.01**	**0.68 ± 0.02**	**0.58 ± 0.03**

heights present means from five-fold cross-validation of the f1-score metric (preferred for unbalanced data), whereas error bars reflect standard deviation.

The following dependencies can be observed in Fig. 2:

- The lowest identification rates are observed for the Magnitude transform [5] and for the Hoang transform [7] cases. Both algorithms used a dimensionality reduction step (Fig. 1). In contrast, more successful metrics was observed for algorithm that preserve the dimensions of the signal (Global transform [6], and Gadaleta [8]). it can be hypothesize that significant reduction in the length of the signal (feature vector) negatively affects the identification performance.
- Global transform [6] usually produced better results than Magnitude transform [5] and Hoang transform [7]. However, in exceptions in this area can be found (Grass surface and Thigh Left sensor).
- Observing all eight cases of different stringencies and six types of different sensor placements, it can be seen that the Gadaleta et al. method [8] provided generally the highest recognition metrics. There are no specific cases of substrate type or sensor location where competing methods have gained a significant advantage.

A natural continuation of the experiments is to verify replacing the SVM classifier with a Deep Learning network and abandoning manual feature selection will modify the results. Figure 3 shows identification results in the F1-score metric obtained by LSTM classifier.

The following dependencies can be observed in Fig. 3:

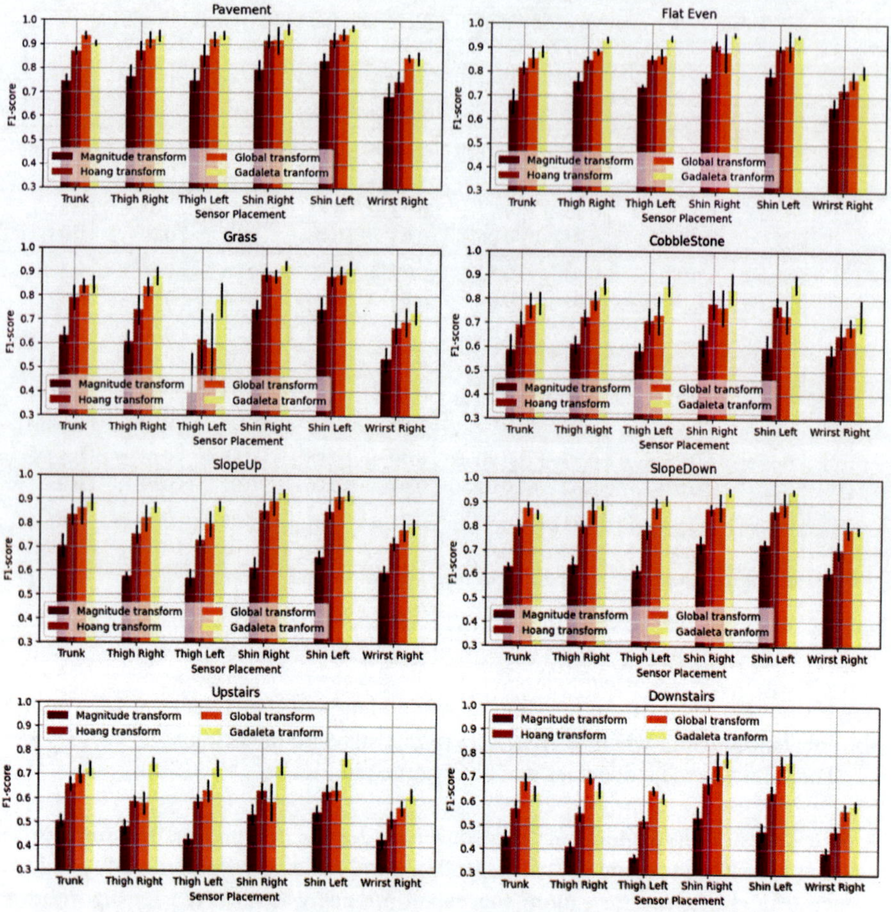

Fig. 2. F1-score identification metric obtained by SVM classifiers

- Generally, the worst identification score is observed when using the signal magnitude transformation [5]. There are of course some exceptions (the cobblestone surface and the wrist sensor). It should be noted, that in comparison to Fig. 2, the degradation in identification score relative to other competing methods is not as remarkable.
- For uneven surfaces such as cobblestone or grass, Global Transform [6] method provided a lower identification score than the Hoang et al. transformation [7].
- Generalizing, the method of Gadaleta et al. [8] again reached the highest identification scores, for all eight surface types and all six sensor placements. Although, for surfaces such as Grass, Pavement, SopeUp and SlopeDown, Hoang et al. transformation can also be recommended.

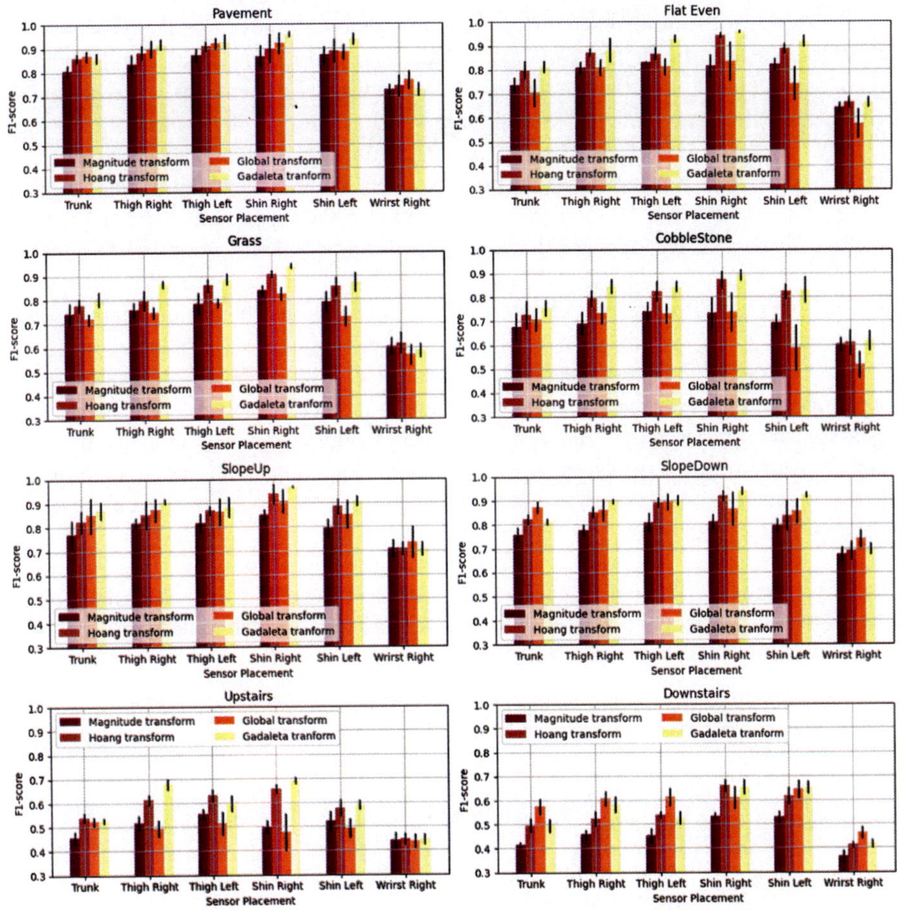

Fig. 3. F1-score identification metric obtained by SVM networks

5 Conclusions

The purpose of this study is to investigate the effect of using orientation invariant gait matching algorithms on the performance of IMU-based biometric systems. The research compares four popular signal preprocessing transformations Magnitude [5], Hoang et al. [7], Global transform [6] and Gadaleta et al. [7]. Experiments were conducted for a publicly available database of 30 participants and four different substrate types [11], using an SVM classifier and a deep neural network. To the best of the authors' knowledge, no studies have been published on the association between gait surface type and signal preprocessing methods.

Referring to Figs. 1 and 2, it can be seen that the method of Gadaleta et al. [8] obtained the highest identification scores. This algorithm can be considered as the preferred one for systems based on gait samples collected on an unknown surface.

It is also worth noting that there is a difference between the classical approach and the deep learning approach. The Global Transform [6] and Hoang Transform [7] methods are closely related, with the second having an additional dimensionality reduction step (Fig. 1). In classical applications, the Global Transform [6] obtains better scores than Hoang's method [7] (Fig. 2). The higher dimensionality of the signals leads to a longer feature vector, which eventually has a positive effect on the classification scores. For deep learning methods (Fig. 3), the relationship between the identification results for these methods is inverse. Hoang algorithm [8] achieves higher classification results compared to the global transform method [7]. It is assumed that the reduction in dimensionality had a beneficial effect on the network training process in the case of a rather limited learning set.

Regardless of the preprocessing algorithm used, the highest classification results were obtained for even surfaces (pavement and flat level), acceptable for surfaces such as grass, cobblestone, slope, and insufficient for ascending and descending stairs (Fig. 2, Fig. 3). Moreover, the results of the presented work demonstrate that the classification performed using a sensor located on the wrist provided worse results in comparison to sensors in other locations.

Acknowledgment. This work was supported by grant WZ/WI-IIT/4/2020 from Białystok University of Technology and funded with resources for research by the Ministry of Science and Higher Education in Poland.

References

1. Zou, Q., et al.: Deep Learning-Based Gait Recognition Using Smartphones in the Wild. arXiv:1811.00338 (2020)
2. Sprager, S., Juric, M.B.: Inertial Sensor-based gait recognition: a review. Sensors (Basel). **15**(9), 22089–220127 (2015)
3. Wan, Ch., Wang, L., Phoha V.V.:A survey on gait recognition. ACM Comput. Surv. **51**(5), 35 (2019). https://doi.org/10.1145/3230633
4. Giorgi, G., Martinelli, F., Saracino, A., Sheikhalishahi, M.: Try walking in my shoes, if you can: accurate gait recognition through deep learning. In: Tonetta, S., Schoitsch, E., Bitsch, F. (eds.) Computer Safety, Reliability, and Security. LNCS, vol. 10489, pp. 384–395. Springer, Cham (2017). https://doi.org/10.1007/978-3-319-66284-8_32
5. Siirtola, P., Röning, J.: Recognizing human activities user-independently on smartphones based on accelerometer data, Int. J. Interact. Multim. Artif. Intell, pp. 38–45 (2012)
6. Comotti, D., et al.: Inertial based hand position tracking for future applications in rehabilitation environments. In: 2015 6th International Workshop on Advances in Sensors and Interfaces (IWASI) (2015)
7. Hoang, T., Choi, D., Nguyen, T.: On the instability of sensor orientation in gait verification on mobile phone. In: Proceedings of 12th International Joint Conference on e-Business Telecommunication (ICETE), vol. 4, pp. 148–159 (2015)
8. Gadaleta, M., Rossi, M.: IDNet: smartphone-based gait recognition with convolutional neural networks. Pattern Recogn. **74**, 25–37 (2018)
9. Subramanian, R., et al.: Orientation invariant gait matching algorithm based on the Kabsch alignment. In: IEEE International Conference on Identity, Security and Behavior Analysis, pp. 1–8 (2015)

10. Subramanian, R., Sarkar, S.: Evaluation of algorithms for orientation invariant inertial gait matching. IEEE Trans. Inf. Foren. Secur. **14**, 304–318 (2018). https://doi.org/10.1109/TIFS.2018.2850032
11. Luo, Y., et al.: A database of human gait performance on irregular and uneven surfaces collected by wearable sensors. Sci. Data **2020**, 7 (2019)
12. Walendziuk, W., Idźkowski, A., Sawicki, A.: Estimation of the object orientation and location with the use of MEMS sensors. In: Photonics Applications in Astronomy, Communications, Industry and High-Energy Physics Experiments 2015, SPIE Proceedings Series, 2015, vol. 9662, pp. 1–8 (2015)
13. Sawicki, A., Saeed, K.: Application of LSTM networks for human gait-based identification. In: Zamojski, W., Mazurkiewicz, J., Sugier, J., Walkowiak, T., Kacprzyk, J. (eds.) Theory and Engineering of Dependable Computer Systems and Networks. AISC, vol. 1389, pp. 402–412. Springer, Cham (2021). https://doi.org/10.1007/978-3-030-76773-0_39
14. Zhang, Y., et al.: Accelerometer-based gait recognition by sparse representation of signature points with clusters. IEEE Trans. Cybern. **45**(9), 1864–1875 (2015)
15. Kuncan, F., Kaya, Y., Tekin, R., Kuncan, M.: A new approach for physical human activity recognition based on co-occurrence matrices. J. Supercomput. **78**(1), 1048–1070 (2021). https://doi.org/10.1007/s11227-021-03921-2

Ant Colony Optimization Algorithm for Object Identification in Multi-cameras Video Tracking Systems

Krzysztof Schiff(✉)

Department of Automatic Control and Computer Science, Cracow University of Technology, 31-155 Cracow, Poland
kschiff@pk.edu.pl

Abstract. The multi-dimensional assignment is one of methods used to associate measurement data with tracks in the analysis of image frames in vision systems. This is a very difficult problem and belongs to the hard NP class. Since accurate methods are very time consuming, there is a need for develop other methods to obtain solutions to this problem in a reasonable time. Among them there is the ant colony optimization heuristic (ACO). ACO has already been proposed for this problem, but only with a static function of desire. The article presents the dynamic function of desire together with data structures and discusses the results of the conducted experiments. The ant algorithm with the dynamic function of desire shows its superiority over the ant algorithm with the static function of desire because it gives a better assignment each time the algorithm is run. The proposed ant algorithm with the new function of desire is not time consuming and can be implemented in multi-cameras vision systems for objects tracking.

Keywords: Ant colony optimization algorithm · Multi-camera vision tracking systems · Multi assignment problem

1 Introduction

Tracking objects is one of major problems in multi-cameras vision systems. The association between the lists of measurements and the list of tracks is formulated as a global discrete optimization problem, subject to certain constraints, the purpose of which is to minimize the overall cost of association. Recently the General Maximum Minimal Clique Problem (GMMCP) trace tool was proposed for this issue [1, 2]. The problem of assigning measurements to objects can be mathematically defined as the problem of multidimensional assignment. The multidimensional data assignment problem is modeled by the graph $G = (V, E, W)$. This graph should be divided into n maximal cliques C_i $(i = 1,\ldots,n)$, where n is the number of objects. There is one object for each clique C_i. If there are m cameras, then each of these cliques C_i should have m vertices, $v_i \in V$ and $i = 1,..m$. Each edge $e_{ij} \in E$ has own weight $w_{ij} \in W$ and each edge e_{ij} connects two vertices v_i and v_j, $i = 1,\ldots, n, j = 1,\ldots,m$. Each clique C_i has $n^*(m-1)$ edges so each clique has a sum of its own edge weights. The data association problem is transformed to

W. Zamojski et al. (Eds.): DepCoS-RELCOMEX 2022, LNNS 484, pp. 276–286, 2022.
https://doi.org/10.1007/978-3-031-06746-4_27

a 0–1 optimization problem where the total distance/benefit of assigning targets to measurements is minimized/maximized [3, 14]. The multi-dimensional assignment problem is NP difficult [4–6]. In order to solve this problem many approximation algorithms have been developed [7–14] There are algorithms based on the network flow model [15–21]. There are also algorithms based on binary integer programming or minimal clique optimization [22–25], minimum cost disjoint path [26, 27] or neural network [28, 29], particle swarm optimization [30] or even genetic algorithm [31]. A very good overview of multi-target tracking methods based on motion is presented in [32, 33].

The optimization problem of determining minimum cliques in target tracking is one of the many combinatorial problems and was investigated in articles [22, 34–38]. The solution to the minimum clique optimization problem is the set of cliques covering the graph $G = (V, E, W)$ with the minimum sum of their edge weights. The ant colony optimization heuristics is applicable to the multi-dimensional assignment problem [39]. The ACO method was used in papers [40, 41]: the static desire function with local cost was used in [40] and the static desire function with global cost [41]. The same desire function was used in [42] as in [40], but in order to improve quality of obtained object trajectories, the Kalman filter and an artificial neural network was also used.

This work is organized as follows. Section 2 describes the problem and ant colony algorithm. The experimental results and their discussion are presented in Sect. 3, and conclusions are given in Sect. 4.

2 Mathematical Formalism

The multi-dimensional assignment problem (MAP) aims to minimize the overall cost of assignment when matching elements from $N = \{N_1, ., N_n, n > 2\}$ disjoint sets of equal size m. The multi-dimensional assignment problem (MAP) is usually presented as the following binary integer problem (Eqs. (1–5)):

$$(MAP) : min \sum\nolimits_{i1 \in N1} \sum\nolimits_{i2 \in N2} \ldots\ldots, \sum\nolimits_{in \in Nn} c_{i1, i2, \ldots, in} X_{i1, i2, \ldots, in} \quad (1)$$

$$s.t. \sum\nolimits_{i2 \in N2} \sum\nolimits_{i3 \in N3}, \ldots\ldots, \sum\nolimits_{in \in Nn} X_{i1, i2, \ldots, in} = 1, i_1 \in N_1 \quad (2)$$

$$\sum\nolimits_{i1 \in N1}, , \sum\nolimits_{is-1 \in Ns-1} \sum\nolimits_{is+1 \in Ns+1}, , \sum\nolimits_{in \in Nn} X_{i1, i2,, in} = 1, i_s \in N_s, s = 2, ., n-1 \quad (3)$$

$$\sum\nolimits_{i1 \in N1} \sum\nolimits_{i2 \in N2}, \ldots\ldots, \sum\nolimits_{in-1 \in Nn-1} X_{i1, i2, \ldots, in} = 1, i_n \in N_n, \quad (4)$$

$$X_{i1, i2, \ldots, in} \in \{0, 1\}, \mathrm{i_s} \in \mathrm{N_s},$$

$$X_{i1, i2, \ldots, in} \in \{0, 1\}, \mathrm{i_s} \in \mathrm{N_s}, s = 1, \ldots, \mathrm{n} \quad (5)$$

where for every n-tuple $(i1, i2, \ldots.in) \in N1 \times N2 \times, \ldots, \times Nn$, variable xi1, i2,..,in takes the value of one, if the elements of the given n-tuple belong to this same assignment and zero otherwise. The total assignment cost (Eq. (1)) is computed as the cost of matching elements from different sets *Ni*. As an example, an assignment which select n elements (*i1*, *i2*, ..., *in*) to be grouped together would have cost of *ci1, i2, .., in*.

Most of MAP can be associated with a weighted n-partite graph, in which the elements are represented by the vertices of the graph, each of the edge is assignment of two elements within the same n-tuple and weights on the edges account for the corresponding assignment costs. N-tuples form cliques in this paper.

For the case of the cliques, a feasible assignment includes all possible connections within the elements of each tuple and thus the cost of the tuple $(i_1,i_2,..,i_n) \in N_1 \times N_2, \ldots, \times N_n$ is defined as the sum of all assignment cost (Eqs. (6)):

$$c_{i1,\, i2,\, \ldots,\, in} = \sum\nolimits_{s=1}^{n} \sum\nolimits_{t=s+1}^{n} c_{isit} \tag{6}$$

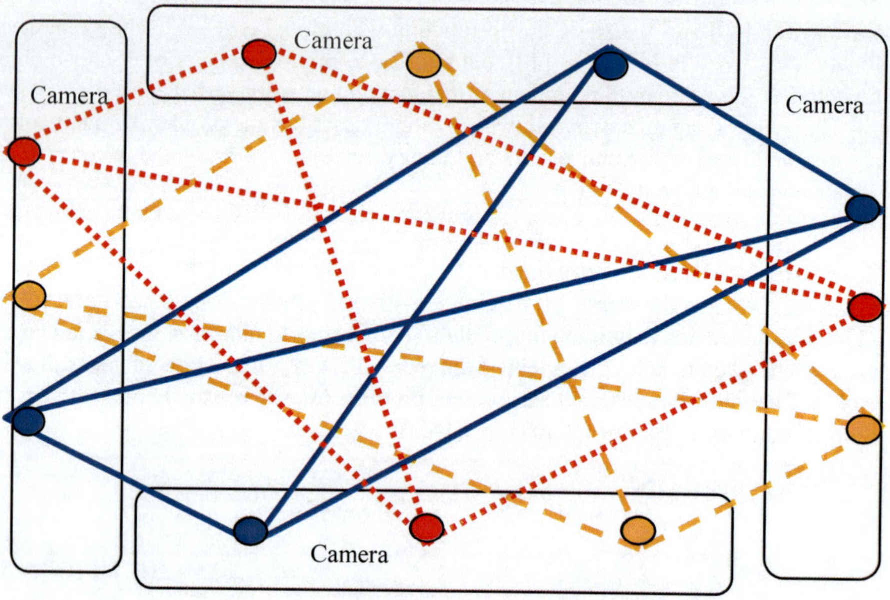

Fig. 1. The 4 cameras and the 3 objects (cliques) detected by the ant algorithm for the MAP

3 Graphical Representation of the Problem

In order to graphically present the MAP for object tracking Fig. 1 has been prepared. We have 4 cameras that allowed here to detect 3 objects. We see objects in the form of cliques. So we have a graph whose vertices represent the measurements. Cliques connect vertices with each other, that is, measurements pointing to the same object. The clique is therefore representative of the object monitored by the cameras. So if we have 4 cameras as in Fig. 1, we can assign 4 vertices (measurements) to each other. Each measurement coming from a different camera. In this way the vertices assigned to each other form cliques. Each clique is related to an object and trajectory. Now at the beginning we have a graph with vertices and with edges between each pair of vertices and weights assigned to each edge of the graph. From all these edges, those that form the cliques, must be selected and the sum of their all weights is the greatest/smallest. The remaining edges of the graph between each pair of vertices (measurements) are not shown in Fig. 1. We can solve the MAP by determining the cliques one by one in such a way that the sum of the weights of their edges was the largest/ smallest.

4 Ant Colony Optimization

4.1 Pseudo-code of the Ant Algorithm

In the algorithm ants work one by one and in the loop "cycle" they finding solutions to the problem. Each ant creates n cliques of dimension m, which corresponds to n objects in m cameras. Each ant takes off from each object (measurement) from the first camera. They are not randomly selected but successively from the first to the last object (measurement) at the first camera. It means that after selecting the first measurement from the first camera, a clique is created and again for the next measurement from the first camera another clique is created by the same ant. The pseudo-code of the ant algorithm is given in Table 1.

Table 1. Pseudo-code of the ACO algorithm.

```
begin
  while (exist cycle) do
     while (exist any ant, which has not worked) do
       while (a solution has not been completed) do
       choose a next vertex to a constructed solution with a probability pi;
        update neighborhood of current state;
       end
     update a best solution if a better solution has been found;
     end
   update a global best solution if a better solution has been found;
   use an evaporation mechanism τ(i) = ρτ(i) ;
   update a pheromone trails τ(i) = τ(i) + Δτ;
  end
end.
```

The ant includes in the created clique each vertex from the adjacent vertices with the probability p_{ij} (Eqs. (7)), so when included in the clique a vertex i the ant goes to a vertex j, when it is selected with a probability p_{ij} and include it in the created clique. Then ant goes to the vertex so included and selects the next vertex from its adjacent vertices, and so on, until it creates a clique with the dimension m. The ant finishes its work when it creates n cliques of dimension m.

Given the pheromone level and the heuristic information for measurements j to be assigned to created clique t to be $\tau_j(t)$ and $\eta_j(t)$, assuming that Ω represents the set of valid measurements for clique t, then the probability of assigning measuring j to clique t is found as follow [13, 43], thus an ant goes from a vertex i to a vertex j with probability p_{ij}, (Eqs. (7)):

$$p_{j/clique(t)} = \frac{(\tau_j)^{\alpha}(\eta_j)^{\beta}}{\sum_{j\in\Omega}(\tau_j)^{\alpha}(\eta_j)^{\beta}} \tag{7}$$

Now the sum of the weights of all the edges used by the ant to create the cliques is calculated. When the sum of the weights is smaller than previously determined by other ants working in a given cycle, it becomes the best solution found in a given cycle. After all ants have finished their work in a given cycle, the best solution from this cycle is compared with the best solution from the previous cycles and the better is remembered.

An evaporation mechanism is used at the end of each cycle. An additional amount of pheromone $\Delta\tau$ appropriate to the quality of the solution is applied to the vertices of the cliques of the solution, (Eqs. (8)):

$$\Delta\tau = \frac{1}{1 - \frac{c_{best} - c_b}{c_{best}}} \tag{8}$$

where: c_{best} – this is the best cost of any cycle to date,
C_b – this is the best cost in the current cycle.

4.2 The Dynamic Desire Function

The aim of the desire function is a better selection of measurements from cameras to the created solution, and these are selected so that the cost of a MAP is as low as possible. The solution to the MAP is a set of cliques with the lowest possible cost. Ultimately, the desire function allows us to solve the problem at a lower cost than if the desire function was not in the algorithm. The higher the value of desire function, the more useful measurement j is getting the best solution.

The desire function can be static or dynamic. The static desired function is computed before two main loops of the ant algorithm start working. The dynamic desire function is computed while each ant is working, and its value is constantly changes each time a vertex is include in the clique created. Each time an ant creates a clique, it increase by one vertex. The vertex represents the measurement from the camera. The remaining vertices from not yet visited cameras have edges that connect them to the vertices of the created clique. The sum of the weights of these edges is the cost. The dynamic desire function is the reciprocal of this cost [13, 39, 43, 44], (Eqs. (9)). Thus, each object (measurement) from unvisited cameras has a newly calculated desire function. Each ant visits each camera only once while creating a clique. When the ant finishes creating a new clique, it starts creating a new clique by visiting all cameras again, but can no longer visit the vertices (measurements) used to build previous cliques by itself.

$$\eta_j(t) = \frac{1}{c_j(t)} \tag{9}$$

4.3 Considering Data Structures

Data structures for pheromone traces and for costs used in the ant algorithm with static function of desire were not described in [40, 41]. In this paper d(d−1)/2 two-dimensional tables for costs and m two-dimensional tables for pheromones were used. Each camera has own set of measurements. Each of measurements concerns each separate object. There are multiple weighted edges between each pair of cameras, and more specifically between pairs of measurements form different cameras. Since there are d(d−1)/2 pairs of cameras, there are d(d−1)/2 two-dimensional tables whose elements are edge weights. Each ant create m cliques. For all cameras with measurements there are an one-dimensional table for indicating already visited camera by the ant and it is used when ant constructs a clique. For each clique there is one two-dimensional table of pheromone traces. The first dimension if for the number of measurements and the second dimension is for the number of cameras. Each ant takes one measurement from each of the cameras, so it takes m measurement from all cameras together and creates a clique. All the edges of this clique are removed from the two-dimensional table with edges, so they are no longer available for further selection and thus the ant cannot use them when it creates next cliques. Each ant when make a selection uses the two-dimensional table of pheromones and the two-dimensional table of costs (first dimension – cameras, second dimension- measurements). The elements of this cost table can be the weights of the connecting edges already included in the created clique of measurement with other available measurements from the cameras not yet visited, so these costs are not calculated while the algorithm is running and therefore they are static and are used in the ant algorithm with static function of desire. The elements of this table of costs could be the sums of edge weights. These sums are obtained from the edge weights connecting the available measurements of still unvisited cameras with all measurements already included into the clique being built, and are calculated for each of the available measurements. Because each time the measurement is included in the created clique, the set of measurements that make up this clique changes, each time the sum of weights for each available measurement must be recalculated. Each time the measurement is included in the created clique, the sum of the weights is recalculated and these values of the sums of weights change dynamically during the algorithm's operation, so the desire function based on the sum of weights is used in the ant algorithm with the dynamic desire function.

5 Experiments

We are most interested in how the algorithms will behave when the number of cameras d or the number of monitored objects (measurements) m increases, when number of cycles lc, number of ants lm and evaporation rate ρ will be constant.

The formula for the probability of assigning the another measurement to a clique includes in the numerator and denominator: the dynamic desire function and the level of pheromone (Eq. 7). We also see the coefficients α and β.

The test results are presented in the Tables 1–6. These are the averages of 10 runs of the algorithms with an ant count of 10, an evaporation factor of 0.95 and a number of cycles of 30. At each run of the algorithms, the ant algorithm with dynamic desire function turned out to better in operation than the algorithm with static desire function, so the difference in score always had the same sign. Random edge weights were generated from 1 to 100. The results in the Tables 1–3 were obtained with the number of objects (measurements) equal to 30 and only the number of cameras was changed. The results in the Tables 4–6 were obtained with the number of cameras equal to 25 and only the number of objects (measurements) was changed. We can see that the increase in the number of cameras and tracked objects (measurements) does not affect the results (Tables 1–6) and only the coefficients α and β from the equation for probability affect the obtained results (Eq. 7, Tables 3 and 6). Increasing the value of α and β coefficients means intensifying the influence of the desire function and the amount of pheromone on the results obtained.

The main focus of the research was to use the dynamic desire function versus the static desire function. With coefficients $\alpha = 1$ and $\beta = 1$ better results were obtained for the static desire function, similarly for $\alpha = 2$ and $\beta = 2$, and with $\alpha = 3$ and $\beta = 3$ it turned out that the dynamic desire function works better than the static desire function, both when the number of cameras and the number of objects (measurements) changed. In this way the dynamic desire function showed its superiority over the static desire function (Tables 3 and 6).

Table 2. Average of the sum of the cliques weights at $\alpha = \beta = 1$

Number of cameras	25	20	15	10	5
Static heuristics	427 913, 6	266 847, 6	143 499, 2	58 353,0	10 874, 1
Dynamic heuristics	434 846, 6	271 892, 8	147 197, 1	60 426, 5	11 253, 4

Tests were also carried out with the changing parameter ρ and constant parameters *m, d, lm, lc*, and with the changing parameter lm and constant parameters ρ, *m, d, lc* and with changing parameter *lc* and constant parameters ρ, *m, d, lm*. The results for each run of the algorithm always showed that the algorithm with the dynamic desire function was better than the algorithm with the static desire function ($\alpha = 3$ and $\beta = 3$). Here, only the relevant test results for emphasizing the superiority of the ant algorithm with the dynamic desire function over the algorithm with the static desire function are presented.

Table 3. Average of the sum of the cliques weights at $\alpha = \beta = 2$

Number of cameras	25	20	15	10	5
Static heuristics	418 051, 7	259 363, 8	138 131, 2	54 946, 3	9 523, 7
Dynamic heuristics	422 990, 1	262 857, 3	140 175, 8	55 738, 0	9 586, 1

Table 4. Average of the sum of the cliques weights at $\alpha = \beta = 3$

Number of cameras	25	20	15	10	5
Static heuristics	416 377, 0	257 625, 3	137 191, 3	54 245, 6	9 111, 3
Dynamic heuristics	409 969, 8	253 988, 1	133 713, 0	51 743, 3	8 464, 3

Table 5. Average of the sum of the cliques weights at $\alpha = \beta = 1$

Number of measurement	30	25	20	15	10
Static heuristics	427 913, 6	356 223, 6	284 802, 8	213 545, 2	141 983, 9
Dynamic heuristics	434 846, 6	362 566, 7	289 605, 0	215 808, 6	143 931, 0

Table 6. Average of the sum of the cliques weights at $\alpha = \beta = 2$

Number of measurement	30	25	20	15	10
Static heuristics	418 051, 7	348 338, 3	279 165, 3	209 082, 5	139 906, 6
Dynamic heuristics	422 990, 1	352 447, 9	282 035, 9	211 268, 9	140 861, 1

Table 7. Average of the sum of the cliques weights at $\alpha = \beta = 3$

Number of measurement	30	25	20	15	10
Static heuristics	416 377, 0	347 452, 5	277 633, 8	208 457, 2	139 211, 0
Dynamic heuristics	409 969, 8	341 465, 0	273 689, 3	205 401, 2	137 806, 6

6 Conclusions

The introduction of the dynamic desire function to the ant colony optimization algorithm gave better results compared to the ant algorithm with the static desire function. A lower total weight of the cliques edges that make up the solution to the multidimensional assignment problem was obtained. The effect is noticeable at $\alpha = \beta = 3$. The data structures for the ant algorithm and the dynamic desire function proposed and presented in this article allowed for the development of a new algorithm. There is no way to accurately compare the developed algorithm with the one presented in [40, 41] because in [40, 41] it was not stated what data structures were used, and they could have been different. This paper presents the results of the ant algorithm to identify objects in multi-cameras vision systems for a much larger number of objects and with a much larger number of cameras than presented in [40, 41]. The ant algorithms presented in [40, 41] have so far been the only ant algorithms developed for the problem of object identification in multi-camera video tracking systems, so the ant algorithm presented in this paper with the dynamic desire function is an expression of progress in scientific work on this problem. This article presents a new ant algorithm with a dynamic desire function, while the usefulness of the ant algorithm for the identification of objects in multi-camera vision systems, and their practical and successful implementation has already been presented in works [40, 41].

References

1. Dehgan, A., Assari, S.M., Shah, M.: GMMCP tracker: globally optimal generalized maximum multi clique problem for multiple object tracking. Conf. Comput. Vision Pattern Recogn. **1**, 4091–4099 (2015)
2. Wen, L., Lei, Z., Lyu, S., Li, S.Z., Yang, M.: Exploiting hierarchical dense structures on hypergraphs for multi-object tracking. IEEE Trans. Pattern Anal. Mach. Intell. **38**(10), 1983–1996 (2016)
3. Wang, H., Kirubarajan, T., Bar-Shalom, Y.: Precision large scale air traffic surveillance using IMM/assignment estimators. Trans. Aerosp. Electron. Syst. **35**, 255–266 (1999)
4. Deb, S., Yeddanapudi, M., PattiPati, K., Bar-Shalom, T.: A generalized S-D assignment algorithm for multi-sensor-multitarget state estimation. IEEE Trans. Aerosp. Electron. Syst. **33**, 523–538 (1999)
5. Feremans, C., Labbe, M., Laportee, G.: Generalized network design problem. Eur. J. Oper. Res. **148**, 1–13 (2003)
6. Koster, A.M.C.A., Hoesel, S.P.M., Kolen, A.W.J.: The partial constraint satisfaction problem: facets and lifting theorems. Oper. Res. Lett. **23**, 89–97 (1998)
7. Sinha, A.; Kirubarajan, T.: A randomized heuristic approach for multidimensional association in target tracking. In: Proceedings SPIE edited by Drummo nd, O.E., vol. 5428, pp. 202–210 (2004)
8. Zamir, A.R., Dehgan, A., Shah, M.: GMPC – tracker: global multi-object tracking using generalized minimum clique problem. In: Computer Vision – European Conference on Computer Vision, pp. 343–356 (2012)
9. Andriyenko, A., Schindler, K.: Multi-target tracking by continuous energy minimization. In: Conference on Computer Vision and Pattern Recognition, pp. 1265–1272 (2011)
10. Andriyenko, A.; Schindler, K.; Roth, S. Discrete-continuous optimization for multi-target tracking. In: Conference on Computer Vision and Pattern Recognition, pp. 1926–1933 (2012)
11. Gabrovsek, B., Novak, T., Povh, J., RupnikPoklukar, D., Zerovnik, J.: Multiple Hungarian method for k-assignment problem. Mathematics **8**, 1–18 (2020)
12. Knigawka, L.: Exploitation of Minimum Clique Graphs for Multi-Object Tracking, Bachelor thesis, Warsaw University Of technology (2021)
13. Bozdogan, A.O., Yilmaz, A.E., Efe, M.: Performance analysis of swarm optimization approaches for the generalized assignment problem in muli-target tracking applications. Turkish J. Elect. Eng. Comput. Sci. **18**, 1059–1076 (2010)
14. Walteros, J.L., Vogiatzis, C., Pasiliao, E.L., Pardalos, P.M.: Integer programming models for the multidimensional assignment problem with star costs. Eur. J. Oper. Res. **235**, 553–568 (2014)
15. Liang, Y., Lu, X., He, Z., Zheng, Y.: Multiple object tracking by reliable tracklets. SIViP **13**(4), 823–831 (2019). https://doi.org/10.1007/s11760-019-01418-3
16. Berclaz, J., Fleuret, F., Turetken, E., Fua, P.: Multiple object tracking using k-shortest paths optimization. IEEE Trans. Pattern Anal. Mach. Intell. **33**, 1806–1819 (2011)
17. Zhang, L., Li, Y., Nevatia, R.: Global data association for multi-object tracking using network flows. In: Proceedings of the 2008 IEEE Conference on Computer Vision and Pattern Recognition, pp. 1–8 (2008)

18. Pirsiavash, H., Ramanan, D., Fowlkes, C.C.: Globally-optimal greedy algorithms for tracking a variable number of objects. In: Proceedings of the 2011 IEEE Conference on Computer Vision and Pattern Recognition, pp. 1201–1208 (2011)
19. Gaidon, A., Wang, Q., Cabon, Y., Vig, E.: Virtual worlds as proxy for multi-object tracking analysis. In: Proceedings of IEEE Computing Society, Conference on Computing and Pattern Recognition, pp. 4340–4349 (2016)
20. Lenz, P.; Geiger, A.: Urtasun, R. FollowMe: efficient online min-cost flow tracking with bounded memory and computation. In: Proceedings of IEEE International Conference on Computer Vision, 4364–4372 (2015)
21. Chari, V., Lacoste-Julien, S., Laptev, I., Sivic, J.: On pairwise costs for network flow multi-object tracking. In: Proceedings of IEEE Conference on Computer Vision and Pattern Recognition, pp. 5537–5545 (2015)
22. Ristani, E., Tomasi, C.: Tracking multiple people online and in real time. In: Cremers, D., Reid, I., Saito, H., Yang, M.-H. (eds.) ACCV 2014. LNCS, vol. 9007, pp. 444–459. Springer, Cham (2015). https://doi.org/10.1007/978-3-319-16814-2_29
23. Kumar, R., Charpiat, G., Thonnat, M.: Multiple object tracking by efficient graph partitioning. In: Cremers, D., Reid, I., Saito, H., Yang, M.-H. (eds.) ACCV 2014. LNCS, vol. 9006, pp. 445–460. Springer, Cham (2015). https://doi.org/10.1007/978-3-319-16817-3_29
24. Tang, S., Andres, B., Andriluka, M., Schiele, B.: Subgraph decomposition for multi-target tracking. In: Conference on Computer Vision and Pattern Recognition, pp. 5033–5041 (2015)
25. Morefield, C.L.: Application of 0–1 integer programming to multi-target tracking problems. IEEE Trans. Autom. Control **22**, 302–312 (1971)
26. Poore, A.: Multidimensional assignment and multi-target tracking. In: Cox, I.J., Hansen, P., Julesz, B., (Eds.) Partitioning Data Sets, pp. 169–196. American Mathematical Society (1995)
27. Poore, A.B.: Multidimensional assignment formulation of data association problems arising from multitarget tracking and multisensor data fusion. Comput. Optim. App. **3**, 27–57 (1994)
28. Yoon, K., Kim, D.Y., Yoon, Y.C., Jeon, M.: Data association for multi-object tracking via deep neural networks. Sensors **19**(3), 1–17 (2019)
29. Lee, B., Erdenee, E., Jin, S., Rhee, P.K.: Efficient object detection using convolutional neural network-based hierarchical feature modeling. Image Video Proc. **10**(8), 1503–1510 (2016). https://doi.org/10.1007/s11760-016-0962-x
30. Kennedy, J., Eberhart, R.: Particle swarm optimization. In: Proceedings of IEEE International Conference on Neural Networks, vol. 4, 1942–1948 (1995)
31. Chen, G., Hong, L.: A genetic based multi-dimensional data association algorithm for multi sensor multi target tracking. Math. Comput. Models **26**, 57–69 (1997)
32. Qiu, C., Zhang, Z., Lu, H., Luo, H.: A survey of motion-based multi-target tracking methods. Progr. Electro-Magn. Res. **62**, 195–223 (2015)
33. Luo, W., Xing, J., Milan, A., Zhang, X., Liu, W., Kim, T.K.: Multiple object tracking: a literature review. Artif. Intell. **293**, 1–23 (2021)
34. Ferrari, V., Tuytelaars, T., Gool, L.: Real-time affine region tracking and coplanar grouping. In: Computer Vision and Pattern Recognition (2001)
35. Marın-Jimenez, M., Zisserman, A., Eichner, M., Ferrari, V.: Detecting people looking at each other in videos. Int. J. Comput. Vision **106**, 282–296 (2014)
36. Zamir, A.R., Dehghan, A., Shah, M.: GMCP-tracker: global multi-object tracking using generalized minimum clique graphs. In: Fitzgibbon, A., Lazebnik, S., Perona, P., Sato, Y., Schmid, C. (eds.) ECCV 2012. LNCS, pp. 343–356. Springer, Heidelberg (2012). https://doi.org/10.1007/978-3-642-33709-3_25
37. Liu, W., Camps, O., Sznaier, M.: Multi-camera multi-object tracking. In: Computer Vision and Pattern Recognition, pp. 1–7 (2017)

38. Kumar, R., Charpiat, G., Thonnat, M.: Multiple object tracking by efficient graph partitioning. In: Cremers, D., Reid, I., Saito, H., Yang, M.-H. (eds.) ACCV 2014. LNCS, vol. 9006, pp. 445–460. Springer, Cham (2015). https://doi.org/10.1007/978-3-319-16817-3_29
39. Dorigo, M., Stutzle, S.: Ant Colony Optimization. MIT Press, Cambridge (2004)
40. Bozdogan, A.O., Efe, M.: Improved assignment with ant colony optimization for multi-target tracking. Expert Syst. App. **38**, 9172–9178 (2011)
41. Bozdogan, A.O., Efe, M.: Ant colony optimization heuristic for the multidimensional assignment problem in target tracking. In: IEEE National Radar Conference (2008)
42. Joelianto, E., Wiranto, I.: An application of ant colony optimization, Kalman filter and artificial neural network for multiple target tracking problems. Int. J. Artif. Intell. **7**, 384–400 (2011)
43. Stuzle, T.: MAX-MIN Ant system. Future Gener. Comp. Syst. **16**, 889–914 (2000)
44. Blum, C.: Ant colony optimization: Introduction and recent trends. Phys. Life Rev. **2**, 353–373 (2005)

Towards Explainability of Tree-Based Ensemble Models. A Critical Overview

Dominik Sepiolo(✉) and Antoni Ligęza

AGH University of Science and Technology, al. A. Mickiewicza 30, 30-059 Krakow, Poland
{sepiolo,ligeza}@agh.edu.pl

Abstract. Tree-based ensemble models are widely applied in artificial intelligence systems due to their robustness and generality. However, those models are not transparent. For the sake of making systems trustworthy and dependable, multiple explanation techniques are developed.

This paper presents selected explainability techniques for tree-based ensemble models. First, the aspect of black-boxness and the definition of explainability are reported. Then, predominant model-agnostic (LIME, SHAP, counterfactual explanations), as well as model-specific techniques (fusion into a single decision tree, iForest) are described. Moreover, other methods are also briefly mentioned. Finally, a brief summary is presented.

Keywords: Random Forest · Tree-based models · Explainability · Explainable AI · XAI · Interpretable AI · Reliable AI · Trustable AI

1 Introduction

Recently, the *explainability* aspect of complex Machine Learning (ML) models has gained the attention of researchers involved in areas of Computational Intelligence (CI), such as Deep Learning. Contemporary Artificial Intelligence (AI) embraces a variety of complex techniques. In classical, symbolic AI, most of the solutions or decisions can be fully explained, justified and evaluated; some very first three examples at hand may include graph-search algorithms, automated planning, and rule-based expert systems. On the other hand, in several areas of CI and ML, with *Deep Learning* based on large neural nets being a prominent example, the way of the output solutions or decisions are obtained is hard to explain. Hence, the analysis of potential error sources becomes almost impossible. In fact, most of the modern and so effective methods are opaque. In order to make AI systems more reliable, trustable and dependable, a variety of *explainability techniques* are being developed.

Random Forests (RF) – ensembles of decision trees – are among the most popular ML techniques, widely applied in research and industry solutions. Their versatility (application for classification and regression tasks) and robustness, as well as resistance to overfitting, made them highly regarded [19]. Tree-based models can be more accurate than neural networks – their direct competitor

W. Zamojski et al. (Eds.): DepCoS-RELCOMEX 2022, LNNS 484, pp. 287–296, 2022.
https://doi.org/10.1007/978-3-031-06746-4_28

– in many applications [22]. Moreover, it can be easier to interpret and thus understand their output. The open nature of decision tree-based models gives a promise of making a step towards explainable solutions.

The main focus of this paper is on existing explanation techniques that can be applied for tree-based ensemble models. Besides the model-agnostic techniques, promising model-specific methods are described and compared. The contributions of this paper include: (i) the survey and critical evaluation of current research work in the area of XAI (especially in the context of tree-based ensemble models), (ii) indication of possible research gaps and, (iii) pointing to future research directions, i.e. XAI based on open causal and functional models.

The paper is structured as follows: Sect. 2 describes aspects of opaqueness of models and explainability. The model-agnostic explainability techniques are presented in Sect. 3. This is followed in Sect. 4, which presents model-specific explainability methods for tree-based ensemble models. Sect. 5 presents conclusions and future works.

2 Black-Boxes and Explainability

Using simple, interpretable models is widely considered a good practice. However, complex, so-called black-box models (e.g. Deep Neural Networks, Random Forests) gain more and more applications as they provide better accuracy. They are especially useful when a declarative open-box model is missing.

Intuitively, a black-box can be understood as a model, which behavior is not transparent, i.e. it is not understandable by itself. In [1] *simulability* (the ability of a model of being simulated by a human), *decomposability* (the ability to explain each of the parts of a model: input, parameters, and calculation) and *algorithmic transparency* (the ability of the user to understand the process) are denoted as features of transparent, interpretable models.

Due to the fact that frequently used black-box models are not simulable, decomposable and algorithmically transparent, emerging needs of explainability techniques can be observed. There are different descriptions of what explainability means. Miller in [25] states, after Lipton [21], that explanation is post-hoc interpretability. Later, he equates interpretability and explainability. An abstract definition was proposed by Chazette et al. in [5]:

Definition 1. *A system S is explainable with respect to an aspect X of S relative to an addressee A in context C if and only if there is an entity E (the explainer) who, by giving a corpus of information I (the explanation of X), enables A to understand X of S in C.*

Some definitions are complimentary, while others describe different concepts. As Explainable Artificial Intelligence (XAI) is a relatively new research topic, the terminology is not standardized yet [1]. A detailed discussion regarding standards for XAI terminology can be found in [6]. Nevertheless, further research should be carried out in order to create widely recognized standards for terminology.

Different taxonomies can be introduced for explainability techniques [1–3,13, 20,32]. The most popular ones are:

- local vs global explanations,
- model-agnostic vs model-specific explanations.

As this paper is focused on tree-based ensemble models, the second taxonomy is used.

3 Model-Agnostic Methods

Model-agnostic techniques can be applied to any class of machine learning models without assumptions about their structure, which means that only access to the model input and output is required [32]. For that reason, those methods begin increasingly popular in practice [40], as they provide the possibility to compare explanations for different classes of machine learning models. Frequently used model-agnostic techniques for random forest explanations are presented below.

3.1 LIME

The Local Interpretable Model-agnostic Explanations (LIME) method proposed by Ribeiro et al. in [29] is one of the most prominent ones. LIME is an algorithm (and software package) that can explain the predictions of any machine learning model by approximating it locally with an interpretable (usually linear) surrogate model. LIME works on a single prediction level.

Explanations can be expressed as follows:

$$explanation(x) = \arg\min_{g \in G} L\left(f, g, \pi_x\right) + \Omega(g) \tag{1}$$

The explanation for instance x is the model g minimizing loss L, which measures how close the explanation is to the prediction of the original model f, while keeping low complexity $\Omega(g)$. G is the set of possible explanations. The proximity measure π_x defines the size of the neighborhood of instance x. In practice, the user has to determine the complexity, while LIME optimizes only the loss [26].

While theoretical analysis performed by Garreau and von Luxburg [11] confirmed that LIME might forget some important features and the surrogate model might not always be faithful, new variants of LIME were developed (e.g. [33,40,41]) in order to overcome those obstacles.

3.2 Shapley Values

SHapley Additive exPlanations (SHAP) were introduced by Lundberg et al. [24] as a unified measure of feature importance. SHAP explains a particular instance of an output by computing the contribution of each input feature to the final prediction. The properties of the SHAP explanations are:

- *Local accuracy:* an explanation for approximated model and the original model match,

- *Missingness:* features missing in the original input have no impact on the output,
- *Consistency:* if a feature increases or stays the same regardless of the other inputs, the attribution of that input should not decrease.

The SHAP explanation is expressed as a linear function of binary variables:

$$g(z') = \phi_0 + \sum_{j=1}^{M} \phi_j z'_j \tag{2}$$

where $z' \in \{0,1\}^M$, M is the number of simplified input features, and $\phi_i \in \mathbb{R}$.

There is extensive research regarding improvements of SHAP for tree-based models. While classical SHAP can explain the output of any machine learning model, TreeSHAP is an algorithm to compute SHAP values for trees and tree ensembles reducing the computational complexity of exact SHAP value computation from exponential to low order polynomial [23]. Such optimization is possible owing to the structure of tree-based models and properties of Shapley values, mainly additivity. Further research resulted in TreeExplainer, a method that enables the tractable computation of optimal local explanations [22].

3.3 Counterfactuals

Counterfactual explanations (CFEs) have been researched for a long time in different research fields (like philosophy, psychology, and social sciences) [37]. Although in the context of computer science and artificial intelligence, counterfactuals can be applied in many ways (for example to supplement incomplete data or improve the training process of generative adversarial networks) their most noticeable use is in respect of generation of explanations for opaque models [4]. The growing popularity of counterfactual explanations may be caused by the fact that they resemble human reasoning [4,10].

Counterfactual explanations are generated on a single prediction level. Given a datapoint and its prediction, CFE explain a prediction by calculating a change (usually minimal) in the datapoint that would cause a change in the response of the model (if an input datapoint were x' instead of x, then the model output would be y' instead of y) [36]. However, it might not be easy to identify the best counterfactual, as there is possibly limitless number of potential x' datapoints.

While the naive approach to generate CFEs was searching by trial and error, Wachter et al. proposed in [38] an efficient computation method by minimizing the following objective function:

$$\arg\min_{x'} \max_{\lambda} \lambda \cdot (\hat{f}(x') - y')^2 + d(x, x') \tag{3}$$

where d is a distance function that measures how far the counterfactual x' and the original datapoint x are from one another. Maximization over λ is done by iteratively solving for x' and increasing λ until a sufficiently close solution is found.

As counterfactuals became a popular explanation technique, there is research focused on counterfactuals for tree-based models. Parmentier and Vidal propose in [27] an efficient, scalable mathematical model to search for counterfactual explanations in tree ensembles with a number of binary variables that is logarithmic in the number of vertices, and therefore scales to large trees while retaining its ability to find an optimal solution. A method to extract a set of counterfactuals from RF, Random Forest Optimal Counterfactual Set Extractor (RF – OSCE) was presented by Fernández et al. in [10]. The technique is based on fusion of tree predictors into a single decision tree (counterfactuals exported from individual trees might be incompatible) and returns a counterfactual set that contains the optimal counterfactual.

3.4 Other Techniques

Besides the methods mentioned above, there are also numerous model-agnostic techniques that could be successfully applied for generation of explanations. Some of the most important ones are:

- Out-of-bag Variable Importance [14],
- Rules Extraction [15],
- Partial Dependence Plot [12],
- Model Explanation System (MES) [34],
- Scoped Rules (Anchors) [30].

4 Model-Specific Methods

Model-specific explainability techniques can be applied only to one class of models or a family of similar model types. Those methods are based on assumptions regarding the model structure or other specific features.

4.1 Transformation into a Decision Tree

First approaches to combine multiple models into one interpretable model were proposed in the twentieth century (for example, using meta learning [9]). This approach can be utilized for tree-based ensemble models in order to generate one glass-box decision tree.

The procedure to build a decision tree that mimics the performance of random forest by studying the asymptotic behavior of splits was proposed by Zhou and Hooker in [43]. First, an initial number of sample points from the random forest is generated at each node. Next, possible splits are compared based on this set. After that, a decision is made, whether to choose the one with the smallest Gini Index with certain confidence or request more sample points. In the latter case, more sample points are generated incrementally, until the confidence assumed for the split is reached. The procedure is repeated until the best split for the node is found. Then, the algorithm is applied to the child nodes.

GENetic Extraction of a Single, Interpretable Model (GENESIM), an algorithm that transforms an ensemble of decision trees into a single decision tree, was proposed by Vandewiele et al. in [35]. In the initialization phase, data is divided into new training and validation sets, then decision trees are created and added to the population. Next, the accuracy is evaluated. In the selection phase, individuals, that will get recombined in the next phase, are selected during the tournament. In each iteration, an individual has a certain probability to be mutated. In the replacement phase they are sorted by their fitness and a certain amount of the best ones is chosen. Those steps are repeated iteratively. The resulting model of GENESIM is interpretable, as it has very low complexity.

Another method, Forest Based Tree (FBS), was recently proposed by Sagi and Rokach in [31]. Although the technique was proposed in order to apply FBS instead of ensemble models, it can be used for generation of explanations for the random forest. The first step is to prune iteratively the decision forest, maximizing AUC of the forest until no improvement is achieved. Then, the pruned forest is divided into a set of rule conjunctions, where each rule conjunction corresponds to a possible output of the decision forest. After that, a single decision tree is constructed. The root node, that contains the entire set of conjunctions, is defined. The node is then split into two conjunction sets. The splitting criterion is the highest information gain of the split. Splitting a node stops when the entropy of the conjunction set equals 0. The resulting tree is interpretable.

4.2 iForest

iForest is an interactive visualization tool proposed by Zhao et al. [42]. It mainly focuses on interpreting random forests, however, it can be applied to other tree-based ensemble models.

Three goals of iForest explanations are to: reveal the relationships between input features and predictions, uncover the underlying working mechanism, and provide case-based reasoning. The goals are fulfilled by some analytical tasks:

- Encoding the feature importance and partial dependence information,
- Encoding the split point distribution of each feature,
- Encoding the prediction results and summarizing the similarities of decision paths,
- Review structures of the decision paths,
- Identification of the training data clusters and outliers,
- Encoding the training data value distribution,
- Supporting the interactive model inspection.

Besides data overview, which allows training data and prediction results exploration, and feature importance view, the Decision Path Projection method is proposed. This method provides an overview of all decision paths for prediction based on their similarities. Moreover, the summary of critical feature ranges of decision paths and detailed information regarding decision paths layer by layer are provided.

4.3 Other Techniques

There are other explainability methods that have been developed for tree-based ensemble models.

User-Centered Approach in Enhancing Random Forest Explainability (RFEX) was proposed by Petkovic et al. in [28]. In a series of steps, RFEX produces a summary report, that can be easily understood by human users. The report contains information such as a RF base accuracy, Feature Rank, f-score, and one-page report summary.

The inTrees framework was developed by Deng in [8]. This method, based on rule extraction, provides a simplified tree ensemble learner (STEL). It can be applied to both classification and regression problems, and is applicable to many types of tree ensembles.

A framework to post-process any tree-based ensemble classifier to extract an optimal actionable plan that can change a given input to the desired class with a minimum cost using linear programming was proposed by Cui et al. in [7]. However, it can only be applied when the class label is binary.

5 Conclusions and Future Works

This paper presents an attempt to describe predominant techniques used for generation of explanations for tree-based ensemble models. Selected methods, that have been described in this paper, are presented and compared in Table 1.

Table 1. Selected explainability techniques for tree-based ensemble models.

Technique	Local/Global	Model-agnostic/ Model-specific	Explanation type	Reference
LIME	Local	Model-agnostic	Model simplification	[29]
SHAP	Local (can be combined into global)	Model-agnostic	Feature contribution	[24]
Counterfactual explanations	Local	Model-agnostic	Explanation by example	[37]
Transformation into a Decision Tree	Global	Model-specific	Model simplification	[9,35,43]
iForest	Global	Model-specific	Visualization, feature contribution	[42]

Most commonly used model-agnostic techniques, such as LIME or SHAP, generate local explanations. However, there are also global methods, that can be applied to any class of machine learning models. While many model-agnostic frameworks can be successfully applied for generation of explanations, some of them have also extensions for tree-based ensemble models, that can perform more efficiently and/or accurately.

Despite the fact that model-specific explainability techniques for tree-based ensemble models seem to be less explored topic, there are numerous publications proposing various methods. Many approaches involve fusion of the ensemble model into a single, interpretable decision tree.

The main contribution of this paper is a critical, but also constructive, overview of existing explainability techniques that can be applied for tree-based ensemble models. The aforementioned explainability methods are being developed for the same purpose: in order to enable responsible, trustable AI solutions. However, they provide different kinds of explanations, therefore can be applied for different cases and target audiences. This paper might be also considered as a step towards the debate regarding an analysis of the quality of explanations as well as standardization of XAI terminology.

Further discussion of computational complexity and discrepancies of explanations generated using presented techniques could be performed in future works. One of the considered alternatives is to proceed towards *causal model discovery*, a declarative model underlying the superficial tree-based one obtained with ML. A review of existing approaches is presented in [39]. A promising idea is the one of *Causal Decision Trees* [16]. Some initial study of application of Constraint Programming for model discovery was put forward in [17] and followed in [18].

References

1. Barredo Arrieta, A., et al.: Explainable artificial intelligence (XAI): concepts, taxonomies, opportunities and challenges toward responsible AI. Inf. Fusion **58**, 82–115 (2020)
2. Biecek, P., Burzykowski, T.: Explanatory Model Analysis. Chapman and Hall/CRC, New York (2021)
3. Burkart, N., Huber, M.F.: A survey on the explainability of supervised machine learning. J. Artif. Intell. Res. **70**, 245–317 (2021)
4. Byrne, R.M.J.: Counterfactuals in explainable artificial intelligence (XAI): Evidence from human reasoning. In: IJCAI (2019)
5. Chazette, L., Brunotte, W., Speith, T.: Exploring explainability: a definition, a model, and a knowledge catalogue. In: 2021 IEEE 29th International Requirements Engineering Conference (RE), pp. 197–208 (2021)
6. Clinciu, M.A., Hastie, H.: A survey of explainable AI terminology. In: Proceedings of the 1st Workshop on Interactive Natural Language Technology for Explainable Artificial Intelligence (NL4XAI 2019), pp. 8–13. Association for Computational Linguistics (2019)
7. Cui, Z., Chen, W., He, Y., Chen, Y.: Optimal action extraction for random forests and boosted trees. In: Proceedings of the 21th ACM SIGKDD International Conference on Knowledge Discovery and Data Mining, pp. 179–188. KDD 2015, Association for Computing Machinery, New York (2015)

8. Deng, H.: Interpreting tree ensembles with inTrees. Int. J. Data Sci. Anal. **7**(4), 277–287 (2019)
9. Domingos, P.: Knowledge acquisition from examples via multiple models. In: Proceedings of the Fourteenth International Conference on Machine Learning, pp. 98–106. Morgan Kaufmann, San Francisco (1997)
10. Fernández, R.R., de Diego, I.M., Aceña, V., Fernández-Isabel, A., Moguerza, J.M.: Random forest explainability using counterfactual sets. Inf. Fusion **63**, 196–207 (2020)
11. Garreau, D., von Luxburg, U.: Explaining the explainer: a first theoretical analysis of lime. In: Proceedings of the 23rd International Conference on Artificial Intelligence and Statistics (AISTATS). Proceedings of Machine Learning Research, vol. 108, pp. 1287–1296. PMLR (2020)
12. Greenwell, B.M., Boehmke, B.C., McCarthy, A.J.: A simple and effective model-based variable importance measure. ArXiv https://arxiv.org/abs/1805.04755 (2018)
13. Guidotti, R., Monreale, A., Ruggieri, S., Turini, F., Giannotti, F., Pedreschi, D.: A survey of methods for explaining black box models. ACM Comput. Surv. **51**(5) (2018)
14. Hooker, G., Mentch, L., Zhou, S.: Unrestricted permutation forces extrapolation: variable importance requires at least one more model, or there is no free variable importance. Stat. Comput. **31**(6) (2021)
15. Huysmans, J., Baesens, B., Vanthienen, J.: Using rule extraction to improve the comprehensibility of predictive models. Behav. Exp. Econ. (2006)
16. Li, J., Ma, S., Le, T., Liu, L., Liu, J.: Causal decision trees. IEEE Trans. Knowl. Data Eng. **29**(2), 257–271 (2017)
17. Ligęza, A.: An experiment in causal structure discovery. a constraint programming approach. In: Kryszkiewicz, M., Appice, A., Ślęzak, D., Rybinski, H., Skowron, A., Raś, Z.W. (eds.) ISMIS 2017. LNCS (LNAI), vol. 10352, pp. 261–268. Springer, Cham (2017). https://doi.org/10.1007/978-3-319-60438-1_26
18. Ligęza, A., et al.: Explainable artificial intelligence. model discovery with constraint programming. In: Stettinger, M., Leitner, G., Felfernig, A., Ras, Z.W. (eds.) ISMIS 2020. SCI, vol. 949, pp. 171–191. Springer, Cham (2021). https://doi.org/10.1007/978-3-030-67148-8_13
19. Ligęza, A., Kluza, K., Jemioło, P., Sepioło, D., Wiśniewski, P., Jobczyk, K.: Evaluation of selected artificial intelligence technologies for innovative business intelligence applications. In: Borzemski, L., Selvaraj, H., Świątek, J. (eds.) ICSEng 2021. LNNS, vol. 364, pp. 111–126. Springer, Cham (2022). https://doi.org/10.1007/978-3-030-92604-5_11
20. Linardatos, P., Papastefanopoulos, V., Kotsiantis, S.: Explainable AI: a review of machine learning interpretability methods. Entropy **23**(1) (2021)
21. Lipton, Z.C.: The mythos of model interpretability. Queue **16**, 31–57 (2018)
22. Lundberg, S.M., Erion, G., Chen, H., DeGrave, A., Prutkin, J.M., Nair, B., Katz, R., Himmelfarb, J., Bansal, N., Lee, S.I.: From local explanations to global understanding with explainable AI for trees. Nat. Mach. Intell. **2**(1), 56–67 (2020)
23. Lundberg, S.M., Erion, G.G., Lee, S.I.: Consistent individualized feature attribution for tree ensembles. ArXiv https://arxiv.org/abs/1802.03888 (2018)
24. Lundberg, S.M., Lee, S.I.: A unified approach to interpreting model predictions. In: Guyon, I., Luxburg, U.V., Bengio, S., Wallach, H., Fergus, R., Vishwanathan, S., Garnett, R. (eds.) Advances in Neural Information Processing Systems. vol. 30. Curran Associates, Inc. (2017)

25. Miller, T.: Explanation in artificial intelligence: insights from the social sciences. Artif. Intell. **267**, 1–38 (2019)
26. Molnar, C.: Interpretable machine learning. Lulu.com (2020)
27. Parmentier, A., Vidal, T.: Optimal counterfactual explanations in tree ensembles. In: International Conference on Machine Learning (2021)
28. Petkovic, D., Altman, R., Wong, M., Vigil, A.: Improving the explainability of random forest classifier - user centered approach. In: Biocomputing 2018. World Scientific (2017)
29. Ribeiro, M.T., Singh, S., Guestrin, C.: "Why should i trust you?": Explaining the predictions of any classifier. In: Proceedings of the 22nd ACM SIGKDD International Conference on Knowledge Discovery and Data Mining, pp. 1135-1144. KDD 2016, Association for Computing Machinery, New York (2016)
30. Ribeiro, M.T., Singh, S., Guestrin, C.: Anchors: high-precision model-agnostic explanations. In: Proceedings of the AAAI Conference on Artificial Intelligence, vol. 32 (2018)
31. Sagi, O., Rokach, L.: Explainable decision forest: transforming a decision forest into an interpretable tree. Inf. Fusion **61**, 124–138 (2020)
32. Schwalbe, G., Finzel, B.: XAI method properties: a (meta-)study. ArXiv https://arxiv.org/abs/2105.07190 (2021)
33. Shi, S., Zhang, X., Fan, W.: A modified perturbed sampling method for local interpretable model-agnostic explanation. CoRR **abs/2002.07434** (2020). https://arxiv.org/abs/2002.07434
34. Turner, R.: A model explanation system. In: 2016 IEEE 26th International Workshop on Machine Learning for Signal Processing (MLSP), pp. 1–6 (2016)
35. Vandewiele, G., Lannoye, K., Janssens, O., Ongenae, F., De Turck, F., Van Hoecke, S.: A genetic algorithm for interpretable model extraction from decision tree ensembles. In: Kang, U., Lim, E.-P., Yu, J.X., Moon, Y.-S. (eds.) PAKDD 2017. LNCS (LNAI), vol. 10526, pp. 104–115. Springer, Cham (2017). https://doi.org/10.1007/978-3-319-67274-8_10
36. Verma, S., Dickerson, J., Hines, K.: Counterfactual explanations for machine learning: challenges revisited (2021)
37. Verma, S., Dickerson, J.P., Hines, K.E.: Counterfactual explanations for machine learning: a review. ArXiv https://arxiv.org/abs/2010.10596 (2020)
38. Wachter, S., Mittelstadt, B., Russell, C.: Counterfactual explanations without opening the black box: automated decisions and the GDPR. Harvard J. Law Technol. **31**, 841–887 (2018)
39. Yu, K., Li, J., Liu, L.: A review on algorithms for constraint-based causal discovery (2016)
40. Zafar, M.R., Khan, N.: Deterministic local interpretable model-agnostic explanations for stable explainability. Mac. Learn. Knowl. Extract. **3**(3), 525–541 (2021)
41. Zhao, X., Huang, W., Huang, X., Robu, V., Flynn, D.: Baylime: Bayesian local interpretable model-agnostic explanations. In: de Campos, C., Maathuis, M.H. (eds.) Proceedings of the Thirty-Seventh Conference on Uncertainty in Artificial Intelligence. Proceedings of Machine Learning Research, vol. 161, pp. 887–896. PMLR (2021)
42. Zhao, X., Wu, Y., Lee, Cui, W.: iforest: Interpreting random forests via visual analytics. IEEE Trans. Visual. Comput. Graph. **25**, 407–416 (2019)
43. Zhou, Y., Hooker, G.: Interpreting models via single tree approximation (2016)

Identification of Information Needs for Risk Analysis Processes in Inland Navigation

Emilia T. Skupień(✉) and Agnieszka A. Tubis

Wroclaw University of Science and Technology, 27 Wyspiańskiego Street, Wroclaw, Poland
{emilia.skupien,agnieszka.tubis}@pwr.edu.pl

Abstract. Identification of information needs is the primary step preceding the creation of databases for risk assessment needsd. Decision-makers' information needs determine the scope and characteristics of the collected data for analysis, but also the method of their processing and the frequency of updating. Each branch of transport has its own specificity. Due to this fact, for risk assessment, data are collected with a standard scope for all modes of transport and a group of data typical for a given means of transport. Meanwhile, numerous publications on risk in water transport mainly refer to maritime transport. However, river transport has different characteristics, which should be taken into account when assessing the risk associated with cargo transport. The objective of the article is to identify these needs, taking into account the specificity of river transport, and present, using the example of a selected data set, how they are used in the analytical process. The research presented in the article was conducted in four stages. The results show the identified information needs of decision-makers based on the direct interviews conducted and the sources of obtaining these data. Then, on the selected example, the difficulties related to the preparation of primary data are presented for the needs for risk analysis in river transport. The results of the analysis presented in the article are a fragment of the research by the authors on building a risk assessment and management system in inland waterway cargo transport.

Keywords: Risk assessment · Information needs · Inland waterways transportation

1 Introduction

River transport is considered one of the safest modes of transport, mainly due to the low rate of accidents registered in this transport system. However, its implementation is very strongly dependent not only on the transport infrastructure used but, above all, on the prevailing weather conditions, particularly those affecting the hydrological conditions of waterways. Due to the high variability of these conditions, the results of risk analyzes are of great importance in assessing the reliability of inland transport. This risk is assessed in terms of the probability and consequences of an undesirable event that will adversely affect the performance of cargo transport. However, the effectiveness of the conducted inference depends mainly on the quality of the data and the methods of their use in the

W. Zamojski et al. (Eds.): DepCoS-RELCOMEX 2022, LNNS 484, pp. 297–307, 2022.
https://doi.org/10.1007/978-3-031-06746-4_29

analytical process. A literature review shows that the number of publications aimed at risk assessment in inland waterway transport is limited [1]. For this reason, this area should be constantly developed, particularly in light of the European Union guidelines on the use of green modes of transport in the main parts of cargo transport [2].

The quality of inference obtained based on the risk assessment is determined mainly by the quality of the data used in the analytical process, including, in particular, its completeness and timeliness. For this reason, a key element in building a knowledge base for risk analysis is the correct identification of the information needs of decision-makers responsible for the reliable implementation of the transport process. Data on adverse events in shipping cargo handling are currently difficult to obtain. Disturbances affecting the correct implementation of the transport process are not always registered; they are registered only in the case of large losses. Due to the low percentage of inland navigation in freight handling, there are currently no data collection requirements for the safety risk assessment of transport. Therefore, this area requires research aimed at identifying the information needs of various stakeholder groups in the river transport system and developing guidelines for the collection and distribution of the necessary data. For this reason, the aim of the article is to initially identify the information needs regarding the factors influencing the occurrence of various operational risk groups, taking into account the specificity of river transport, to determine their availability and the need for updating. Then, on the example of a selected group of data, the difficulties related to the preparation and analysis of available data for the purposes of risk assessment will be presented.

The structure of the article assumes that a literature review is presented from the studied area in Sect. 2. In Sect. 3, the adopted research methodology was discussed, according to which the research described in the article was carried out. In Sect. 4, the obtained research results will be presented and discussed. Finally, Sect. 5 presents the final conclusions and summary.

2 Literature Review

The method of defining the risk influences the scope of the analyses performed and the interpretation of the results obtained [3]. Depending on current decision-making needs, the risk can be equated [4]:

- only with the occurring hazard and denote an undesirable event generating losses or
- with the occurring threat and the effect of loss, as well as the existing opportunities and the related effects in the form of profit.

A detailed review of the literature presented in [4] proves that in the case of transport risk analysis, risk is usually equated to the occurrence of adverse events, the effects of which are negatively evaluated by decision-makers. This dominant approach is even manifested in the way some authors define risk. An example would be the risk defined by [5], who described it as "the probability of accidents occurring". A similar approach to risk in transport is also present in research conducted jointly by leading research centers in Poland dealing with transport systems. An example is the ZEUS project, whose participants in their publications promoted a definition that describes risk as "a combination of the probability of threat activation in an adverse event and the resulting damage [6].

Based on the review of the literature on risk in water transport, it can be observed that most publications in this area mainly focus on the risk associated with sea transport [4]. This is mainly because maritime transport plays a crucial role in the functioning of global supply chains and therefore influences global economic development. However, at the same time, its functioning generates the risk of catastrophic accidents, the severe effects of which are assessed in human life and health units, financial losses, and a negative impact on the environment [7]. Meanwhile, the analysis of the adverse event comparator, which is the subject of the risk analysis for both modes of transport of water, shows that both the characteristics of the events that occur and the frequency and effects of their occurrence are entirely different [8]. In the case of inland transport, the most significant consequences of accidents concern losses related mainly to transported goods and damage to vessels [9]. At the same time, the subject of risk analysis is the assessment of transport safety; therefore, research focuses primarily on events such as [10]: accident, collision, contact (striking any fixed or floating objects other than those included in collision or grounding), grounding (being aground or impact / touching the shore or sea bottom or underwater objects), fire, and explosion. Among the main reasons for the appearance of adverse events in water transport are indicated [11]:

- accidents caused by inadequate human failure,
- accidents intentionally caused by man,
- accidents due to technical failures,
- accidents due to poor weather.

However, as Kaup [12] notes, in the case of river transport, it is equally important to take into account infrastructure objects located above the waterway, e.g., high voltage lines, pipelines. They should be kept in good technical condition because, for example, in unfavorable weather conditions, they can potentially pose a threat to the units flowing underneath them.

One of the critical elements of the environment that significantly affects the level of risk in water transport is the prevailing weather conditions, particularly the hydrological situation on the waterways. Their impact on safety was assessed, among other things, in the studies described in [9]. Therefore, the analysis of historical data on these conditions is fundamental in the framework of the knowledge base created for risk analysis. However, the influence of the human factor on the adverse events that occur is not less critical. Relevant data in this regard should be reported in the information systems of selected stakeholders in the water system. In the absence of internal reporting of these events, the analysis of offenses registered by the Inland Navigation Office may be helpful. An example of such analysis can be found, e.g., in [13].

Therefore, it can be concluded that the scope of information collected for the purposes of risk in water transport has its specificity, which should be considered in the knowledge bases created. Therefore, the identification of decision-makers information needs is the basis for defining the reporting system requirements and the methods of analyzing the shared data. Although the literature contains numerous publications and reports on creating knowledge bases for risk analysis needs (including [14]). In the opinion of the authors, there is no research that takes into account the specificity of river transport. The results presented below may at least partially fill this gap.

3 Methodology

The research aimed to identify the information needs to assess the risk of water cargo transport and to analyze the usefulness of the available data to estimate the probability and effects of the identified events. It should be emphasized that the conducted risk assessment is carried out according to the needs of the river freight transport participants. For this reason, the information collected corresponds to the information needs of this group of stakeholders of the analyzed transport system. Therefore, the adopted methodology includes four basic research steps, which are presented in Fig. 1.

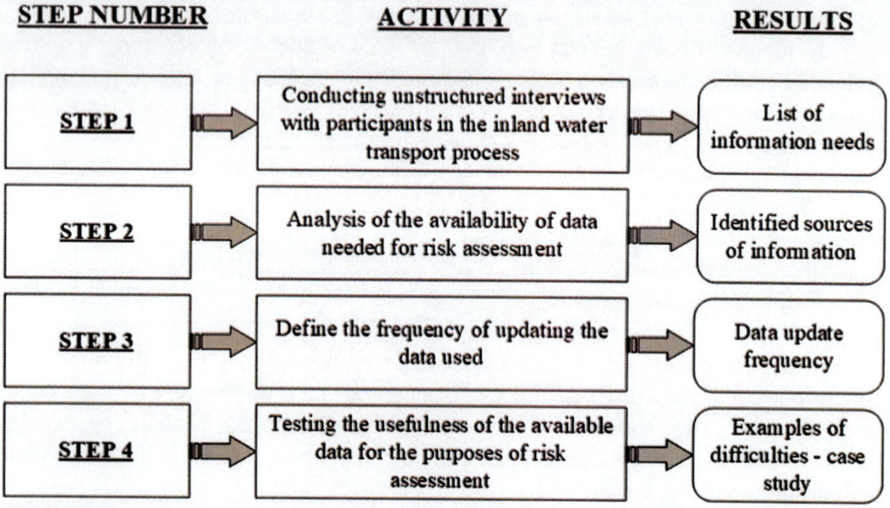

Fig. 1. Research steps

First, unstructured face-to-face interviews were conducted. The unstructured nature of the interview included the lack of a specific survey form and the implementation of the interview in the form of brainstorming based on the respondent's knowledge and experience. The interviews were attended by eight respondents who represented the community of ship captains and freight forwarders handling cargo transported by water in Poland. Based on the obtained answers, a list of information requirements needed for risk assessment by the participants of the transport process was created.

The main objective of this research was to identify what type of information must be gathered for risk analysis. However, identification of adverse events and estimation of the probability and consequences of their occurrence were not the main goals of the analysis at this stage.

For further analysis of this article, information needs were selected that were repeated in the results of individual interviews (the number of indications greater than 2). For this information, research was carried out to determine its availability in the form of legal documents, reports, data from industry portals, waterway administrators, institute of meteorology and water management, regional waterway authority, and navigation announcements. Then each need is assigned an appropriate source of information that should feed the knowledge base for risk assessment. The following research step determined the required frequency of updating the data in the databases. This frequency was determined on the basis of the following:

- changes in applicable legal provisions,
- changes in the available transport infrastructure,
- changes in navigational conditions,
- changes in weather conditions,
- changes in the requirements of contractors.

At the same time, the analysis of the data used for risk assessment has shown that not all information can be used in the original version available from the indicated sources. Some of the data requires appropriate processing, taking into account the specific conditions of water transport. For this reason, a case study is presented at the end of the article. Its purpose is to outline the difficulties related to the processing of available quantitative data for risk assessment in inland transport.

4 Results and Discussion

4.1 Identified Information Needs

The quality of the risk assessments performed largely depends on the scope, timeliness, and detail of the data supplying the analytical process. Therefore, the key issue is to determine the information needs necessary in the process of identifying potential adverse events and assessing the probability and consequences of their occurrence. Based on interviews with shippers and captains of ships, the information needs of decision-makers responsible for the organization of freight transport by inland transport were identified. The information needs reported by the study participants were limited to issues related to the identification of adverse events and the accompanying risk, which should be taken into account in order to better organize inland waterway transport. According to the methodology presented in Sect. 3, this information has been supplemented with additional parameters that are presented in Table 1.

Based on the data collected in Table 1, it can be concluded that the information reported as needed can be divided into groups related to its source, frequency of checking it, and what it concerns.

Legal requirements and point infrastructure matters need to be checked less frequently than once a year. Formal requirements must be checked for every transportation process. Operation information should be checked on a regular basis. But there is also a group of data checked every day. These are data connected to water, because it changes very often. These data are needed on the time of transportation, but what was underlined

Table 1. Parameters of information needed for risk analysis processes in river transport

Information group	Type of information	Source of information	Frequency of information checking
Applicable legal regulations	• Ship regulations, • Crew regulations, • Local regulations, • Documents related to a specific type of shipment	• EU regulations, • Polish regulations, • Local regulations	Once a year
Formal requirements	• Transport documents, • Contractors' requirements	• Shipping documents from the contractor	Every transportation
Infrastructure	• Waterway parameters, • Locks parameters, • Crossing infrastructure, • Night mooring, bunkering, waste disposal, drinking water collection, leaving the car ashore, • Transhipment infrastructure	• Facility administrator, • Waterway administrator, • Navigation information, • RIS	Every transportation, at least every 3 years
Operational conditions of the transportation	• Precipitation, temperature, fog, wind, • Transit depth, • Information on the customer's requirements (e.g. cleaning of loading compartments)	• Weather forecast, • Navigation announcements, • Contractor information	Every day Customer's requirements – every transportation
Exclusions	• Information on periodic shutdowns, renovations, closings, and accidents	• Navigation announcements	Every day
Current operating information	• Operating parameters, • Information on possible damage to the ship and/or equipment	• Equipment available on board	On a regular basis

by the users of the inland navigation system in Poland was that in the matter of risk, the most important is whether the navigation conditions will allow them to transport the goods and how likely it will change.

Collecting all that data influences transportation safety and feasibility. The data have various impact, and their analysis differ, but all should be taken into account. The process of data analysis was shown in an example. The case study was based on transit depth of a waterway.

4.2 Difficulties in Preparing Data for Risk Assessment – Case Study

When inland waterway transport is performed, one of the sources that need to be checked every day are the navigation announcements given by the waterway administrator. It may state if the waterway is closed or the transit depth for the managed sections.

But there are also data, the analysis of which is difficult because they cannot be interpreted only in the original version. The preparation of these data for risk analysis is difficult in terms of inland navigation. Data must be marked with the specificity of river transport, taking into account the parameters and characteristics of a specific section of the river.

The case study is based on Oder waterway in Poland, managed by the Regional Water Management Authority. Table 2 shows an example of the reasons and the number of failures and difficulties that caused the closure of the waterway section in 2018, 2019, and 2020.

Table 2. Failures that caused the closure of the Oder waterway in the Wroclaw area and the number of days with closing.

Reason of waterway closing	2018	2019	2020
A flood wave	0	1	4
Ice phenomena on the river	1	1	0
Fluctuating water level	1	3	0
Damage to the lock gates	1	0	2
Problems with damming water	0	4	3
Renovations	5	6	2
Timed shutdown without failure	1	2	0

As may be seen, during the navigation season (270 days a year) for about 10 days navigating was not possible at all. It is not so often, but, to be sure, navigation announcement must be checked. Another important thing to mention - Table 2 concerns the canalized section, so is the focus on problems other than water level or ice phenomena, because it is a problem mostly on free flowing rivers. It is also worth noting that on some European rivers, days of closing the waterway due to ice phenomena changed places with days of closing due to low water level caused by drought [15].

Another reason to check the announcements is the fact that they include the depth of the transit waterway. It is very important in Poland, as the foreflowing rivers fluctuate a lot. The water level is monitored by the Institute of Meteorology and Water Management, and on the basis of these data, the Waterway Management Authority calculates the transit depth. The announced transit depth guarantees that the ship of that immersion can safely navigate truth in this section.

Ensuring an adequate depth of the waterway is highly important because a level too low may endanger transportation execution. Another thing is that if the ship is loaded on some level, it cannot be unloaded during the voyage, and for example the route from Wrocław to Szczecin lasts 3 days. The other important fact is that if one starts the voyage, wants to finish it without interrupting, stopping for several number of days. So, one thing is the transit depth and the other is the duration of this depth. And the duration of the transit depth is the subject of risk analysis. It may influence the decision to use inland navigation at all.

To analyze the data, the water gauge was chosen. The Oder river has the lowest level in the free flowing section, near the mouth of the Nysa Luzycka river. Therefore, it determines the draft of the ship across the entire Odra River. Data were collected from the Institute of Meteorology and Water Management. The analysed period included 9 years. The indication of the water gauge for each day was calculated on the depth of the transit. It was not possible to use the data from the Regional Water Management announcements because it includes only working days.

Figure 2 shows the graph of the depth of transit in the Oder River over several years (level 0 corresponds to the days when shipping was not allowed). It was decided to show it this way to emphasize the fact that there were closures of the waterway. The chosen analysed periods result of specific events - reconstruction of the water junction in Wroclaw and reconstruction of hydrotechnical devices which could distort the results of the analyses.

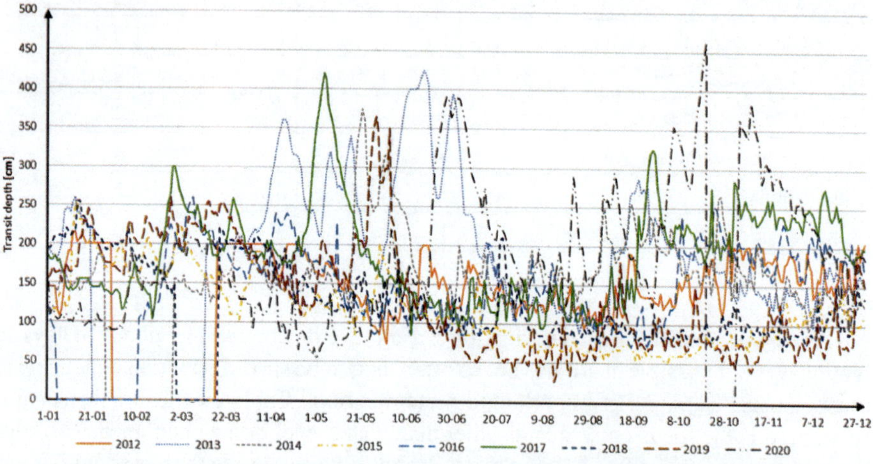

Fig. 2. Transit depth on the free flowing section of the Oder River in several years.

The data shown in Fig. 2 clearly shows that the water level changes frequently, so it is not easy to predict whether the water level that the ship owner needs will occur the day it is needed and will last as many days as planned. This is the reason why monitoring of waterway transit depth is needed not only every day of voyage, but it may be helpful to assess the risk of voyage feasibility.

For the purpose of this analysis, data was collected. Calculations take into account depths for a given number of consecutive days. So that the voyage could take place without the need to stop. To calculate the probability of the occurrence of an appropriate transit depth for the desired number of days, the depth of the waterway and its duration were assumed as independent random variables.

The normal distribution was fitted to the transit depth (first random variable). Its mean was $\mu = 126.276$ cm and the standard deviation was $\sigma = 51.140$ cm. This means that according to this distribution, the expected depth values of the waterways fluctuated in the range from $1.26–0.51 = 0.75$ m to $1.26 + 0.51 = 1.77$ m. For a depth in the range of 0.00–3.68 m (theoretical values assumed by the normal distribution), the distribution function was determined, i.e. the probability that the water level will be lower than the specified one. By subtracting this value from one, the probability of the expected depth or greater is obtained.

In a similar manner, the distribution of the lengths of occurrence of individual depths was fitted to the log-normal distribution. Its mean was $\mu = 0.732$ days and the standard deviation was $\sigma = 0.671$. By counting the distribution function for a given number of days, from the range of 1–365 days (the value used in the calculations for the log-normal distribution), the probability of the transit depth occurring for a given number of days was calculated.

Examples of data obtained for the described distributions are presented in Table 3.

Table 3. The probability of a specified transit depth and the probability of the depth occurring over a specified number of days.

Transit depth [cm]	Probability that the level will be ≥ from the given	Number of days	Probability that the number of days will be ≥ from the given
90	0.841135591	1	1
120	0.567417284	3	0.839770039
150	0.325054144	6	0.599305705
180	0.140410879	9	0.476944131

The probability of maintaining a given transit depth for a given number of days is the product of both indicated probabilities. Examples of calculated probabilities are shown in Table 4.

Presented calculations indicate that carry on the transportation of a ship immersed more than 150 cm is burdened with high uncertainty of performance. Therefore, the next step of this research will be calculating the probability for narrowed time periods, e.g. individual quarters.

Table 4. Sample data of the probability of a specific transit depth occurring for 3 days

Transit depth [cm]	Number of days	Probability of reaching the set depth for 3 days
90	3	0.707107712
120	3	0.477274623
150	3	0.273769532
180	3	0.118730096

5 Summary

The research carried out by the authors of the article shows that there is currently a significant research gap in the literature on risk assessment and management systems, taking into account the specificity of river transport. In the last decade, many publications on the risk of water transport have appeared, but their primary focus is primarily on maritime transport. Meanwhile, the specificity of river transport means that applying the assumptions of the analysis formulated for sea transport does not always allow conducting a research procedure that is appropriate for the needs.

An important element of the functioning of effective risk management systems is a sufficiently prepared and used knowledge base about the factors that influence the risk of specific adverse events. For this reason, the article presents the results of research on the identification of information needs, the sources of their acquisition, and the difficulties in preparing data useful for the analytical process. The proposed scope of the data obtained for the risk analysis needs is practical. It should be used by shippers and shipowners' employees responsible for organizing transporting cargo by water. At the same time, it should be emphasized that the results presented in the article are part of the research currently carried out by the authors. This research aims to develop a risk assessment method that considers the specificity of the water-based cargo transport process.

References

1. Skupień, E., Tubis, A.: The use of linguistic variables and the FMEA analysis in risk assessment in inland navigation. TransNav Int. J. Marine Nav. Saf. Sea Transp. **12**(1), 143–148 (2018)
2. Roadmap to a Single European: Transport Area – Towards a competitive and resource efficient transport system. COM (2011)
3. Aven, T.: Risk assessment and risk management: review of recent advances on their foundation. Eur. J. Oper. Res. **253**, 1–13 (2016)
4. Tubis, A.: Method of Operational Risk Management in Road Transport. Publishing House of the Wrocław University of Technology, Wrocław (2018)
5. Hauer, E.: Traffic conflicts and exposure. Accid. Anal. Prev. **14**(5), 359–364 (1982)
6. Krystek, R. (ed.): Integrated Transport Safety System. Volume 2. Conditions for the Development of Transport Safety Systems Integration, Wydawnictwo Komunikacji i Łączności, Warszawa (2009)

7. Wang, S., Yin, J., Khan, R.U.: The multi-state maritime transportation system risk assessment and safety analysis. Sustainability **12**(14), 5728 (2020)
8. Tubis, A., Skupień, A., Rydlewski, M: Water transport risk - comparative analysis. In: 8th Carpathian Logistics Congress, CLC 2018: Conference Proceedings, December 3rd–5th 2018, Prague, pp. 879–884. Tanger, Czech Republic. Ostrava (2019)
9. Łozowicka, D., Kaup, M.: Safety aspects of inland water transport in Poland. TTS **12**, 2016–2020 (2015)
10. Li, S., Meng, Q., Qu, X.: an overview of maritimewaterway quantitative risk assessment models. Risk Anal. **32**(3), 496–512 (2012)
11. Neumann, T.: Telematic support in improving safety of maritime transport. Trans. Nav. Int. J. Mar. Navig. Saf. Sea Transp. **12**(2), (2018)
12. Kaup, M.: Selected problems of inland transport of oversize cargo. Logistyka. **4**, 3981–3990 (2015)
13. Skupień, E., Tubis, A., Rydlewski, M.: The analysis of offenses in Inland Navigation in Poland 2013–2017. Int. J. Marine Nav. Saf. Sea Transp. **13**(1), 151–156 (2019)
14. A guide to assessing your risk data aggregation strategies. https://www2.deloitte.com/content/dam/Deloitte/ca/Documents/risk/ca-en-15-2733H-risk-data-aggregation-reporting-BCBS-239.PDF. Accessed 20 Dec 2021
15. Bačkalić, T., Maslarić, M., Skupień E.: Analysis of navigation accessibility and fairway availability: a case study the middle Danube river. In: 10th International Scientific Conference, VII International Symposium of Young Researcher, pp. 53–62. Transport Problems, Katowice (2018)

Using Fault Injection for the Training of Functions to Detect Soft Errors of DNNs in Automotive Vehicles

Peng Su and DeJiu Chen(✉)

KTH Royal Institute of Technology, 100 44 Stockholm, Sweden
{pensu,chendj}@kth.se

Abstract. Advanced functions based on Deep Neural Networks (DNN) have been widely used in automotive vehicles for the perception of operational conditions. To be able to fully exploit the potential benefits of higher levels of automated driving, the trustworthiness of such functions has to be properly ensured. This remains a challenging task for the industry as traditional approaches to system verification and validation, fault-tolerance design, become insufficient, due to the fact that many of these functions are inherently contextual and probabilistic in operation and failure. This paper presents a data centric approach to the fault characterization and data generation for the training of monitoring functions to detect soft errors of DNN functions during operation. In particular, a Fault Injection (FI) method has been developed to systematically inject both layer- and neuron-wise faults into the neural networks, including bit-flip, stuck-at, etc. The impacts of injected faults are then quantified via a probabilistic criterion based on Kullback-Leibler Divergence. We demonstrate the proposed approach based on the tests with an Alexnet.

Keywords: Neural Networks · Anomaly · Soft errors · Fault injection · Kullback-Leibler Divergence · Machine learning

1 Introduction

Deep neural networks (DNNs) are learning-enabled components (LEC) [4] currently being widely used in advanced functions for operation perception in Automotive Vehicles (AVs). Recent studies show that operational anomalies associated with some environmental conditions as well as some key parameters of a DNN structure (e.g., weights) can alter the computational performance drastically [19]. Such anomalies, if improperly managed, would become system faults, causing undesired system errors and safety hazards.

There are several fundamental questions that have to be carefully treated for ensuring the trustworthiness of DNN based system functions [5], relating to: 1). What are the possible operational faults of a DNN in AVs? 2). How much could each operational fault affect the functionality and performance of a DNN? 3). How to effectively detect or tolerate the errors and failures caused

W. Zamojski et al. (Eds.): DepCoS-RELCOMEX 2022, LNNS 484, pp. 308–318, 2022.
https://doi.org/10.1007/978-3-031-06746-4_30

by an operational fault? For example, for the design of safety mechanisms, it is important to know what error behaviors by means of operational signals and data would be given by possible hardware faults and how such error behaviors could be distinguished effectively from other behaviors. Such information is hard to reveal by analytical approaches, given the complexities of DNNs and their implementations in AVs. Conventional approaches to system dependability and trustworthiness rely heavily on system verification and validation, fault tolerance design (e.g. Triple Module redundancy (TMR) [24]). However, for DNN components, such conventional approaches become often insufficient due to the fundamental challenge of state space coverage for neural networks [26].

In this work, we present a data centric approach to the fault characterization and data generation for the training of embedded monitoring functions that help to detect soft errors of DNN functions during operation based on Fault Injection (FI). The paper is organized as follows: In Sect. 2, we discuss the related work. In Sect. 3, we present our approach by explaining the key concepts of faults, and the methods applied for fault injection and error detection. The approach is applied for case study using *Alexnet* and the results are presented in Sect. 4. Finally, we conclude with Sect. 5 regarding the effectiveness and future work.

2 Background and Related Work

In recent years, various DNN based functions have been introduced into AVs for the perception of environmental conditions during operation [8]. The design is characterised by a directed acyclic graph consisting of multiple computation layers [18]. As one key step, the training of DNN involves using optimization for a minimization of the discrepancy between the ground-truth data and the network output, normally based on a Back Propagation (BP) algorithm [17].

During the operation of AVs, the performance of DNN based perception functions is however strongly affected by the run-time environments and would exhibit error conditions that in turn would lead to system hazards. For example, both the corruption of input pixels [9] and the change of the brightness of the image may leads to misclassification of DNNs [7]. One category of such error conditions is referred to as *soft errors* [3]. By definition, a soft error is an anomalous state or error condition not caused by any primary design faults, but by faults from the run-time environment due to hardware aging, memory corruption, thermal disturbances, and many other anomalies of the software and hardware platforms [26]. According to the temporal characteristics, a fault can be *transient* (i.e. occurring occasionally and persisting only for a short period) or *permanent* (i.e. persisting over time once occurred).

Fault Injection (FI) is a testing technique which aids in understanding how a system behaves when stressed in unusual ways [21]. For example, it could be used to falsify some input data to verify the functionality and robustness of a system function. For neural networks, traditional fault injection methods [12,15] could suffer from performance issues due to the inherent complexity of DNNs.

Currently, there are a few fault injection tools dedicated to the design of neural networks. InjectFI [2] is a fault injector for *Tensorflow* for identifying the the effects of random hardware faults on neural networks. To inject faults, the tool splits original network model into different parts. The approach is delimited by only supporting the injections to the output data of specific layers. TensorFI [20] is another fault injector for *Tensorflow*. It provides a more fine-grained support by first replicating some network nodes (layers), adding new operations masking the original inputs/outputs, and then substituting the original ones in the network. NvBit [27] is a compiler-based fault injector that aims at corrupting the register states to optimize the hardware resilience and redundancy. One possible delimitation of all these fault injectors would be the lack of explicit support for investigating the error behaviors of specific neurons, which is necessary to understand how faults propagate through specific neurons. The identification of such a more fine-grained fault or failure logic plays an important role for the design of fault-tolerance mechanism, such as by redundant neurons [14].

For anomaly detection, it is necessary to first identify the nominal operational ranges of some related data in target DNN and then quantify the observed discrepancies [6]. One approach, referred to as *Symptom-based-detector* (SBD), is based on outlier detection [19,26]. The main process is to first collect a set of data from a few of layers with fault-free operations, then to identify the corresponding maximum and minimum boundaries as well as the thresholds for anomalous conditions. In operation, once the actually observed data by corresponding layers or neurons are beyond the predefined thresholds, the target system would be agged as anomalous. A similar approach is to identify the maximum-minimum conditions for a few layers collectively [11]. As DNN is inherently probabilistic in computation, our approach investigate in particular the usage of probabilistic criteria for effective error detection. In this work, we consider especially a criterion based on the well-known Kullback-Leibler Divergence [10], widely used in statistics as a measure between two probability density functions.

3 Fault Injection and Training for Error Detection

Our approach to a FI based training of functions for error detection consists of the following four steps:

Step 1: Basic System Modeling. This step is taken to identify the computational characteristics of a target neural network. Basically, in a DNN, the atomic computational unit is given by *neuron*, which is essentially a non-linear function defined as follows:

$$\textbf{Basic Activation Function (ACT) of Neuron: } f(ax+b) \tag{1}$$

where, x refers to the input to the neuron, a and b are the weight and bias parameters for adjusting the outputs. A transfer function of any neuron n in layer m can be denoted as follows:

$$\textbf{Activation Function (ACT) in Network: } y_n^m = f(a_n^m x_n^m + b_n^m) \quad (2)$$

Step 2: Configuration of Fault Injection. At this step, the faults to be injected in a neural network are identified and configured. See Table 1 below for an overview of the fault types currently being supported.

Table 1. Common fault types being considered for DNNs.

Fault type	Description
Stuck-at	A permanent omission (i.e. *Stuck-at* Low) or commission fault (i.e. *Stuck-at* High) on signals due to software or hardware anomaly
Noise	A transient fault on signal level due to external interference
Bit-flip	A transient fault on bit level due to external interference

The configuration of FI is supported by the following parameters for each fault injection case $Fault^i$:

$$\textbf{Fault Injection Case: } Fault^i = (Eve^i, Loc^i) \quad (3)$$

where Eve^i refers to the fault event, for which the temporal property is further characterized by a binary variable $\mathcal{C}_i$ for deterministic occurrence, and by a random variable $\mathcal{D}_i$ for probabilistic occurrence. For example, a stuck-at fault could occur deterministically (i.e. always occurring or not occurring during FI) for randomly (i.e. occurring or not occurring in a probabilistic way). Moreover, the probabilistic case with $\mathcal{D}_i$ is defined by a set of k distribution models $\mathbf{D}_k$. For example, we denote a specific Gaussian Distribution model with $\mathbf{D}_1$.

Step 3: Configuration of Fault Injection Scenarios. Our approach to the fault injection targets explicitly the activation functions (ACT) of Neurons as well as to the outputs of different layers (Eqs. 1 and 2). An overview of the procedure is shown in Fig. 1. In general, we allow faults to occur by any parameter (weights, bias, input, and outputs) of a *neuron* and at an arbitrary layer. Each fault injection scenario stipulates the configurations of different fault injection cases, the injection procedures as well as the assessment.

$$\textbf{Fault Injection Scenario: } \mathcal{S}^i = (\mathcal{M}^i, Fault^i); \quad (4)$$

$$\mathcal{M}^i \in (Mode\_A, Mode\_B) \quad (5)$$

Each fault injection scenario $\mathcal{S}^i$ stipulates the configurations of different fault injection cases $Fault^i$ with assigned injection modes $\mathcal{M}^i$, the injection procedures as well as the assessment. There are basically two modes of injection:

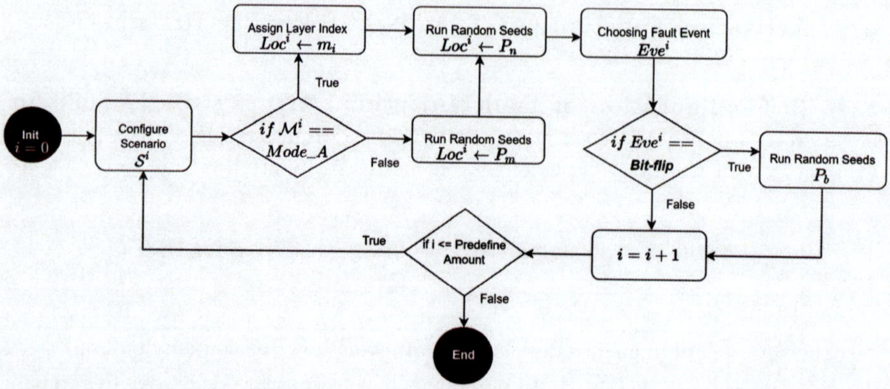

Fig. 1. A functional overview of the fault injection procedure based UML Activity Diagram. After the initialization, a FI scenario is configured by the selections of fault events and locations based on the preferred FI modes.

- $Mode\_A$: Injecting fault in neurons in one or more locations of specified layers k, which is assigned by users;
- $Mode\_B$: Injecting fault in neurons randomly in any location of the entire network.

While the $Mode\_B$ constitutes the basis for state space exploration, the $Mode\_A$ provides dedicated support for identifying specific fault models. That is, by $Mode\_A$, the configuration of injection layer m is decided by the system designer. By $Mode\_B$, the injection is defined based on a probabilistic selection. These two modes of FI complement each other, while enhancing the flexibility of fault injection in comparison to the related approaches introduced in Sect. 2. Technically, the target layers and neurons are determined in $Mode\_B$ by using two random seeds P_m and P_n. For the faults of the type *Bit_flip*, a random seed P_b is defined for selecting the reverting bits from a pre-defined length of bitstream, which is normally *Float 32* [9,13].

Step 4: Impact Assessment with KL Criterion. For the assessment of system-wide effects of faults being injected, a probabilistic criterion based on Kullback-Leibler Divergence [10] is used. This criterion (D_{KL}) quantifies the discrepancy of two probability density functions $f(x)$ and $g(x)$ of a specific random variable x as follows:

$$D_{KL}(f||g) = -\int f(x) \ln \frac{f(x)}{g(x)} dx \tag{6}$$

In our case, the probability density functions $f(x)$ and $g(x)$ are used to characterize the samples from the injected fault (i.e. $f(d_l)^{Fault^i}$) with a specify type $Fault^i$ (Eq. 3) as well as the samples by nominal operations $g(D^{Nominal})$. Here, $(d_l)^{Fault^i}$ refers to one sample l from a collection of k samples $\{d_1, d_2, ..., d_k\}$

by k simulation rounds for a fault type, and $D_{nominal}$ refers to all the samples from the simulation of nominal operational conditions. We assume both models follow *Gaussian* distribution and k is large enough by *Monte Carla* simulation. The samples are also assumed to be independent identically distributed random variables (i.i.d.). Accordingly, the observe KL discrepancy for each fault i (i.e. $D_{KL}^{Fault^i}(f||g)$) can be expressed by the following equation [10]:

$$D_{KL}^{Fault^i}(f||g) = -\frac{1}{k}\sum_{l=1}^{k} \ln \frac{f({d_l}^{Fault^i})}{g(D^{Nominal})} \tag{7}$$

Compared to a *symptom-based detection criterion*, this KL based approach would provide a better support for quantifying the differences among the error behaviors given by the FI samples of random variables. Due to the inherent probabilistic nature of DNNs, their fault emissions are also fundamentally probabilistic, making any deterministic classification less effective. For example, although a symptom-based method could be useful for the assessment for neural networks [11], it is difficult to distinguish among the impacts of various fault types.

For a specific fault being injected, the error emissions at different neurons and layers differ among specific DNNs. If a fault is not vanishing (i.e. being tolerated), it propagates across the system and the impacts should be captured by the discrepancy based on the KL criterion. Currently, the assessment is supported by two algorithms: K-nearest-neighbor (KNN) [1] and Multilayer Perception (MLP) [23]. To evaluate the performance of a specific fault assessment method, we use the following accuracy metric:

$$Accuracy = \frac{\hat{\alpha}}{\alpha} \times 100\% \tag{8}$$

where $\hat{\alpha}$ refers to the faults with their impacts being detected by a assessment method, α refers to the number of faults being injected.

4 Case Study and Results

We demonstrate the proposed approach based on a case study[1] with *Alexnet* [16], which contains 5 convolutional layers and 3 dense layers with Relu functions. The software implementation is based on *Pytorch* [22]. The Code Segment 1 below shows the configuration of fault scenario.

```
# Define some classes to decompose the neural network
net = torch.load(modelPath)
layer_info = Map_Layer(net)
neuron_data = Map_Tensor(net)
# Map_Layer and Map_Tensor are functions to load the detail of Alexnet
def Mode_Type(mode_type,layer_info,*args):
    if mode_type =='A':
# Preprocess the *arg
        layer_index = *arg
    else:
```

[1] The code is available in https://gits-15.sys.kth.se/pensu/Fault-Injection-Alexnet.

```
        layer_index = Layer_Proportion(layer_info)
    return layer_index
# The random function, defined in another class, is dependent on the
    neurons amount of each layer.
def Inject_Fault(neuron_data,fault_amount, fault_event, mode_type):
    count = 0
# Define the parameters to start the framework.
    for i in range (0,fault_amount):
        if count> fault_amount:
            break
        else:
# Fault scenario contains a mode_type, a fault_case (fault_event,
    fault_location)
            fault_location = Mode_Type(mode_type)
            fault_case = np.array(fault_amount, fault_event,fault_location)
            fault_network = Fault_type.Fault_Tensor(neuron_data, fault_case
    )
# Configure the fault injection scenario. The neuron injector is defined by
    another class Fault_type
            count = count + 1
        return fault_network
# Save the fault neuron network model, data process to the neuron network
```

Code Segment 1: The Python pseudo code configuring a specific fault scenario.

The amount of faults injected in the neural networks have been decided both randomly and statically chosen. The results are shown in Fig. 2. In particular, for the faults being injected, it is observed that their significance levels range from high to low with: *Bit-flip* → *Stuck-at High* → *Noise* → *Stuck-at Low*.

The faults of *Bit-flip* at the bit level propagate more significantly than all other fault types. The reason for this is that a bit-flip fault can make the original value shift significantly, in accordance to the numerical significance of bit position for the fault injection. Meanwhile, it is shown that the faults of *Stuck-at* Low have the least propagation. One reason for this is that the *stuck-at* Low faults are in effect similar to *dropout* [25], which is a technique widely used for preventing a DNN from over-fitting during training. This means a DNN can always have a certain degree of inherent robustness or tolerance regarding such faults. Similarly, during the training of DNNs, it is also common to introduce random disturbances to improve the robustness. This means that a DNN can also have certain degree of inherent robustness or tolerance regarding the *Noise* faults. Of course, the actual degree depends on whether the *Noise* faults of concern are taken into consideration during the training.

For the fault injection scenarios of *Mode_A* and *Mode_B*, the corresponding fault amounts chosen for the case study are 300 and 500. The labels of dataset collected in advanced when injecting faults. We use KNN and MLP as solvers to classify the results given by KL divergences. We process these tests 10 times to get the average accuracy. The results can be found in Table 2. Our monitor, especially using MLP as the solver, is high efficient to detect bit flip, stuck-at, and random faults. Since these faults leads to discrepancy of the distribution.

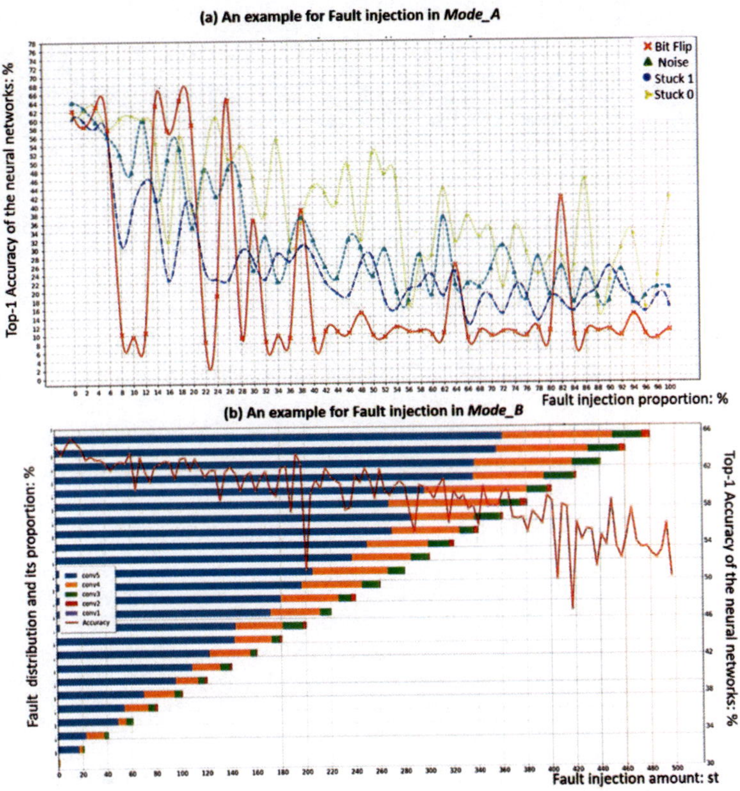

Fig. 2. The results of FI scenarios with the injection modes *Mode_A* and *Mode_B*. The color bar in Fig. (b) shows the faults' quantity of each layer selected for the simulation rounds. For example, the shortest bar at bottom of Fig. (b) shows the injection of 20 faults in the entire network with the portion on different layers as indicated by the colorbar, which results in 66% accuracy as indicated by the Accuracy curve. Similarly, when 500 faults are injected with the same portion on the layers, the accuracy is 53.5%.

Table 2. Results of fault classification based on KNN and MLP.

Fault amounts	Fault type	MLP based classification	KNN based classification
300	*Stuck-at* Low	64.50%	62.30%
	Stuck-at High	84.60%	61.60%
	Noise	71.40%	61.20%
	Bit-flip	91.00%	72.00%
500	*Stuck-at* Low	72.50%	64.20%
	Stuck-at High	90.20%	67.60%
	Noise	72.20%	59.20%
	Bit-flip	91.40%	72.00%

5 Conclusion

This paper presents a FI method for a data centric approach to the treatment of soft errors of DNNs in AVs. Through a case study, the capabilities of this approach, regarding its support for a neuron-wise fault injection and a probabilistic criterion based impact assessment, have been demonstrated. The adoption of *KL Divergence* provides a more effective method for the discrepancy quantification. Future studies would enhance the support for fault classification and assessment. An inclusion of *hypothesis test* could allow a confidence based fault detection and a richer set of failure modes. Normally, when faults are detected, DNN redesign or optimization would be used to improve the robustness. For the AVs, one complexity is due to the incompleteness of data for learning as emergent operational situations are not unavoidable. This means we plan to use suitable probabilistic models or NNs for encoding and generating operational data. The approach proposed by this paper aims to support active monitoring as a safety mechanism instead of passive hardware redundancy. This would allow an effective detection of common failure modes of DNNs due to same training faults.

Acknowldgement. This work is supported by 1. KTH Royal Institute of Technology with the industrial research project ADinSOS (2019065006); and 2. the Swedish government agency for innovation systems (VINNOVA) with the cooperative research project Trust-E (Ref: 2020-05117) within the programme EUREKA EURIPIDES.

References

1. Altman, N.S.: An introduction to kernel and nearest-neighbor nonparametric regression. Am. Stat. **46**(3), 175–185 (1992)
2. Beyer, M., et al.: Fault injectors for tensorFow: evaluation of the impact of random hardware faults on deep CNNs. arXiv preprint arXiv:2012.07037 (2020)
3. Borkar, S.: Designing reliable systems from unreliable components: the challenges of transistor variability and degradation. IEEE Micro **25**(6), 10–16 (2005)
4. Cai, F., Koutsoukos, X.: Real-time out-of-distribution detection in learning-enabled cyber-physical systems. In: 2020 ACM/IEEE 11th International Conf. on Cyber-Physical Systems (ICCPS), pp. 174–183. IEEE (2020)
5. Chen, D.J., Lu, Z.: A model-based approach to dynamic self-assessment for automated performance and safety awareness of cyber-physical systems. In: Bozzano, M., Papadopoulos, Y. (eds.) IMBSA 2017. LNCS, vol. 10437, pp. 227–240. Springer, Cham (2017). https://doi.org/10.1007/978-3-319-64119-5_15
6. Ding, K., Ding, S., Morozov, A., Fabarisov, T., Janschek, K.: On-line error detection and mitigation for time-series data of cyber-physical systems using deep learning based methods. In: 2019 15th European Dependable Computing Conference (EDCC), pp. 7–14. IEEE (2019)
7. Dreossi, T., Jha, S., Seshia, S.A.: Semantic adversarial deep learning. In: Chockler, H., Weissenbacher, G. (eds.) CAV 2018. LNCS, vol. 10981, pp. 3–26. Springer, Cham (2018). https://doi.org/10.1007/978-3-319-96145-3_1

8. Grigorescu, S., Trasnea, B., Cocias, T., Macesanu, G.: A survey of deep learning techniques for autonomous driving. J. Field Robot. **37**(3), 362–386 (2020)
9. Henriksson, J., Berger, C., Ursing, S.: Understanding the impact of edge cases from occluded pedestrians for ml systems. In: 2021 47th Euromicro Conference on Software Engineering and Advanced Applications (SEAA), pp. 316–325. IEEE (2021)
10. Hershey, J.R., Olsen, P.A.: Approximating the Kullback Leibler divergence between gaussian mixture models. In: 2007 IEEE International Conference on Acoustics, Speech and Signal Processing-ICASSP2007. vol. 4, pp. IV–317. IEEE (2007)
11. Hoang, L.H., Hanif, M.A., Shafique, M.: FT-ClipAct: resilience analysis of deep neural networks and improving their fault tolerance using clipped activation. In: 2020 Design, Automation & Test in Europe Conference & Exhibition (DATE), pp. 1241–1246. IEEE (2020)
12. Hsueh, M.C., Tsai, T.K., Iyer, R.K.: Fault injection techniques and tools. Computer **30**(4), 75–82 (1997)
13. Kalamkar, D., et al.: A study of bfloat16 for deep learning training. arXiv preprint arXiv:1905.12322 (2019)
14. Khunasaraphan, C., Vanapipat, K., Lursinsap, C.: Weight shifting techniques for self-recovery neural networks. IEEE Trans. Neural Netw. **5**(4), 651–658 (1994)
15. Klees, G., Ruef, A., Cooper, B., Wei, S., Hicks, M.: Evaluating fuzz testing. In: Proceedings of the of the 2018 ACM SIGSAC Conference on Computer and Communications Security, pp. 2123–2138 (2018)
16. Krizhevsky, A., Sutskever, I., Hinton, G.E.: ImageNet classification with deep convolutional neural networks. Adv. Neural Inf. Process. Syst. **25**, 1097–1105 (2012)
17. LeCun, Y., Bengio, Y., Hinton, G.: Deep learning. Nature **521**(7553), 436–444 (2015)
18. LeCun, Y., Kavukcuoglu, K., Farabet, C.: Convolutional networks and applications in vision. In: Proceedings of 2010 IEEE International SYM on Circuits and Systems, pp. 253–256. IEEE (2010)
19. Li, G., et al.: Understanding error propagation in deep learning neural network (dnn) accelerators and applications. In: Proceedings of the International Conf. for High Performance Computing, Networking, Storage and Analysis, pp. 1–12 (2017)
20. Li, G., Pattabiraman, K., DeBardeleben, N.: TensorFi: a configurable fault injector for tensorFlow applications. In: 2018 IEEE International SYM on Software Reliability Engineering Workshops (ISSREW), pp. 313–320. IEEE (2018)
21. Moradi, M., Van Acker, B., Vanherpen, K., Denil, J.: Model-implemented hybrid fault injection for simulink (tool demonstrations). In: Chamberlain, R., Taha, W., Törngren, M. (eds.) CyPhy/WESE -2018. LNCS, vol. 11615, pp. 71–90. Springer, Cham (2019). https://doi.org/10.1007/978-3-030-23703-5_4
22. Paszke, A., et al.: Pytorch: an imperative style, high-performance deep learning library. Adv. Neural Inf. Process. Syst. **32**, 8026–8037 (2019)
23. Ramchoun, H., Idrissi, M.J., Ghanou, Y., Ettaouil, M.: Multilayer perceptron: architecture optimization and training with mixed activation functions. In: Proceedings of the 2nd International Conference on Big Data, Cloud and Applications, pp. 1–6 (2017)
24. Shooman, M.L.: Reliability of Computer Systems and Networks: Fault Tolerance, Analysis, and Design. John Wiley & Sons, New York (2003)

25. Srivastava, N., Hinton, G., Krizhevsky, A., Sutskever, I., Salakhutdinov, R.: Dropout: a simple way to prevent neural networks from overfitting. J. Mach. Learn. Res. **15**(1), 1929–1958 (2014)
26. Torres-Huitzil, C., Girau, B.: Fault and error tolerance in neural networks: a review. IEEE Access **5**, 17322–17341 (2017)
27. Villa, O., Stephenson, M., Nellans, D., Keckler, S.W.: NVBit: a dynamic binary instrumentation framework for NVIDIA GPUs. In: Proceedings of the 52nd Annual IEEE/ACM International SYM on Microarchitecture, pp. 372–383 (2019)

FPGA Implementations of BLAKE3 Compression Function with Intra-Round Pipelining

Jarosław Sugier(✉)

Department of Computer Engineering, Wrocław University of Science and Technology, 11/17 Z. Janiszewskiego Street, 50-372 Wrocław, Poland
jaroslaw.sugier@pwr.edu.pl

Abstract. BLAKE3 is the latest evolution of the cryptographic hash function BLAKE – a cipher which, although was not selected as the new SHA-3 standard, after the NIST contest has become an accepted method of choice in contemporary software cryptographic systems. This work explores selected FPGA organizations of the BLAKE3 core– its compression function – realized in Xilinx Spartan-7 devices and evaluates benefits of the new modification from the point of view of efficient hardware implementation which has never been a strong point of this algorithm. In total 6 architectures are proposed and evaluated: the basic iterative one with one round instance, modifications with the round split into two and four pipeline stages and three possible versions with six stages. The results allow for evaluation of various speed vs. size trade-offs of the proposed organizations. Additionally, the paper includes a comparison with equivalent implementations of the predecessor BLAKE2 as well as SHA3-256 (KECCAK) standard.

Keywords: Cryptographic hash function · Hardware implementation · BLAKE3 · SHA3

1 Introduction

Hash functions suitable for efficient computer application yet with good cryptographic strength have been studied since the second half of the XX century. In 1995 US National Institute of Standards and Technology (NIST) normalized a group of such algorithms in a standard entitled Secure Hash Algorithm (SHA) and this norm has been universally accepted as the basic reference. Although in 2002 an extended SHA-2 standard was announced, constant increase in available computational power eventually made some of its algorithms vulnerable to brute force attacks. In November 2007 NIST announced an open contest for development of the next SHA standard emphasizing the need for an entirely new approach in hash design in order to compete with recent advances in cryptanalysis. BLAKE was one of the submitted contenders and was among the best 5 algorithms selected for the final consideration. The new SHA-3 standard was published in 2015 and by NIST decision it was based on a different submission – the KECCAK algorithm.

W. Zamojski et al. (Eds.): DepCoS-RELCOMEX 2022, LNNS 484, pp. 319–330, 2022.
https://doi.org/10.1007/978-3-031-06746-4_31

Although BLAKE ultimately lost this contest the cipher was repeatedly acclaimed in all stages of evaluation for excellent cryptographic strength and great performance in software implementations. For its advantages (and also due to slower operation of KECCAK in software) it quickly became popular in cryptographic community. Already in 2012 the authors modified the original version submitted to the contest in [1] and announced BLAKE2 [2] which quickly found application in various data processing or communication systems, from RAR file archive format through Linux kernel utilities to Argon2 key derivation function. BLAKE3 was proposed in 2020 in [5], with further modifications which simplified the processing and led to additional increase in effective hashing speed. The newest version instantly has become a candidate for application e.g. in blockchain transaction protocols [6] or automatic key generation [3].

Efficiency of hardware implementation was one of the main conditions which were required in the SHA-3 contest [4] and it always remains an important characteristic of any cryptographic method. The purpose of this works is to evaluate BLAKE3 with regard to this aspect, in particular: a) to compare it with its predecessor (ver. 2) and with its main competitor - the SHA3-256 algorithm (KECCAK), b) to test how its overall throughput can be improved by pipelining. Of many possible organizations of the cipher in hardware the study included the standard iterative one (with one round instance) and pipelined variants with the round divided into 2, 4 and 6 pipeline stages. Rest of the text is organized as follows. In the next chapter the BLAKE3 algorithm is described within the scope relevant to our analysis, then ch. 3 introduces the selected architectures and presents results of their implementations, ch. 4 is devoted to their evaluation, and the last chapter ends the presentation with conclusions.

2 The BLAKE3 Algorithm

2.1 Processing Scheme

The input message is first split into 1024-byte data chunks which are arranged as leaves in a binary tree. Each chunk can be compressed independently and the final value comes from combining intermediate hashes in the internal nodes up to the root of the tree. Such a computational flow calls for application of parallel processing of nodes – e.g. pipelining which is proposed in this paper – because in other schemes where data chunks need to be hashed sequentially parallelism can be applied only for multiple and independent input streams. Compression of each chunk is done sequentially in 64B message blocks; for a chunk consisting of blocks $M_1, M_2 \ldots M_n$ ($n = 16$ unless the chunk is the last one) the 256b hash value h is computed as:

$$h = \mathrm{IV}$$

$$\text{for} \quad i = 1 \cdots n \quad \text{do} : h = \ \mathrm{F}(h, M_i);$$

where IV is some constant initialization vector and F() is the compression function which implements the essential cryptographic processing of the algorithm. Implementations of F() in FPGA arrays are the subject of this work.

2.2 Compression Function

The function operates on internal state V consisting of sixteen words $v_0 \dots v_{15}$ (the term "word" will from now on denote a 32-bit vector). Initially state V (512b) is filled with the input hash h (256b), IV constants (128b) and other parameters (128b) with block counter and domain separation flags indicating, among others, different phases of tree processing. The state is then transformed in $N_R = 7$ rounds with each round recomputing every state word twice by so called quarter-round functions G_i. Operating on groups of four words, the first set of quarter-rounds is applied as:

$$G_0(v_0, v_4, v_8, v_{12});\ G_1(v_1, v_5, v_9, v_{13});\ G_2(v_2, v_6, v_{10}, v_{14});\ G_3(v_3, v_7, v_{11}, v_{15}); \quad (1)$$

and then the second:

$$G_4(v_0, v_5, v_{10}, v_{15});\ G_5(v_1, v_6, v_{11}, v_{12});\ G_6(v_2, v_7, v_8, v_{13});\ G_7(v_3, v_4, v_9, v_{14}). \quad (2)$$

The quarter-round functions make up the lowest level of processing; operation of $G_i(a, b, c, d)$ is defined as the following sequence:

$$\begin{aligned} a &= a + b + m_{2i}; \\ d &= (d \oplus a) >> 16; \\ c &= c + d; \\ b &= (b \oplus c) >> 12; \\ a &= a + b + m_{2i+1}; \\ d &= (d \oplus a) >> 8; \\ c &= c + d; \\ b &= (b \oplus c) >> 7; \end{aligned} \quad (3)$$

where words m_i come from the message block M ($M = m_0 \dots m_{15}$), and the symbols denote the following operations on words:

- $\oplus$ - bitwise XOR (sum mod 2),
- $+$ - addition mod 2^{32} (i.e. regular 32b addition with carry out ignored),
- $>>$ - vector rotation to the right by a given number of positions.

Processing inside every G_i instance is identical and its ordinal number i selects only a different pair of message words m_{2i} & m_{2i+1} loaded n (3).

After each round the message words are renumbered according to some random but constant permutation defined by authors in specification. It is worth noting that the words are actually additional parameters to the G_i functions (two per each function, with all 16 used in every round) and they must be submitted to them during calculations. This is in contrast with most of other algorithms (SHA3 included) where the hashed message is used only for initialization of the state and can be discarded afterwards. BLAKE, in turn, has never used the message in state initialization.

The chaining hash value $h' = h'_0 \dots h'_7$ – the result of F() – is extracted from the state after the last round just by xor'ing the v_i words:

$$h'_i = v_i \oplus v_{i+8}, \ i = 0 \cdots 7 \quad (4)$$

2.3 Comparison with the Second Version

The authors proposed the third version as the evolution of BLAKE2. Excluding new organization of the tree and introduction of additional modes (e.g. keyed hashing and extendable output), the list of changes which are related to the scope of this work contains the following modifications:

1 removal of 10 permutations σ_r in Eq. (3) which in BLAKE2 selected different words m_i for each round,
2 simplification of Eq. (4) which previously included also the input h,
3 reduction in number of rounds N_R from 10 to 7.

Ad. 1 In BLAKE2 the pair of message words passed to the quarter-rounds were different in each round; relevant Eq. (3) were in the form

$$a = a + b + m_{\sigma r(2i)} \; (\cdots) \; a = a + b + m_{\sigma r(2i+1)} \tag{5}$$

i.e. in the round no. r the r-th permutation σ_r selected the words. In hardware this led to very extensive (and expensive) multiplexing: along with a stored message each G_i module required two 32b-wide 16:1 multiplexers for propagating the words according to σ_r. Such a message schedule can be seen in the left part of Fig. 1 which will be used for illustration of the discussion in the following chapter. In our previous work in [8] this issue was analyzed and, as a solution, replacement of the multiplexers with dedicated memory modules was proposed. Resultant reductions in array utilization were even by half on some FPGA platforms, although on the cost of external memory blocks and additional time for their initialization. In BLAKE3 this expensive mechanism is replaced with re-numbering m_i after each round according to just one, constant permutation; as a result, each G_i function always uses message words $2i$ & $2i + 1$. This is *much* simpler in hardware implementation: the registers with m_i words are re-loaded with new values after each round and permuting them is obtained completely in routing, without absorbing any logic. Compared to BLAKE2's 16:1 multiplexers of total width 16 × 32b this is a substantial saving as also can be seen Fig. 1.

Ad. 2 In BLAKE2 the final compression result (Eq. (4)) was computed as

$$h'_i = h_i \oplus v_i \oplus v_{i-8}, \; i = 0 \cdots 7 \tag{6}$$

This meant that the input h had to be registered and used in h' calculation which required128b of memory (registers) plus an extra path of the same width to the module output. Both of these are removed in BLAKE3, further simplifying the hardware (compare Fig. 1). The new algorithm uses the h_i values like in Eq. (6) only in extended output mode but this mode of operation is out of scope of this work.

Ad. 3 Reduction in number of rounds, although publicized by the authors in the first place as the modification which leads directly to decrease of hashing time, has virtually no impact on hardware size nor clocking speed and brings only trivial change in the control counter. For the subject of this study the first two points are much more important.

3 Hardware Implementations

In this work we propose six different architectures for the BLAKE3 compression function which aim at reaching possibly maximum throughput with limited hardware occupancy. Taking a typical iterative organization with one complete round instantiated in hardware and the state propagated through this instance N_R times in a loop, the higher-throughput variants were derived through pipelining the round and processing multiple data chunks in parallel. In this paper the basic iterative (non-pipelined) organization will be denoted as "X1" and the pipelined variants as "Pk", with k indicating total number of pipeline stages inside the round (thus also the number of parallel data streams which are hashed concurrently). The set of pipelined variants consisted of P2, P4 and three versions of P6 architectures, denoted as P6a, P6b and P6c.

Additionally, for comparison another two algorithms were implemented on the same device platform with identical tools: the previous version BLAKE2 and the main rival and alternative – the permutation function KECCAK of the SHA3-256 standard, both in the basic iterative organization X1.

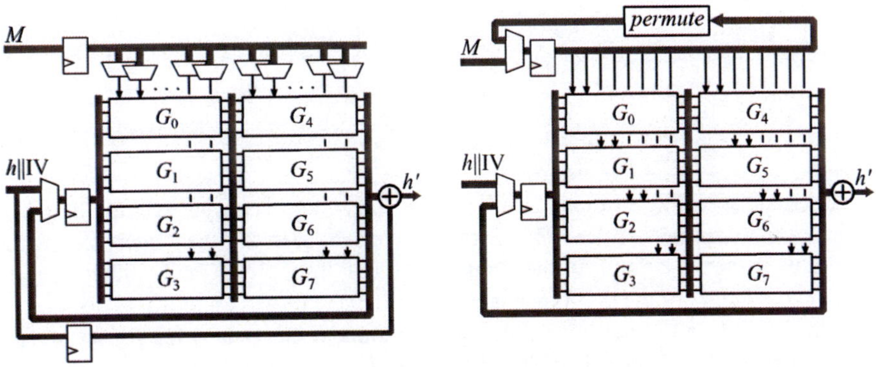

Fig. 1. Organization of the basic iterative architecture in BLAKE2 (left) and BLAKE3 (right).

3.1 The Architectures

Diagrams of the two latest BLAKE variants in their X1 architectures are presented in Fig. 1 and they illustrate modifications brought by BLAKE3 at the hardware level. Sharing the same scheme with 8 quarter-round instances (organized in a cascade 4 + 4) and iterative loopback in the state processing, the diagrams underline differences in message schedules. As it was already signaled in point 2.3, elimination of the 16 message multiplexers which selected message words for the G_i modules is a significant simplification, particularly that they were replaced with just the permutation block which is implemented completely in routing. Removal of the h register and its propagation path to the output xor block (Eq. (6) vs. (4)) should also be noted.

Starting from the basic X1 organization, the pipelined BLAKE3 variants were created through internal division of the round unit into multiple pipeline stages. It was obvious that a natural boundary for such division was the crossing in the cascade of the two successive G_i blocks – and this is the structure of the P2 variant. The remaining architectures P4 and P6 required further splitting of the G_i inner logic into, respectively, two or three pipeline stages.

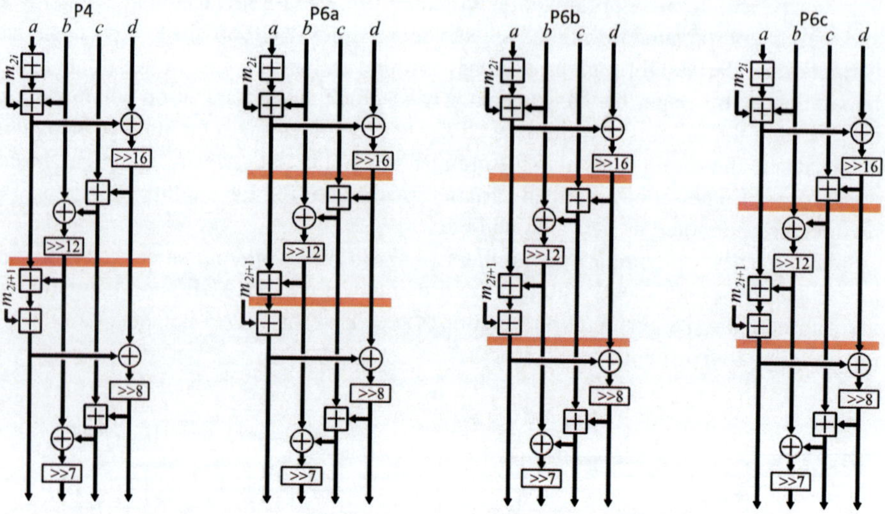

Fig. 2. Proposed stage boundaries inside G_i modules with internal pipelining.

Internal pipelining of the quarter-round block should consider its inner data paths which resulted from Eq. (3). Looking at Fig. 2 it can be seen that the longest propagation path goes from the a input to the b output passing through all the 6 adders. Because those 32b adders are the components with the longest propagation it was decided first to split the path in the balanced $3 + 3$ proportion and this is how the P4 organization was created. Keeping the rule of equal number of adders in each stage, variant P6a was proposed with two adders per stage ($2 + 2 + 2$). Because in this division the second boundary separated the two consecutive adders in the a lane and this could potentially impair optimizations, it was decided to include also another two P6 variants which divided the path in proportions $2 + 3 + 1$ (P6b) and $3 + 2 + 1$ (P6c). This kept both adders together in the second pipeline stage but led to imbalanced loads between the stages.

3.2 Implementation Results

To create practical hardware implementations all 8 ciphers units were equipped with basic input/output buffers which provided means for iterative loading of the input messages and unloading the results. The buffers, although consumed a significant number of register resources, had little impact on realization of actual cryptographic cores which were primarily (and heavily) oriented on combinatorial logic. The designs were then synthesized and implemented by Xilinx Vivado software for devices from the Spartan-7 family [10]: all BLAKE3 designs were fit in the smallest xc7s6-2 chip while the BLAKE2 and KECCAK X1 cases – being too large – had to be implemented in the second chip in the lineup, xc7s15-2. The smallest possible devices and the budget-oriented family of the 7 Series were chosen intentionally.

Table 1 presents results produced by Vivado tools after automatic implementation. Performance parameters are estimated from static timing analysis of fully routed designs with constrains selected so that the minimum clocking period was achieved. Size of the designs is given as the number of occupied slices (elementary groups of logic cells) and LUT function generators; in addition, the table lists utilization of some other specific elements: MUX dedicated multiplexers used in extra-wide multiplexing and CARRY primitives essential in multi-bit adders [9]. The longest propagation path which determines minimum clock period is described with two parameters: the number of logic levels in the path (additionally with the number of CARRY primitives among them) and percentage of its delay incurred by routing rather than by actual logic. The speed characteristics in the last three rows of the table include: highest frequency of operation estimated by the implementation tools, subsequent throughput of the round unit and the resulting throughput of the entire hashing operation.

Table 1. Implementations of all BLAKE3 variants analyzed in this paper.

Algorithm architecture	Blake3						Blake2	Keccak
	X1	P2	P4	P6a	P6b	P6c	X1	X1
Size: slices	797	886	1037	1247	1214	1172	1317	1362
Size: LUTs	2668	2676	2774	3471	3446	3488	4593	5298
Primitives: MUX	0	0	0	0	0	0	1536	0
Primitives: CARRY	256	256	256	320	256	256	256	0
Levels of logic (included Carry prim.)	44 (34)	22 (16)	15 (11)	13 (10)	16 (10)	16 (13)	47 (32)	3 (0)
Routing delay	44.6%	44.2%	38.2%	33.9%	36.6%	36.8%	48.9%	80.3%
F_{max} [MHz]	57.1	107.4	174.6	217.6	216.1	199.4	48.6	258.5
Round speed [Gbps]	29.2	55.0	89.4	111.4	110.7	102.1	24.9	281.3
Hashing speed [Gbps]	4.2	7.9	12.8	15.9	15.8	14.6	2.5	11.7

Figure 3 complements these parameters by presenting distribution of LUT2 ÷ LUT6 primitives in all the implementations. In FPGA devices from Xilinx every logic cell includes a function generator in the form of a Look Up Table (LUT, [9]) which in the 7th generation is able to compute any Boolean function of 6 variables. If the tools during the implementation process are capable to decompose the combinatorial parts of design into 6-input functions then the potential of the generators is used up to its maximum. In real-world cases this is never possible in 100% of cases and smaller functions of 5 down to even 1 argument must also be programmed – which leaves parts of the LUTs unused. Therefore, looking at the distribution of LUT2 ÷ LUT6 primitives to which the design was mapped one can judge if its logic can utilize this specific capacity of the array to its fullness.

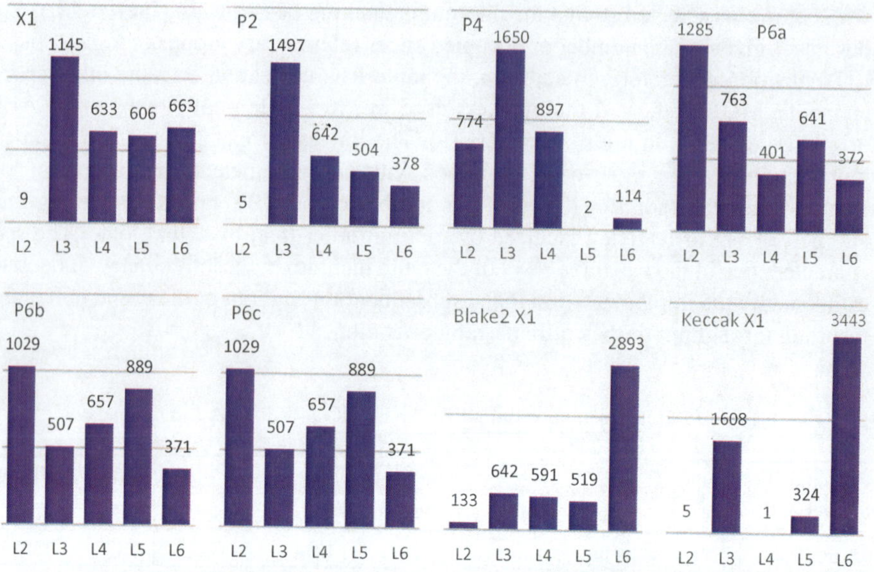

Fig. 3. Distributions of LUT2 ÷ LUT6 primitives in all implementations.

With regard to this particular aspect, among the designs tested in this work the clear winner would be the SHA3 algorithm with 64% of its LUTs using all 6 inputs, vs. 3% ÷ 22% in the BLAKE3 versions. Involved and elaborated transformations of this algorithm operate on individual state bits (and not on words like in the BLAKE family) and, as these results indicate, can be efficiently combined to fit the 6-input generators. The apparently second place of BLAKE2 should be treated with caution because majority of its LUT6 primitives are consumed by multiplexing in the message schedule; when this multiplexing is removed in BLAKE3 the results e.g. in the X1 case are reversed, with most of the stress found on only LUT3 elements. Regarding the BLAKE3 algorithm, just in the pipelined architectures, mainly in the three P6 variants, the use of LUT5 begins to be remarkable but this positive effect is counter-weighted by even stronger increase in LUT2. In particular, peculiar situation is observed for the P4 architecture which makes very little use of 6- and 5-input generators. It could be implemented with not much worse effect using only 4-input LUTs which have been available since the first FPGA generations 30 years ago.

4 Evaluation

4.1 Three Ciphers in the Basic Iterative Implementation

For better visualization the size and speed results from Table 1 are graphically presented in Fig. 4 where BLAKE3, its predecessor and KECCAK (SHA3), all in X1 organizations, are present as the top and two bottom bars.

Comparing BLAKE3 and BLAKE2 (in their X1 versions) the results indicate significant improvements: the modifications reduced size to 61% in slice and to 58% in LUT utilization. Such an impressive progress results primarily from changes in the message schedule as it was discussed in points 2.3 and 3.1. After reorganization the implementation finally needs no MUX primitives which in the first two versions were involved in multiplexing of the message words; Table 1 lists 1536 their instances in B2 X1 vs. zero in all B3 versions, whereas the differences in characteristics like levels of logic, carry elements and routing delay are practically insignificant. What is also important, simplifications in hardware led to 17% increase in frequency of operation which – combined with reduction in number of rounds – gave a remarkable growth in BLAKE3 final hashing throughput by 68%.

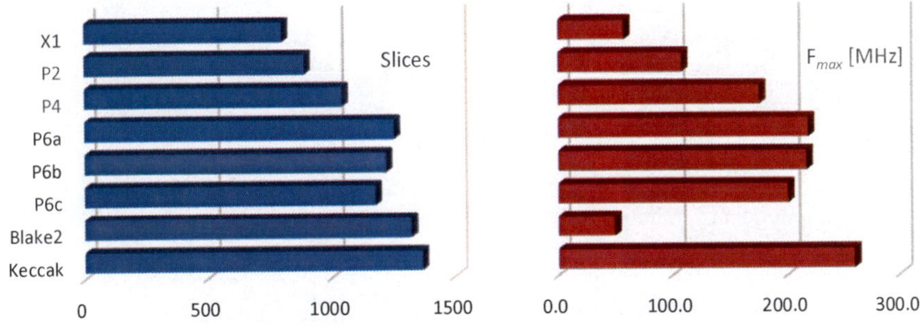

Fig. 4. Size and speed of the evaluated designs.

This improvement changes to some extent BLAKE3 relation with KECCAK. While BLAKE2 was approximately of the same size as the SHA3 permutation function, the new version is by 41% smaller and its inferiority in the processing speed – it still turned out to be nearly 3 times slower in the final hashing throughput – at least is somewhat compensated with improved size/speed productivity. But the problems with BLAKE3 inferior speed still prevail in this comparison. KECCAK transformations were outstandingly packed into the 6-input LUT generators with only 3 levels of logic in the longest path, albeit the extended size of the module boosted the delay generated by routing up to 80% – the value about twice as large as in the BLAKE implementations (still also an effect of a low base of this percentage). Nevertheless, SHA3 round instance can operate with much higher frequencies (259 MHz vs. 57 MHz). The difference is even greater in raw round throughput as it is amplified by KECCAK's exceptionally wide state (1600b vs 512b) but significantly higher number of rounds (24 vs. 7) makes its final hashing throughput "only" 2.8 times better.

4.2 The Pipelined Organizations

When converting a X1 design into a pipelined version with k stages, it is expected that dividing propagation paths into k segments will reduce the minimum clock period also k times thus the maximum frequency of operation will be multiplied by the same factor (because of components with register propagation and setup times which are not divided the actual effect will be slightly worse). As for the size after the pipelining, it is expected that: a) the combinatorial functions will effectively remain the same hence LUT utilization should not increase, and b) the extra pipeline registers – that must be added to the design – should be absorbed by the logic cells which so far used only LUT generators (a common case especially in very logic-intensive cryptographic designs) thus occupancy array expressed in slices should also remain on the same level. Actual size can differ from this ideal model because of different optimization and routing conditions after pipelining which may change the implementation due to design specifics and particulars of array resources.

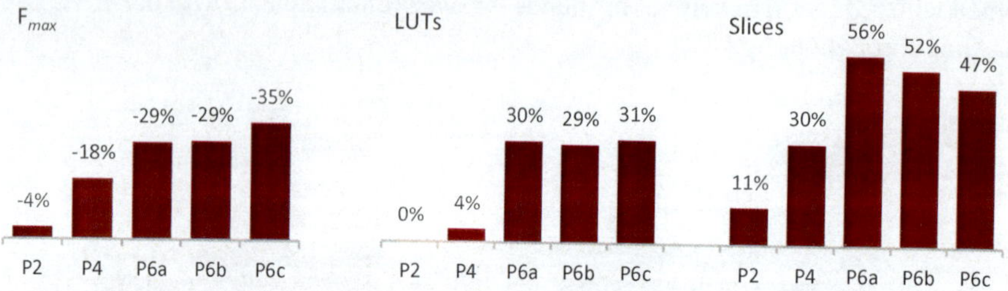

Fig. 5. Differences between actual and expected parameters of the pipelined BLAKE3 variants in: maximum frequency, number of LUTs and number of slices.

Evaluation of actual results versus the expected values obtained as above are shown in Fig. 5. Generally, among the three cases of $k = 2$, 4 and 6 the P2 implementation did manage to achieve the speed close to the anticipated (lower by just 4%) and size in slices larger by 11%, but with deeper pipelining the parameters start to deviate. It can be seen that in the P4 case LUT utilization is actually very good although apparently packing of the pipeline registers could not be as efficient because the growth in slice size begins to be remarkable higher (30%) which is supplemented by 18% drop in speed. Only in the P6 designs, on the other hand, the increase in LUT number is evidently higher which may indicate different effect of synthesis of the paths divided to this level; as a result, both the speed and the slice size are significantly worse than expected. In [7] an analogous analysis was made for Keccak tested in P2 and P3 organizations and similar results were found: in the fastest designs the P2 organization was 19% larger and 5% slower than expected while for P3 the change was by + 26% and –21%, respectively – although in that algorithm the P3 implementation reached the bottom limit of single LUT elements per critical path and there was no room for further subdivisions with higher k factors.

Yet overall, in addition to the above critical comparison to ideal conditions, it should be noted that the pipelining mechanism did work and accomplished its main mission i.e. it increased the final hashing throughput of the X1 implementation, as the results from

Table 1 show: by 88% for P2, ×3 for P4 and ×3.8 for P6a. With 4 stages the algorithm is able to exceed the speed of SHA3 by 9%, and with 6 stages the advantage grows to 36%. The size growth in slices should be judged in the light of these improvements.

As for the three tested options of the P6 organization, the results do not indicate a clear winner. With almost identical LUT utilization, the P6a case turned out to be the fastest one (thus the division of the two adders between two pipeline stages did not turn out to be a problem) and the P6c implementation, while being somewhat slower, showed the most potential for size reductions. The latter still may be the best option in applications which impose a limit on clock frequency lower than cipher's F_{max}. It is symptomatic that two apparently such different divisions of the G_i function into three pipeline stages – with distributions of the adders $2+2+2$ (P6a) and $2+3+1$ (P6b) – led to so similar results.

5 Conclusions

The paper evaluated potential of the latest BLAKE3 hash function in hardware implementations on the FPGA platform. It was shown that the modifications introduced to the second version of the algorithm, although presented by the authors first of all as speed enhancements in software, did changed – and improved – conditions for hardware implementations. Apart from obvious acceleration coming from the reduced number of rounds (which seems to be the crucial modification for software) equally important in our context are the changes in the message schedule and in output XOR computation. As the results showed for the basic iterative organization with one round instance, these modifications reduced size of the implementation by 40%, also leading to 17% higher frequency of operation. The latter improvement combined with number of rounds reduced from 10 to 7 gave a remarkable growth in total hashing throughput by 68%.

The paper also evaluated results of pipelining the iterative architecture with the round divided into 2, 4 or 6 pipeline stages. The mechanism proved its efficiency with increase in throughput by 88% for P2, ×3 for P4 and ×3.8 for P6 organizations. The results for higher number of pipeline stages are not as good as it might be expected but analogous "saturation" of speed for more densely divided pipelines are common also for other algorithms.

Nevertheless, speed in hardware of the newest BLAKE version still remains significantly inferior when compared to the principal alternative – the SHA3 standard. Depending heavily on 32b additions in the core quarter-round functions, the implementations of this cipher must be realized in a different way than they are in KECCAK. In particular, because in contemporary FPGA arrays adders cannot be optimized with enlarged LUT resources, mapping the two algorithms on the programmable array is done with different efficiency and, as a result, the reported advantages of BLAKE3 over SHA3 in software speed – both with CPU and SIMD processors – cannot be repeated in FPGA arrays.

References

1. Aumasson, J.P., Henzen, L., Meier, W., Phan, R.C.-W.: SHA-3 proposal BLAKE, version 1.3 (2010). https://www.aumasson.jp/blake/blake.pdf. Accessed Mar 2022

2. Aumasson, J.-P., Neves, S., Wilcox-O'Hearn, Z., Winnerlein, C.: BLAKE2: simpler, smaller, fast as MD5. In: Jacobson, M., Locasto, M., Mohassel, P., Safavi-Naini, R. (eds.) ACNS 2013. LNCS, vol. 7954, pp. 119–135. Springer, Heidelberg (2013). https://doi.org/10.1007/978-3-642-38980-1_8
3. Ciocan, I.T., Kelesidis, E.A., Maimuţ, D., Morogan, L.: A Modified Argon2i using a tweaked variant of Blake3. In: 2021 26th IEEE Asia-Pacific Conference on Communications (APCC), 11–13 October 2021, Kuala Lumpur, Malaysia, pp. 271–274. IEEE (2021). https://doi.org/10.1109/APCC49754.2021.9609933
4. Gaj, K., Kaps, J.P., Amirineni, V., Rogawski, M., Homsirikamol, E., Brewster, B.Y.: ATHENa – automated tool for hardware EvaluatioN: toward fair and comprehensive benchmarking of cryptographic hardware using FPGAs. In: 20th International Conference on Field Programmable Logic and Applications, Milano, Italy, pp. 414–421 (2010). https://doi.org/10.1109/FPL.2010.86
5. O'Connor, J., Aumasson, J.P., Neves, S., Wilcox-O'Hearn, Z.: BLAKE3: one function, fast everywhere. Real World Crypto 2020 (lightning talk) (2020). https://github.com/BLAKE3-team/BLAKE3-specs/blob/master/blake3.pdf. Accessed Mar 2022
6. Sinha, S., Anand, S., Krishna Prakasha, K.: Improving smart contract transaction performance in hyperledger fabric. In: 2021 Emerging Trends in Industry 4.0 (ETI 4.0), 19–21 May 2021, pp. 1–6. IEEE (2021). https://doi.org/10.1109/ETI4.051663.2021.9619202S
7. Sugier, J.: Intra-round pipelining of KECCAK permutation function in FPGA implementations. In: Zamojski, W., Mazurkiewicz, J., Sugier, J., Walkowiak, T., Kacprzyk, J. (eds.) DepCoS-RELCOMEX 2020. AISC, vol. 1173, pp. 606–615. Springer, Cham (2020). https://doi.org/10.1007/978-3-030-48256-5_59
8. Sugier, J.: Simplifying FPGA implementations of BLAKE hash algorithm with block memory resources. Procedia Eng. **178**, 33–41 (2017). https://doi.org/10.1016/j.proeng.2017.01.057
9. Xilinx, Inc.: 7 Series FPGAs Configurable Logic Block. UG474.PDF. www.xilinx.com. Accessed Mar 2022
10. Xilinx, Inc.: 7 Series FPGAs Data Sheet: Overview. DS180.PDF. www.xilinx.com. Accessed Mar 2022

An Impact of Data Augmentation Techniques on the Robustness of CNNs

Kamil Szyc(✉)

Wrocław University of Science and Technology, Wrocław, Poland
kamil.szyc@pwr.edu.pl

Abstract. Data augmentation is a well-known and willingly used method of Convolutional Neural Network regularization. However, there are no clear recommendations on which augmentation strategy should be used. We performed the comprehensive analysis based on the CIFARs datasets and two models: the ResNet and the MobileNet. We trained models with a total of 21 different augmentation approaches. The analysis was based on non-modified and strongly distorted images in 9 different ways. It allowed us to focus on checking the robustness of the networks based on various scenarios. We've checked the influence of data augmentation on robustness. Finally, we suggested more general guidelines for improving accuracy and overall robustness increase.

Keywords: Data augmentation techniques · Robustness · Convolutional Neural Network · Computer vision

1 Introduction

Deep Neural Networks (DNNs) [1] need thousands of examples of data to learn how to obtain valuable features. Based on these features, they determine the decision boundaries, allowing the separation of known classes from each other. Due to the millions of parameters of the DNN, which need to be set, the vast number of training data is crucial for modern models. Since the cost of preparing data (especially collecting and labeling) is high and time-consuming, ML developers often work with limited data. Due to the above, a data augmentation technique is used.

Augmentation [2] is a well-known technique of regularization for improving deep learning models. Especially for Convolutional Neural Networks (CNNs), which are the focus of this work. The key idea is based on generating new (from the network's perspective) examples by modifying existing ones. The classical approaches are, e.g., horizontal flip, translation, or color shifting of images. These new examples allow the network to fit the parameters better. The goal is to minimize the distance between the training and testing subsets. The train examples should be as comprehensive as possible, in which the augmentation technique is helpful.

W. Zamojski et al. (Eds.): DepCoS-RELCOMEX 2022, LNNS 484, pp. 331–339, 2022.
https://doi.org/10.1007/978-3-031-06746-4_32

Augmentation is a widespread topic in deep learning literature. There are many different techniques and specific applications where the characteristic strategy should be applied. For instance, in [3] authors focus on the influence of color augmentation for skin image analysis, in [4] on the influence of the images color spaces, in [5] on using GAN in augmentation process, in [6,7] of mixing different classes images, or in [8] on automatically search for improved data augmentation policies.

There are no clear guidelines on how to establish a suitable augmentation policy. Too weak augmentation may not change the distribution of the data. On the other hand, too strong may change the input too much, preventing the network from learning the patterns it was meant to learn. Some augmentation techniques may not influence the learning process in a significant way and be a waste of time and resources. This paper focused on the difficulty of choosing the proper strategy. We compared many of them. We based on the CIFAR-10 and the CIFAR-100 datasets[1] and the ResNet-101 [9] and the MobileNetV2 [10] CNN models. Although there are techniques that automatically find proper augmentation policy, e.g., [8], it still focuses only on final accuracy without assuming robustness.

The robustness is the characteristic that describes how effective the network is while being tested on the new independent dataset. This characteristic is broadly defined. It should enclose the factors that may influence of model's final prediction – for example: by checking the influence of existing biases or spurious correlations, by changing the context (i.e., background) or image properties (i.e., image brightness contrast), or by using adversarial attacks or applying distortions to the image.

The classical accuracy of the test data has become a less critical metric nowadays. The researchers focus more and more on safety, trustworthiness, and explainability. For example, the models can extract spurious correlations between image classes and some characteristics of images used for training [11]. It allows generating natural adversarial attacks [12], which can easily fool the network. The network's robustness becomes an important part of modern deep learning problems. In this paper, we have checked the influence of the augmentation of the robustness of the models. Many papers [13,14] deal with this problem too. However, they focus on different distortions methods based on the ℓ_2- and ℓ_∞-norm-bounded perturbations [15].

2 Methodology

Our main goal was to check the influence of data augmentation on the robustness of classification. For this purpose, we trained CNN models, the ResNet-101 on the CIFAR-10 and the MobileNetV2 on the CIFAR-100 datasets. We set the same hyperparameters for all models except the augmentation strategy. We normalized all images to a uniform mean of 0 and a standard deviation of 1. We tried the following augmentation techniques (mostly we used Albumentations [16] library): None, Affine, CLAHE, CoarseDropout, ColorJitter, CropAndPad,

[1] https://www.cs.toronto.edu/~kriz/cifar.html.

CutMix [7], FancyPCA, GaussNoise, GridDistortion, HorizontalFlip, HueSaturationValue, MedianBlur, MixUp [6], RandomBrightnessContrast, RandomGridShuffle, RandomShadow, Solarize, VerticalFlip, Complex 1 (mix of HorizontalFlip, RandomBrightnessContrastm, ShiftScaleRotate, ImageCompression, HueSaturationValue), and Complex 2 (mix of HorizontalFlip, ShiftScaleRotate, Blur, OpticalDistortion, GridDistortion, HueSaturationValue). See the details of each method in the library documentation[2].

The following is the details for each augmentation method. For CropAndPad, we used $px = (-3, 3)$, for CoarseDropout $max\_holes = 5$, $max\_height = 3$, and $max\_width = 3$, for GaussNoise $var\_limit = (15.0, 80.0)$, for MedianBlur $blur\_limit = 3$, for Complex 1 ShiftScaleRotate with $rotate\_limit = 15$, $scale\_limit = 0.10$, ImageCompression with value 80, and for Complex 2 ShiftScaleRotate with $shift\_limit = 0.05$, $scale\_limit = 0.05$, and $rotate\_limit = 5$, Blur with $blur\_limit = 3$. For all others approaches, we used Albumentations's default parameters. For MixUp, we used the publicly available code from the authors of [6] with default configuration, and for CutMix this code[3] with followig parameters $beta = 1.0$ and $num\_mix = 6$.

In Fig. 1 we showed examples, based on the CIFAR-10, of the above augmentation methods. Note that the probability of application here is artificially increased to 1.0(by default, for most methods, the augmentation is applied with 50% probability).

We can group the methods of augmentation in many ways. For instance, by spatial-level transforms (i.e., Affine, CropAndPad, VerticalFlip), by color manipulations (i.e., ColorJitter, GaussNoise, RandomBrightnessContrast, HueSaturationValue), by adding minor distortion (i.e., CoarseDropout, GridDistortion, GaussNoise, GridDistortion, RandomGridShuffle), by mixing different images (CutMix and MixUp), by "weather" augmentation (i.e., RandomShadow, Solarize), by complexity (Complex 1, Complex 2).

Overall, we had 21 CNN models for the ResNet-101 fitted on the CIFAR-10 and 21 models for the MobileNetV2 fitted on the CIFAR-100.

We decided to test the robustness of each augmentation method on complex images. By the complex ones, we mean the strong augmentation (to make them distorted). The goal of that augmentation was that humans could still recognize objects on the images, despite adding severe distortion. We tested the following methods: GaussNoise, Rotate, Blur, Downscale, Cutout, Hue, RandomBrightness, OverlayImage, ToGray. See the examples of the above in Fig. 2. The experiments consisted of a robustness test by examining the final accuracy for validation data entirely modified with the above robust augmentations separately.

We again used the Albumentations library with the following parameters. For GaussNoise $var\_limit = (0.035, 0.035)$, for Rotate one of the $limit = [90, 90]$ or $limit = [270, 270]$, for Blur $blur\_limit = [4, 4]$, for Downscale $scale\_min = 0.4$ and $scale\_max = 0.4$, for Cutout we used twice Cutout methods with diffrents configurations $num\_holes = 5$, $max\_h\_size = 10$, $max\_w\_size = 3$

[2] https://albumentations.ai/docs/api_reference/full_reference/.

[3] https://github.com/ildoonet/cutmix.

and $num\_holes = 5$, $max\_h\_size = 3$, $max\_w\_size = 10$ for Hue one of the $hue\_shift\_limit = (20, 20)$ or $hue\_shift\_limit = (-20, -20)$ both with $at\_shift\_limit = 0$ and $val\_shift\_limit = 0$, for RandomBrightness one of the $limit = (0.5, 0.5)$ or $limit = (-0.5, -0.5)$. For OverlayImage, one of the five emoticons 12px by 12px image was overlaid (about 14% of the image area). For all complex images, the probability of applied augumentation was set to 1.0. Others parameters were set as default.

3 Results

Table 1 in column *None* presents the accuracy on the unchanged test subset. All augmentation techniques significantly increase the final results from the 75% up to 85% (least MedianBlur) or 94% (most Affine and CropAndPad) for the ResNet-101 fitted on the CIFAR-10. Both increased final results for 19p.p. For the CIFAR-100 based on the MobileNetV2 model, the baseline accuracy is 64%. Up to 9methods do not improve results with four (ColorJitter, GaussNoise, HSV, and MedianBlur), which are even worse than without augmentation. The best method achieved up to 9p.p. better results (Affine).

We can see that CropAndPad with Affine returns the best results. The above suggests that essential augmentation is changing the object's position. HorizontalFlip (92% for the CIFAR-10 and 69% for the CIFAR-100) achieved a similarly good result, which confirms above. The methods that change a single pixel's colors do not achieve the best results. For instance, the methods like ColorJitter, FancyPCA, HSV, or RandomBrightnessConstrast have the lowest impact on the final results compared to other augmentations. It suggests that objects are varied naturally enough, and only minor changes in pixels may be applied. The group mixing the two images achieved not bad results, especially cutting (not mixing). The distortion group is in the middle compared to others, except that which distorts single pixels (like GuassianNoise). The complex augmentation achieved only good results for the CIFAR-10.

In the rest of the columns of Table 1, we presented the results of robustness analysis based on the final accuracy on distorted images.

The Hue is the distortion which it is the least effective approach – the drop of the average values was 3% for the CIFAR-10/ResNet-101 and 10% for the CIFAR-100/MobileNetV2. The networks seem to be relatively robust on it. We hypnotize the same as above: the objects in both datasets are varied on pixel-level naturally enough. For example, there are red and blue cars in the dataset. Interestingly, even augmentation by using HSV does not achieve the best results. A similar case is for another pixel-level change is ToGray. Both are extremely easy for humans.

The Downscale is a complex distortion for the humans and CNN models. The result for the None for the CIFAR-10/ResNet-101 is intriguing. It returns a better outcome than any argumentation.

The well-known fact that the networks are not robust for rotation [17] is also observable in our table. All augmentation methods nearly equally failed.

Fig. 1. The examples of augmentations that we used. By default, we applied the augmentation with 50% probability.

The standard deviation for both datasets is only 3%. VerticalFilp helps, but not significantly.

The most variation distractions are Cutout (std equals 18% the CIFAR-10/ResNet-101 and 15% the CIFAR-100/MobileNetV2) and Blur (std equals

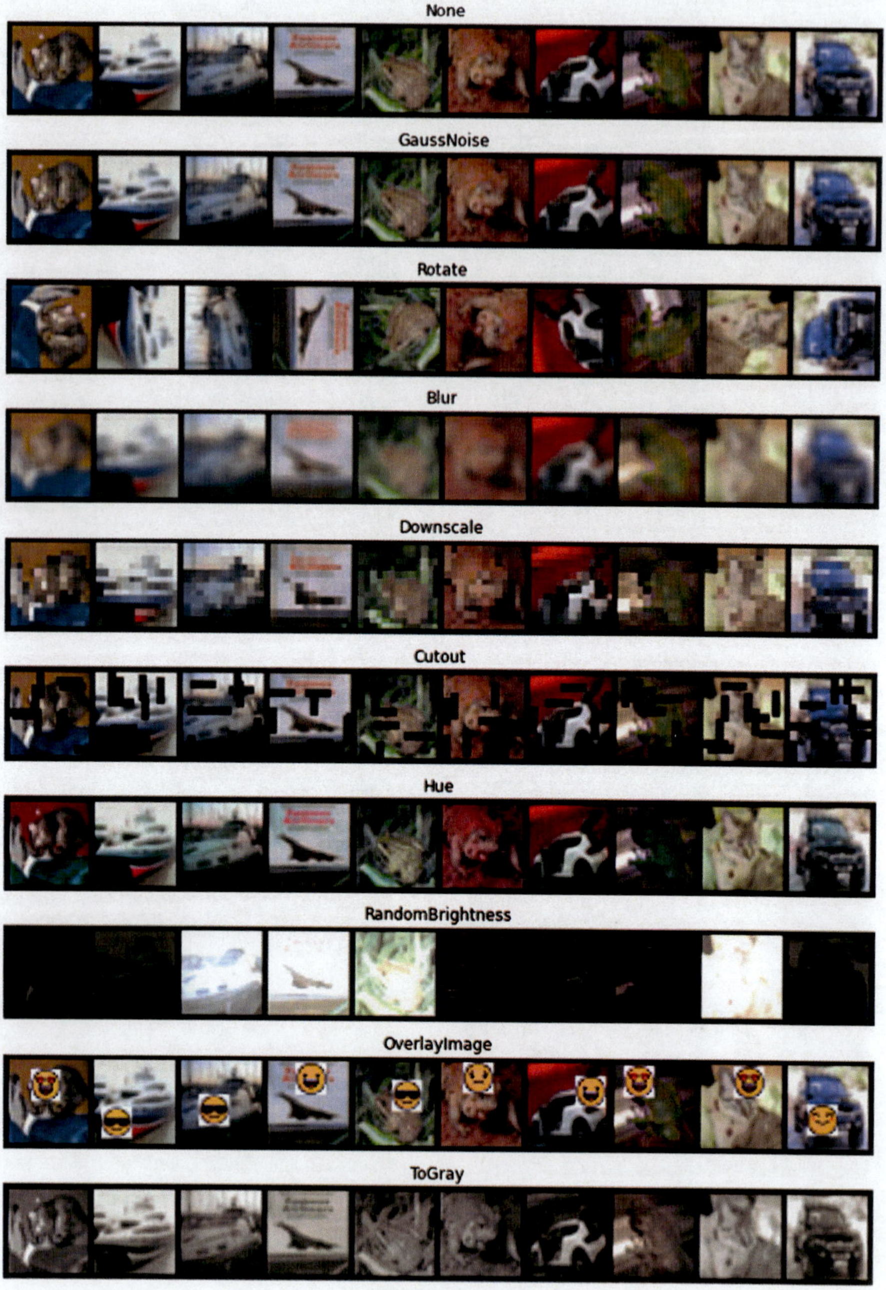

Fig. 2. The example of distorted images used for robustness evaluation. We set parameters so that humans could recognize the correct class of the image in most cases.

Table 1. The robustness analysis based on non-modified and strongly distorted images for the CIFAR-10 and the CIFAR-100 datasets. We presented an evaluation of two different CNN models.

Aug model	None	Blur	Cutout	Downscale	GaussNoise	Hue	OverlayImage	RandomBrightness	Rotate	ToGray	Avg
The ResNet-101 on the CIFAR-10											
None	75%	49%	32%	**61%**	34%	70%	55%	29%	28%	66%	50%
Affine	**94%**	59%	81%	32%	46%	90%	80%	63%	**41%**	**88%**	67%
CLAHE	88%	56%	34%	25%	52%	85%	66%	66%	35%	82%	59%
CoarseDrop	89%	36%	**82%**	24%	49%	85%	71%	58%	34%	85%	61%
ColorJitter	87%	37%	32%	22%	51%	87%	65%	63%	31%	85%	56%
Crop&Pad	**94%**	71%	73%	33%	52%	90%	74%	64%	**41%**	87%	**68%**
CutMix	92%	27%	81%	24%	29%	87%	**87%**	62%	39%	84%	61%
FancyPCA	88%	29%	33%	20%	47%	87%	68%	63%	32%	85%	55%
GaussNoise	88%	43%	33%	28%	51%	85%	67%	56%	32%	83%	57%
GridDist	93%	69%	60%	53%	48%	88%	73%	63%	38%	85%	67%
HorizFlip	92%	32%	39%	29%	52%	89%	73%	64%	37%	87%	59%
HSV	87%	29%	32%	21%	44%	87%	64%	61%	32%	85%	54%
MedianBlur	85%	76%	37%	46%	52%	80%	65%	55%	31%	78%	61%
MixUp	87%	49%	37%	37%	43%	84%	69%	61%	31%	77%	58%
RBrighCon	88%	39%	29%	20%	61%	85%	67%	74%	34%	84%	58%
RGridShuf	91%	24%	69%	21%	37%	86%	80%	61%	40%	85%	59%
RShadow	88%	37%	59%	25%	40%	84%	71%	53%	34%	83%	57%
Solarize	88%	37%	48%	21%	41%	84%	71%	53%	34%	83%	56%
VerticalFlip	88%	27%	32%	28%	39%	83%	66%	56%	40%	81%	54%
Complex 1	93%	67%	42%	38%	**65%**	**92%**	70%	**80%**	34%	**88%**	67%
Complex 2	93%	**89%**	48%	58%	52%	**92%**	70%	63%	34%	85%	**68%**
Avg	89%	47%	48%	32%	47%	86%	70%	60%	35%	83%	60%
The MobileNetV2 on the CIFAR-100											
None	64%	23%	11%	8%	19%	53%	25%	32%	24%	44%	30%
Affine	**71%**	33%	50%	13%	17%	57%	34%	28%	**30%**	45%	**38%**
CLAHE	64%	28%	15%	13%	23%	53%	25%	34%	23%	42%	32%
CoarseDrop	66%	23%	**55%**	11%	20%	54%	27%	32%	24%	44%	36%
ColorJitter	63%	18%	9%	9%	25%	62%	24%	36%	21%	48%	32%
Crop&Pad	70%	43%	43%	13%	18%	57%	31%	32%	29%	45%	**38%**
CutMix	70%	20%	53%	5%	14%	55%	**58%**	31%	**30%**	46%	**38%**
FancyPCA	64%	19%	11%	11%	22%	62%	24%	35%	21%	**50%**	32%
GaussNoise	62%	28%	12%	17%	28%	52%	25%	29%	22%	42%	32%
GridDist	68%	41%	32%	33%	17%	53%	26%	30%	29%	41%	37%
HorizFlip	69%	23%	14%	11%	20%	57%	30%	34%	27%	47%	33%
HSV	63%	18%	10%	10%	22%	62%	26%	33%	21%	49%	31%
MedianBlur	63%	36%	16%	23%	18%	51%	27%	30%	23%	39%	33%
MixUp	68%	29%	16%	19%	21%	57%	33%	35%	25%	46%	35%
RBrighCon	64%	24%	12%	10%	27%	53%	24%	**46%**	23%	43%	33%
RGridShuf	67%	16%	28%	7%	14%	54%	39%	28%	28%	42%	32%
RShadow	65%	23%	35%	8%	19%	53%	38%	29%	25%	43%	34%
Solarize	64%	24%	24%	9%	18%	53%	37%	27%	24%	44%	32%
VerticalFlip	64%	20%	12%	9%	16%	51%	22%	29%	29%	43%	29%
Complex 1	67%	40%	13%	17%	**32%**	**66%**	28%	44%	24%	49%	**38%**
Complex 2	66%	**60%**	20%	**36%**	19%	63%	24%	30%	23%	42%	**38%**
Avg	66%	28%	23%	14%	20%	56%	30%	33%	25%	45%	34%

18% and 10%). For Cutout, the good augmentations are those which are similar to this distortion, so CoarseDropout and CutMix. The Affine is a good alternative method. Interestingly, a similar method - RandomGridShuffle, achieved good but not so good results for Cutout. For Blur, the good methods are also these by using Blur in the training process, so Complex 2 and MedianBlur. The CropAndPad is an alternative method.

There are other noteworthy results. The RandomBrightnessContrastm works well for strong RandomBrightness distortion and only for it. CutMix generally works well for all distortions, especially for OverlayImage. Complex augmentation is a good direction for increasing robustness. However, they are not as significant as we expected. A simple and popular approach - CropAndPad - achieved outstanding results.

4 Summary

We proposed the method of generating distorted images in 9 different ways. Our analysis was performed on the two datasets and two CNN models. We showed the influence of 21 different data augmentation policies for classic accuracy and robustness.

Our analysis leads to some more general conclusions. The most effective augmentation strategy is a modification of the object's position. Modification of pixels' color values do not lead to significant influence on the effectiveness of the network. The idea of mixing images improves robustness (especially by cutting). The variety of augmentations helps achieve more stable results.

This paper focused on the impact of augmentation on the network's robustness. In general, CNNs models failed on distorted images. We showed that even simple to recognize (for humans) distortions, like Rotate or to Gray, lead to a significant deterioration of the results. The challenge for modern CNNs therefore lies in increasing safety, trustworthiness, and explainability.

References

1. Goodfellow, I., Bengio, Y., Courville, A.: Deep Learning. MIT Press, Cambridge (2016)
2. Shorten, C., Khoshgoftaar, T.M.: A survey on image data augmentation for deep learning. J. Big Data **6**, 1–48 (2019)
3. Galdran, A., et al.: Data-driven color augmentation techniques for deep skin image analysis. arXiv preprint:1703.03702 (2017)
4. Szyc, K.: An impact of different images color spaces on the efficiency of convolutional neural networks. In: Zamojski, W., Mazurkiewicz, J., Sugier, J., Walkowiak, T., Kacprzyk, J. (eds.) DepCoS-RELCOMEX 2019. AISC, vol. 987, pp. 506–514. Springer, Cham (2020). https://doi.org/10.1007/978-3-030-19501-4_50
5. Frid-Adar, M., Diamant, I., Klang, E., Amitai, M., Goldberger, J., Greenspan, H.: Gan-based synthetic medical image augmentation for increased cnn performance in liver lesion classification. Neurocomputing **321**, 321–331 (2018)
6. Zhang, H., Cisse, M., Dauphin, Y.N., Lopez-Paz, D.: mixup: Beyond empirical risk minimization. arXiv preprint:1710.09412 (2017)
7. Yun, S., Han, D., Oh, S.J., Chun, S., Choe, J., Yoo, Y.: CutMix: regularization strategy to train strong classifiers with localizable features. In: Proceedings of the IEEE/CVF International Conference on Computer Vision, pp. 6023–6032 (2019)

8. Cubuk, E.D., Zoph, B., Mane, D., Vasudevan, V., Le, Q.V.: AutoAugment: learning augmentation strategies from data. In: Proceedings of the IEEE/CVF Conference on Computer Vision and Pattern Recognition, pp. 113–123 (2019)
9. He, K., Zhang, X., Ren, S., Sun, J.: Deep residual learning for image recognition. In: Proceedings of the IEEE Conference on Computer Vision and Pattern Recognition, pp. 770–778 (2016)
10. Sandler, M., Howard, A., Zhu, M., Zhmoginov, A., Chen, L.-C.: Mobilenetv2: Inverted residuals and linear bottlenecks. In: Proceedings of the IEEE Conference on Computer Vision and Pattern Recognition, pp. 4510–4520 (2018)
11. Szyc, K., Walkowiak, T., Maciejewski, H.: Checking robustness of representations learned by deep neural networks. In: Dong, Y., Kourtellis, N., Hammer, B., Lozano, J.A. (eds.) ECML PKDD 2021, Part V. LNCS (LNAI), vol. 12979, pp. 399–414. Springer, Cham (2021). https://doi.org/10.1007/978-3-030-86517-7_25
12. Hendrycks, D., Zhao, K., Basart, S., Steinhardt, J., Song, D.: Natural adversarial examples. In: Proceedings of the IEEE/CVF Conference on Computer Vision and Pattern Recognition, pp. 15262–15271 (2021)
13. Rebuffi, S.-A., Gowal, S., Calian, D.A., Stimberg, F., Wiles, O., Mann, T.A.: Data augmentation can improve robustness. In: Ranzato, M., Beygelzimer, A., Dauphin, Y., Liang, P.S., Wortman Vaughan, J. (eds.) Advances in Neural Information Processing Systems, vol. 34, pp. 29935–29948. Curran Associates, Inc. (2021). https://proceedings.neurips.cc/paper/2021/file/fb4c48608ce8825b558ccf07169a3421-Paper.pdf
14. Gowal, S., Qin, C., Uesato, J., Mann, T., Kohli, P.: Uncovering the limits of adversarial training against norm-bounded adversarial examples arXiv preprint:2010.03593 (2020)
15. Croce, F., et al.: Robustbench: a standardized adversarial robustness benchmark. arXiv preprint:2010.09670 (2020)
16. Buslaev, A., Iglovikov, V.I., Khvedchenya, E., Parinov, A., Druzhinin, M., Kalinin, A.A.: Albumentations: fast and flexible image augmentations. Information **11**(2), 125 (2020)
17. Lenc, K., Vedaldi, A.: Understanding image representations by measuring their equivariance and equivalence. In: Proceedings of the IEEE Conference on Computer Vision and Pattern Recognition, pp. 991–999 (2015)

A Fault Injection Tool for Identifying Faulty Operations of Control Functions in Automated Driving Systems

Kaveh Nazem Tahmasebi and DeJiu Chen(✉)

KTH Royal Institute of Technology, 100 44 Stockholm, Sweden
{kavent,chendj}@kth.se

Abstract. Along with an increasing usage of intelligent driver-assistance service in automotive vehicles, the guarantee of dependability and trustworthiness of the system functions for operation perception and control becomes particularly critical. To cope with the risks, an investigation for determining how such system functions would be affected by different operational conditions, including the deviations of expected inputs, software, and hardware anomalies, becomes necessary. This paper describes a tool that augments Simulink/MATLAB™ with additional fault models for effective analysis of error behaviors of automated control systems. This constitutes a basis for simulating different fault scenarios, collecting the operational data, and thereby the design of embedded online functions for error detection and treatment. This approach is demonstrated by a case study with an automated emergency braking system.

Keywords: Fault injection (FI) · Simulation · Fault modeling · Method and tool · Embedded control

1 Introduction

In recent years, there has been a trend toward an increasingly more intelligent driver-assistance services in automotive vehicles for cruise control, emergency braking, etc. As such services are inherently safety critical, the guarantee of dependability and trustworthiness of those system functions that provide such services becomes one key engineering task in the system development [1]. To this end, a key step is to reveal the likely effects of possible anomalies of the system functions under different operational conditions, relating to the deviations of expected inputs, software and hardware anomalies [3]. As recommended by ISO26262, FI (Fault Injection) testing is a preferred technique in the safety engineering [2]. However, several challenges remain for engineering practices, mainly due to the existence of a wide range of failure modes with intertwined and often unknown system-wide impacts across the software and hardware solutions. Moreover, for successful adoption of ML (machine-learning) and AI (Artificial

W. Zamojski et al. (Eds.): DepCoS-RELCOMEX 2022, LNNS 484, pp. 340–349, 2022.
https://doi.org/10.1007/978-3-031-06746-4_33

Intelligence) algorithms for operation perception and control, an identification of their corresponding failure modes as well as system-wide impacts in various operational conditions becomes critical in the system development.

This work focuses on the design of a FI tool that augments the most widely used simulation tool for controller design, Simulink/MATLAB^{TM} [4], with additional fault models for the analysis of error behaviors of control functions in automated driving systems. The ultimate goal is to enable more effective identifications of fault models, controller robustness, operational risks and suitable safety mechanisms, etc., early in the system development [1]. For effective fault injection and fault data collection, this approach structures different fault injection scenarios with a set of parameters relating to the occurrences and locations of some specific fault events. This promotes a systematic reasoning of the scope and coverage of testing activities as well as an automated testing procedure.

This paper is organized as follows: Sect. 2 discusses related concepts and technologies. Sect. 3 describes the overall concept and related basic design parameters specifying each fault injection task. Section 4 elaborates on the implementation based on Simulink/MATLAB^{TM}. A case study is presented in Sect. 5. Finally, the conclusion is given in Sect. 6.

2 Related Work

In the development of automotive vehicles, the evaluation of system behaviors together with possible anomalies constitutes a useful method for revealing the critical operational states, including the boundary conditions and necessary barriers for operational safety [5]. Over recent years, several FI tools have been developed, with the scope of support ranging from generic controller design to more specific automated driving systems.

A fault injection tool for functional models of automated control systems in Simulink/MATLAB^{TM} has been described in [6,7]. This tool, referred to as ErrorSim, supports a wide range of fault models. It allows the user to configure a set of fault models for FI simulation, yielding numerical results useful for dependability evaluation. However, the execution of FI scenarios can only be done manually. S-TALIRO [10] is a MATLAB toolbox for the analysis of robustness of control systems. It relies on a stochastic model-checking method for identifying system operational trajectories that may falsify the system requirements given in temporal logic. This tool uses in particular different optimization algorithms (including Cross Entropy Method (CEM) and Simulated Annealing Method (SMA) to search for new operational scenarios for an effective convergence to the expected falsification cases. On the other hand, current S-TALIRO does not provide explicit support for FI. Instead of system models in Simulink/MATLAB^{TM}, AVFI is a tool [8] dedicated to FI for software programs in Python that support the design of perception and other functions of AV (Autonomous Vehicles). AVFI uses CARLA [12], which is an open-source driving simulator, for the simulation of operational environment in order to enable an end-to-end resilience assessment of AV systems. The tool addresses explicitly the support for fault injection to AI algorithms such as deep-learning neural networks. For system functions based on ML and AI algorithms, Kayotee [9] is a tool allowing the injections of faults into the corresponding software and hardware components. In comparison to AVFI, the Kayotee tool would also provide a better support for revealing the actual effects of system-wide error propagation and masking. A comparison of these approaches is given Table 1. Our work aims to complement these approaches by allowing a more effective FI execution as well as a more flexible choice of the FI targets for the development of automated driving systems.

Table 1. A comparison of some FI methods in regard to the richness of fault definition.

	ErrorSim	AVFI	Kayotee
Types of fault	Supporting a wide range of failure modes for system functions in Simulink/MATLAB™	Supporting a wide range of failure modes for Python software programs	Supporting a wide range of failure modes for software and hardware components (incl. CPU, GPU)
Quantification of fault value	Allowing the definition of anomalies of each fault type regarding the nominal value	NA (Not mentioned explicitly in the papers currently being referred)	Allowing a choice of fault values with the options ranging from random to fixed scale values
Occurrence of fault event	Providing a set of stochastic methods for the occurrence definition, using failure probability, mean time to failure (MTTF), and failure rate distribution (e.g. Weibull, gamma, binomial)	NA(Not mentioned explicitly in the papers currently being referred)	Supporting a random occurrence based on the number of (one or more) transient faults per run
Duration of fault event	Allowing a list of options, including just-once, infinite time, and mean time to repair (MTTR)	Allowing transient and permanent persistence of fault events	Allowing only **transient** persistence of fault event
Location of fault injection	Allowing fault injection only at the links of targeted functional blocks	Allowing fault injection at the inputs, outputs, and internal functions of software programs	Allowing fault injection inside software and hardware components

3 Overall Design

By FI, we refer to the technique of introducing faults, characterized formally by some fault models, into a target system, including its operational environment as well as the system itself (i.e. the functional and technical solutions for the operation perception, control, and actuation). The overall design objective is to allow an exploration of the possible impacts of faults, regarding the consequent occurrences of errors, failures, and system hazards in different operational scenarios. This in turn constitutes a solid basis for the identification of the causal chains of *fault-error-failure* for safety analysis (e.g. FTA, FMEA) and the design of safety mechanisms. Considering these, the fault injection tool presented by this paper provides the following key features:

1. Stipulating a set of fault models to be injected, including some well-defined failure modes and related operational assumptions;
2. Supporting a systematic configuration and an automated execution of different FI scenarios, allowing an effective coverage of injection locations and fault behaviors.

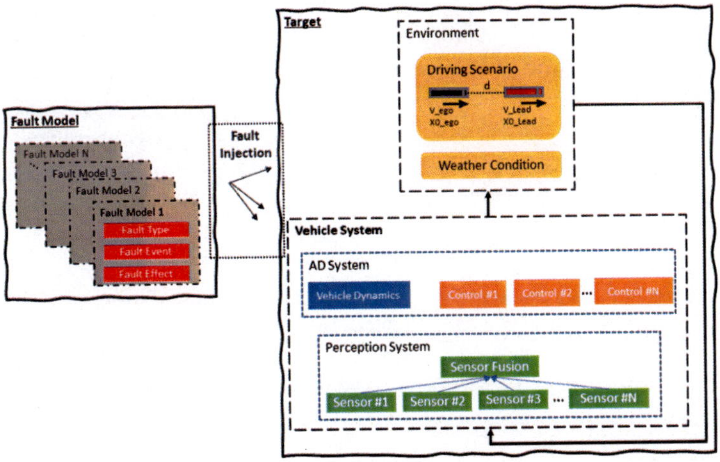

Fig. 1. Overall design concepts for fault injection in an automated driving system.

The overall design concept of our FI tool is illustrated with Fig. 1, with the following modules: i) a vehicle system including a perception (sub)system with some sensor functions for detecting its dynamic operational conditions (e.g. position and velocity) and an AD (sub)system for regulating vehicle dynamics; ii) an environment system generating different driving scenarios including weather and road conditions; and iii) a FI support including some fault models and a fault injection service to inject faults into the target system.

For configuring specific FI scenarios, two sets of parameters, shown in Table 2 and Table 3, have been defined. In particular, with Table 2, a set of configuration parameters have been defined for characterizing the basic fault models, including their types and respective temporal properties. With Table 3, a set of parameters have been defined for characterizing the locations of faults occurred in a target system. On the basis of such parameters, a specific FI scenario can be defined for either a selective fault coverage (e.g. guided by safety experts) or a random fault coverage (i.e. guided by Monte Carlo Simulation).

Table 2. Key parameters in the FI tool for specifying faults and fault events.

FI Parameter	Definition
Fault_Type	Referring to the failure modes by FI, including Omission-, Commission-, Value- and Timing-fault, which are caused by stuck-at, delay, noises, etc
Fault_Value	Referring to the quantification of anomalies underlying a specific failure mode, including min-max value deviation, time offset
Fault_Event_Occurrence	Referring to the characterization of a fault event regarding to the probability of occurrence, which can be either probabilistic (e.g. based on Normal/ Gaussian models) or deterministic at certain time
Fault_Event_Intensity	Referring to the characterization of a fault events regarding the amount of occurrences during a time interval
Fault_Event_Duration	Referring to the duration of operation during which a fault occurs, for which the options include transient, permanent, periodic
Fault_Contextual_Condition	Referring to the contextual conditions (e.g. a specific driving scenario) when a fault event occurs

Table 3. Key parameters in the FI tool for specifying the target locations of fault injection.

FI Parameter:	Definition
Target_Environment	Referring to the environmental conditions chosen as the targets of FI, including weather, road, traffic, etc.
Target_Component	Referring to the functional models, software programs, or hardware modules chosen as the targets of FI
Target_Signal	Referring to the logical or physical signals across system components chosen as the targets of FI

4 Tool Implementation

As mentioned previously, our FI tool augments Simulink/MATLAB^{TM} with additional fault models for the analysis of error behaviors. Technically, it integrates and refines the FIBlock tool presented in [6,7] and allows more effective configuration and execution of complex FI scenarios. The implementation is given by a composition of various control and computation blocks in Simulink as shown in Fig. 2.

The *fault template* and *system template* shown in this figure together allow a parameterized configuration of different fault models. The *fault template* defines a set of configuration blocks. In particular, the blocks of *model_selection_method* in terms of switch gates provide the support for a selection of some atomic

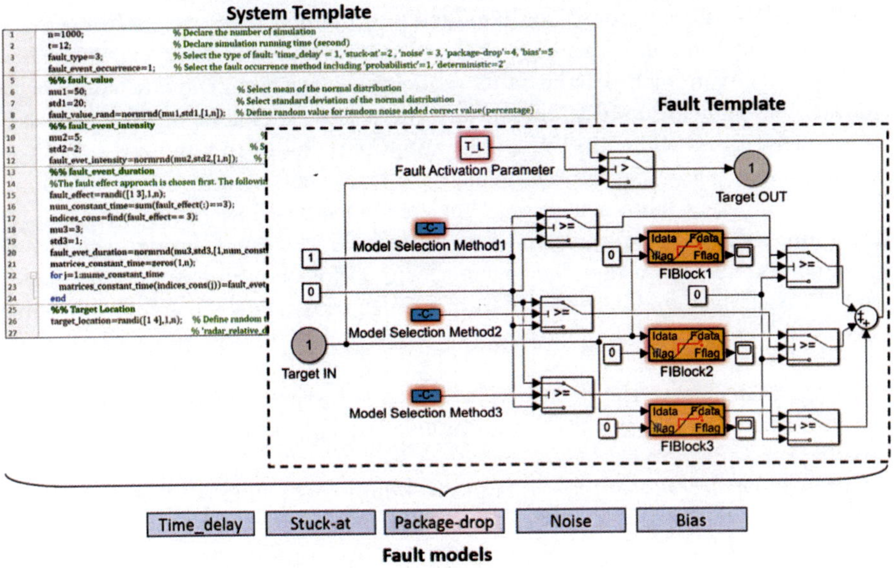

Fig. 2. The support for a two-stage configuration: 1. A **fault template** making it possible to configure different atomic fault models; 2. A **system template** for instantiating atomic fault models and the injection scenarios.

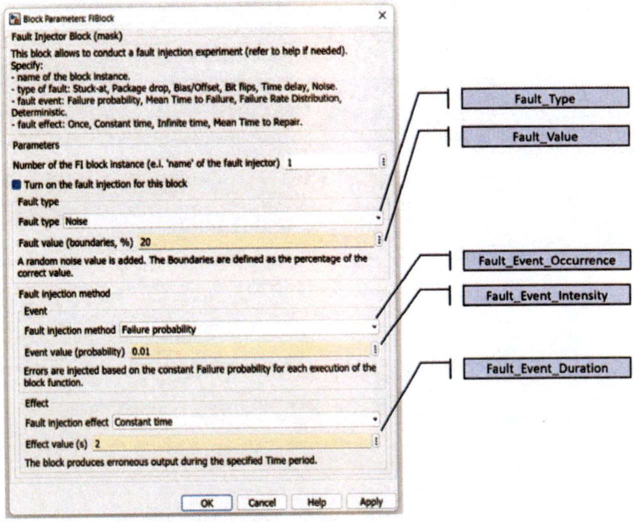

Fig. 3. The definition of an atomic fault model by *FIBlock* [6,7] for a noise of 20%, with probability 0.01 and duration 2 s. The Fault_Event_Intensity can be given either as occurrence frequency or as probabilistic figure depending on the chosen of Fault_Event_Occurrence.

fault models. These atomic fault models are defined using the *FIBlocks* [6,7], with specific configurations of fault type, fault value and fault event as shown in Fig. 3. Given such atomic fault models, the *system template* contains the parameters that are used to instantiate the actual selections of such models for a FI scenario. As shown in Fig. 4, the support includes randomized selection of the values for these parameters, which are then written into the corresponding block parameters (shown in Fig. 3) for each specific FI scenario. For the faults to be injected, the configuration of target locations is done through the location parameters, which are then fed to some switch blocks in Simulink for connecting to the target system. The executions of FI scenarios follow a common procedure, shown in Fig. 5, with the following key steps:

- *Step_1:* Declaring the locations in the target system for the fault injection, such as a camera input port, an internal signal or variable of the sensor fusion block;
- *Step_2:* Declaring the type of faults to be injected, such as time delay, stuck-at;
- *Step_3:* Declaring the event properties for the faults to be injected, such as a probability model for the random occurrence;
- *Step_4:* Declaring the persistence of a fault event, once occurred, during the simulation, such as a transient or permanent.

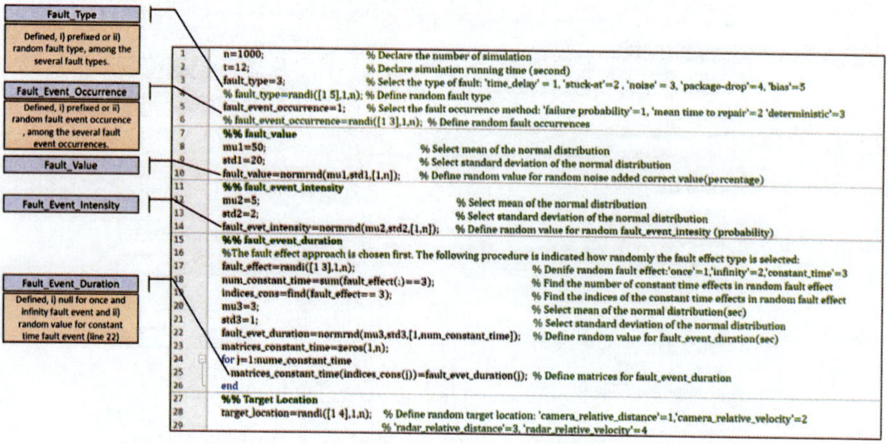

Fig. 4. A FI scenario instantiated using the **system template**.

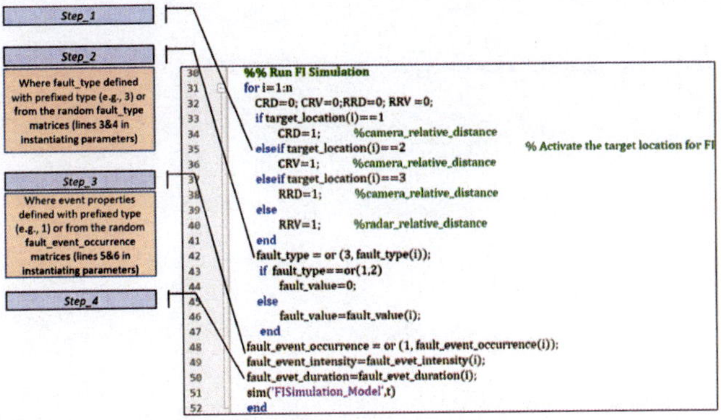

Fig. 5. A FI scenario procedure defined with the **system template**.

5 Demonstration and Results

To demonstrate the capability of this FI support, an autonomous emergency braking (AEB) system originally developed by Mathworks for the design of automated driving functions [11] has been used. The key control functionality is to avoid collision with other front-going vehicles by using cameras and radars for perceiving actual longitudinal conditions, as well as some automated control functions for regulating the vehicle accelerating. The functional model of such a system consists of Simulink blocks for the perception, vehicle control and driving scenarios. We inject faults to the perception system consisting of camera and radar, sensor fusion.

The effect of timing faults injected into the perception system consisting of camera, radar, and sensor fusion is shown in Fig. 6. For each FI scenario, we run the AEB system for 12 s with the same fixed initial conditions: 1. the vehicle under control starting with constant velocity, v_ego = 16.66 (m/s), at the origin, $x_0\_ego = 0$; 2. a front-going vehicle in the environment starting with constant speed, v_lead = 12.5 (m/s) from the x_0_lead =35 m. The impacts of faults are indicated with the Time-to-Collision (TTC) figures. For a Monte Carlo simulation, the faults are defined by a random configuration of transient time delay faults and injected randomly.

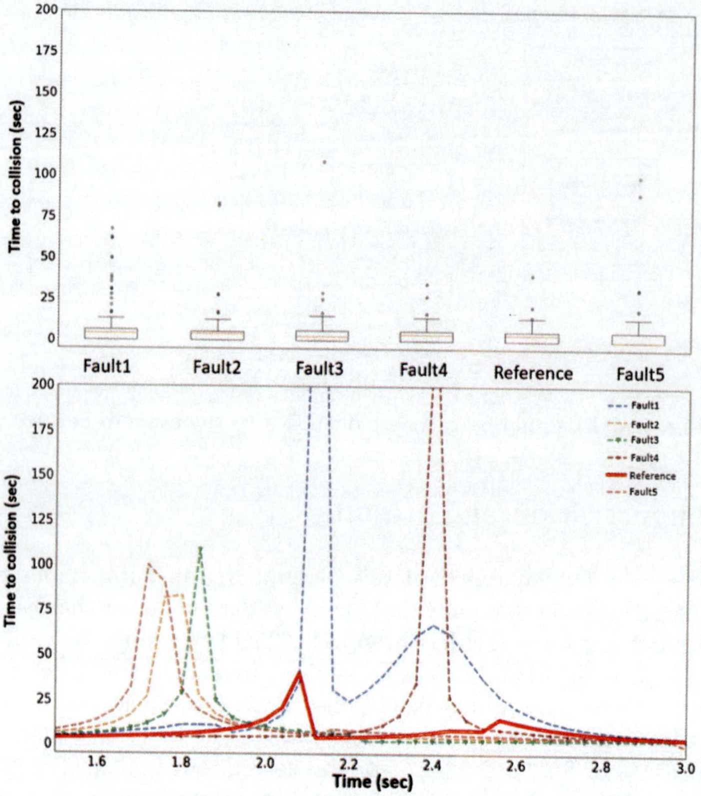

Fig. 6. Time-to-Collision (TTC) behavior based on different *Time Delay* models, comparing the impacts of different faults. The *Reference* refers to the trajectory of TTC value. **Fault 1**: fault injected into camera_relative_distance (CRD) at time 1.1 s for 1.6 s and radar_relative_distance (RRD) at 1.9 s for 1.9 s. **Fault 2**: fault injected into RRD at time 1.3 s for 1.7 s and radar_relative_velocity (RRV) at 0.5 s for 1.4 s. **Fault 3**: fault injected into camera_relative_velocity (CRD) at time 1.6 s for 1.2 s and RRD at 0.4 s for 1.9 s. **Fault 4**: fault injected into RRD at time 0 s for 1.3 s. **Fault 5**: fault injected into RRD at time 1.2 s for 0.8 s.

6 Conclusion

With this paper, we have described a FI tool that extends the support based on some Simulink/MATLAB^{TM} models with fault behaviors for safety analysis. The tool constitutes a basis for simulating different fault scenarios, collecting the operational data, and thereby the design of control functions for error detection and treatment. The main functionalities include a set of predefined fault models and a procedure for configuring and executing various FI scenarios.

As part of the future work, a support for hardware-in-the-loop simulation will be developed, considering in particular the co-simulation with ML/AI functions running with embedded hardware (e.g. GPU). To support the design of safety mechanisms for error detection and fault treatment, a formal analysis for identifying the operational behaviors regarding the hazardous states and the conditions triggering the transitions of such states will be developed. For the simulation of complex vehicle operation scenarios, an integration with the open-source CARLA [12] Simulator system will be investigated.

Acknowledgement. This work is supported by 1. KTH Royal Institute of Technology with the industrial research project ADinSOS (2019065006); and 2. the Swedish government agency for innovation systems (VINNOVA) with the cooperative research project Trust-E (Ref: 2020-05117) within the programme EUREKA EURIPIDES.

References

1. Chen, D., et al.: Design of a knowledge-based strategy for capability-aware treatment of uncertainties of automated driving systems. In: 1st International Workshop on Artificial Intelligence Safety Engineering, WAISE, SAFECOMP, LNCS 11094, pp. 446-457 (2018)
2. ISO 26262 - Road vehicles - Functional safety. https://www.iso.org/home.html
3. Mariani, R.: An overview of autonomous vehicles safety. In: IEEE International Reliability Physics Symposium (IRPS) (2018)
4. MathWorks: Matlab & simulink: Simulink users guide r2020b (2020). https://www.mathworks.com/
5. Avizienis, A., Laprie, J.-C., Randell, B., Landwehr, C.: Basic concepts and taxonomy of dependable and secure computing. IEEE Trans. Depend. Sec. Comput. **1**(1), 11–33 (2004)
6. Saraoğlu, M., Morozov, A., Söylemez, M.T., Janschek, K.: ErrorSim: a tool for error propagation analysis of simulink models. In: Tonetta, S., Schoitsch, E., Bitsch, F. (eds.) SAFECOMP 2017. LNCS, vol. 10488, pp. 245–254. Springer, Cham (2017). https://doi.org/10.1007/978-3-319-66266-4_16
7. Fabarisov, T., et al.: Model-based fault injection experiments for the safety analysis of exoskeleton system. In: Proceedings of the 30th European Safety and Reliability Conference and the 15th Probabilistic Safety Assessment and Management Conference (2020)
8. Jha, J.S., Banerjee, S., Cyriac, J.T., Kalbarczyk, Z.K., Iyer, R.: AVFI: fault injection for autonomous vehicles. In: 2018 48th Annual IEEE/IFIP International Conference on Dependable Systems and Networks Workshops (DSN-W) 2018, pp. 55–56. IEEE, Luxembourg (2018). https://doi.org/10.1109/DSN-W.2018.00027
9. Jha, S., et al.: Kayotee: a fault injection-based system to assess the safety and reliability of autonomous vehicles to faults and errors. ArXiv (2019)
10. Annpureddy, Y., Liu, C., Fainekos, G., Sankaranarayanan, S.: S-TaLiRo: a tool for temporal logic falsification for hybrid systems. In: Abdulla, P.A., Leino, K.R.M. (eds.) TACAS 2011. LNCS, vol. 6605, pp. 254–257. Springer, Heidelberg (2011). https://doi.org/10.1007/978-3-642-19835-9_21
11. MathWorks: automated driving toolbox, autonomous emergency braking with sensor fusion. https://mathworks.com/help/driving/ug/autonomous-emergency-braking-with-sensor-fusion.html
12. Dosovitskiy, A., Ros, G., Codevilla, F., Lopez, A., Koltun, V.: CARLA: an open urban driving simulator. In: Proceedings of the 1st Annual Conference on Robot Learning, pp. 1–16 (2017)

Dual Learning Model for Multiclass Brain Tumor Classification

Rohit Thanki[1](✉) and Sanaa Kaddoura[2]

[1] Data Scientist, Wolfsburg, Germany
rohitthanki9@gmail.com
[2] College of Technological Innovation, Zayed University, Abu Dhabi, UAE

Abstract. A brain tumor occurs in the human body when the brain develops abnormal cells. Tumors are called either benign (noncancerous) or malignant (cancerous). The function of the nervous system is affected by the growth rate and the location of the tumor. The tumor treatment depends on tumor type, size, and location. Artificial intelligence has been widely used to automatically predict various brain tumors using multiple imaging technologies such as magnetic resonance imaging (MRI) and computerized tomography (CT) scan during the last few years. This paper applies a hybrid learning based classifier on an MRI dataset containing benign and malignant images. Moreover, deep learning is also applied to the same dataset. The proposed learning approach's performance is compared to other existing supervised machine learning approaches. The experimental results show that our proposed approach outperforms the existing approaches available in the literature.

Keywords: Brain tumor · Hybrid machine learning · Deep learning · Magnetic resonance imaging · Malignant

1 Introduction

Brain tumors are a collection of abnormal cells in the brain; thus, they are considered life-threatening diseases. Although brain tumors can be benign, they can also be cancerous [1]. Both benign and malignant tumors can cause pressure to the brain skull once they start growing. Therefore, tumors need to be diagnosed in the early stages to increase the percentage of cure. Magnetic resonance imaging (MRI) is considered an excellent technique to reveal the biology of a brain tumor, its cellular structure and functionality, and the necessary treatments for patients at risk. In MRI examination, image segmentation is typically used to measure and visualize the brain structure, investigate changes, inspect disease, and plan treatments or surgical operations. There are typically more than 120 types of tumors [2], including:

- The glioma tumor is a common type of tumor originating in the brain (around 33% of all brain tumors).
- The meningioma tumor is a primary central nervous system tumor (around 30% of all brain tumors).

W. Zamojski et al. (Eds.): DepCoS-RELCOMEX 2022, LNNS 484, pp. 350–360, 2022.
https://doi.org/10.1007/978-3-031-06746-4_34

- The pituitary tumor is a pea-sized endocrine gland at the brain base, behind the nose bridge, and below the hypothalamus (around 10% of all the brain tumors).

The various machine learning approaches, such as Naïve Bayes, neural net- work, and support vector machine, were used to classify tumors. The literature shows several trials to get accurate classifications [3, 4]. In these approaches, the tumor features were manually extracted and dependent on some numerical values, which are not valid and cannot always be reliable. Brain tumor classification using machine learning attracted researchers for a long time. This research is still active to produce the most accurate results. Automatic prediction and various brain tumor images and other medical images play a vital role in a patient's treatment. An early diagnosis of the disease allows the patient body to respond effectively to the treatment, which will enhance people's survival rate. Usually, the process of brain tumor diagnosis is done manually in the clinics. However, automating this task saves the cost and time for the patients. Researchers use many types of images as input to their models, such as [5]: Computed Tomography (CT), MRI (Magnetic Resonance Imaging), Positron Emission Tomography (PET), Single-Photon-Emission Computed Tomography (SPECT), and Magnetic Resonance Spectroscopy (MRS). MRI, which is used in this research, is the most commonly used imaging technique.

Convolution neural networks (CNN) were widely used for classification of brain tumors. The authors in [6–20] are proposed various models based on CNN for classification of brain tumors. These models are tested using various datasets such as BRAT2013, MRBrainS13, BRAT2016, BRAT2017, and BRAT2018, etc. These models are achieved average accuracy around $90 \pm 5\%$ for classification of brain tumors. In [21], the authors proposed a model using the following architectures: GoogLeNet, Alexnet, and VGGNet for classification of brain tumors. This paper presents a new learning model for the classification of multiclass brain tumors. This new model used convolution layers for features extraction from MRI images while machine learning based classifier used for classification of these features. The rest of the paper is organized as follows: Sect. 2 illustrates the methodology of the new machine learning approach adopted in brain tumor classification. The experimental setup and results are represented in Sect. 3, followed by a discussion of the obtained results and comparative analysis of various classifiers in Sect. 4. Finally, Sect. 5 concludes this paper.

2 Methodology

In this paper, we have used approach which contains combination of machine learning and deep learning for multiclass classification of brain tumor. This approach called as hybrid approach (shown in Fig. 1) and model based on this approach is called as dual learning model (shown in Fig. 2). This model uses two types of learning: deep learning and supervised machine learning. The model's input is a brain MRI image. The model's output is the prediction tumor type of input images such as no tumor, glioma tumor, meningioma tumor, and pituitary tumor. In this model, deep learning is used for finding essential features from the input image. These features of the image are then fed to a multiclass classifier. Here, convolutional layers of SqueezeNet [11] are used for features extraction from the brain MRI image. In contrast, different supervised machine learning algorithms (which are worked as classifier in this approach) [12] such as k-nearest neighbor (k-NN), support vector machine (SVM), decision tree (DT), neural network (NN), naïve bayes (NB), and logistic regression (LR) are used for the prediction and classification of tumor class.

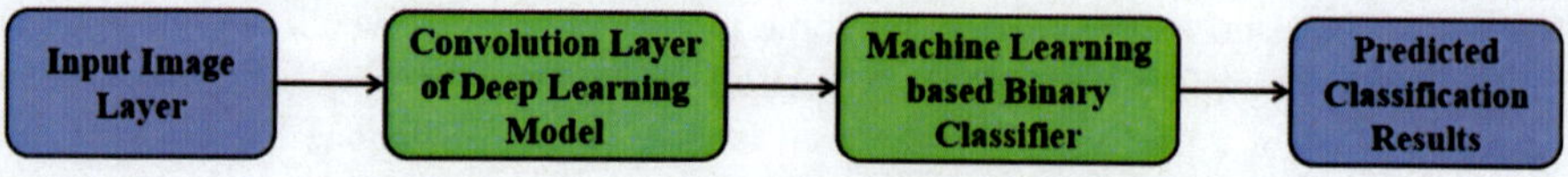

Fig. 1. The Steps of the hybrid learning approach

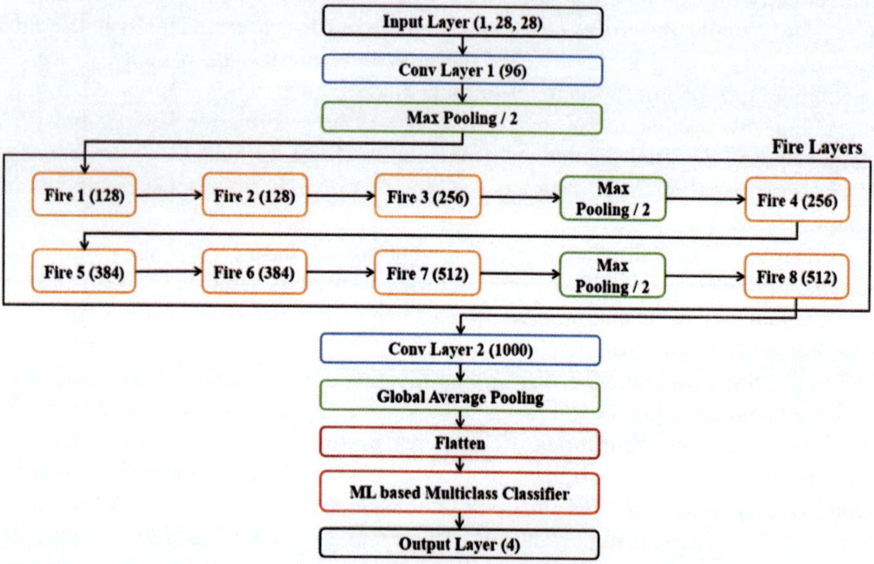

Fig. 2. Dual learning model for multi class tumor classification

2.1 Features Extraction Using Convolutional Layers of SqueezeNet

SqueezeNet model [11] is a deep convolutional neural network (CNN) with compressed architecture, a small number of parameters, and achieved higher accuracy than AlexNet [13] and ImageNet [14] with the same number of parameters. The main advantages of this model have required fewer communication channels for training, is easy to deploy on the cloud server, and can be customized and used on hardware with limited memory. This original SqueezeNet model [11] has 14 layers where contain: Two conventional convolution layers; Eight fire layers; Three max-pooling layers; One global average pooling layer; Softmax. Here, we have used only convolution layers of this model to extract features from the input brain MRI image. After extracting the features from the input brain MRI images, a flatten layer was used, converting features into a one-dimensional layer.

2.2 Machine Learning Based Multiclass Classifier

In machine learning, multiclass classification refers to classifying instances into one of three or more classes. In this case, we need to classify input brain MRI images into one class, such as no tumor, glioma tumor, meningioma tumor, and pituitary tumor. For this purpose, in this model, we have used different conventional classifiers such as k-nearest neighbor (k-NN), support vector machine (SVM), decision tree (DT), naïve bayes (NB), and logistic regression (LR) for classification of multiclass brain tumor images. Here, we have modified each classifier such way that it operates as multi class classifier.

3 Experimental Setup and Results

3.1 Information About Dataset

In this paper, the dataset is taken from online dataset provider website Kaggle and Github [22]. The dataset used in this paper contains approximately 3265 MRI images of different classes of tumors such a "glioma," "meningioma", and "pituitary". The dataset also contains MRI image without tumor as normal image. The dataset is divided into two data samples such as training dataset that contains 2,939 MRI images (90%) and testing dataset that contains 326 MRI images (10%). A sample of MRI images in the dataset is shown in Figs. 2 and 3.

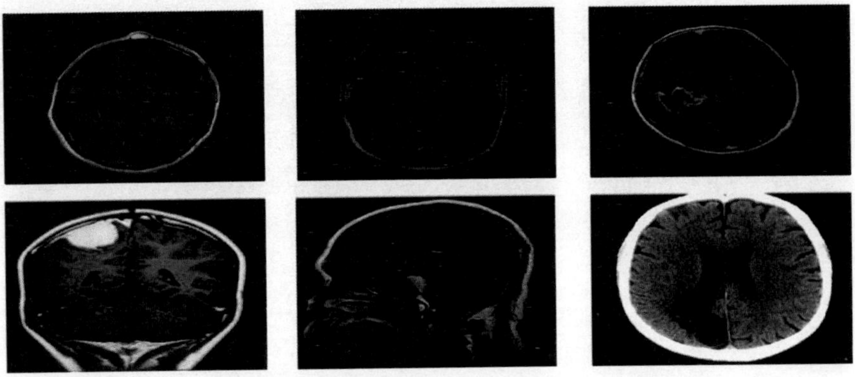

Fig. 3. Sample MRI images from the dataset

3.2 Experimental Results for Training Dataset

After extracting features of MRI images of training dataset using convolutional layers, then passing these extracted features of training dataset to various machine learning classifiers to get experimental results for training dataset and summarized average performance values of various classifiers in the Table 1. The performance of classifiers is validated using performance evaluation metrics such as area under curve (AUC) value of ROC curve, accuracy, F1 score, precision, and recall.

Table 1. Average performance of classifiers for all tumors classes of training dataset

Classifier	AUC	Accuracy	F1 score	Precision	Recall
k-NN	0.993	0.938	0.938	0.938	0.938
DT	0.996	0.967	0.967	0.968	0.967
SVM	0.984	0.897	0.896	0.902	0.897
NN	1.000	1.000	1.000	1.000	1.000
NB	0.888	0.697	0.691	0.699	0.697
LR	0.998	0.983	0.983	0.983	0.983

3.3 Experimental Results for Testing Dataset

After extracting features of MRI images of training dataset using convolutional layers, then passing these extracted features of testing dataset to various machine learning classifiers to get experimental results for testing dataset and summarized average performance values of various classifiers in the Tables 2, 3, 4, 5 and 6. The confusion matrix for each classifier are illustrated in Figs. 4, 5, 6, 7, 8 and 9.

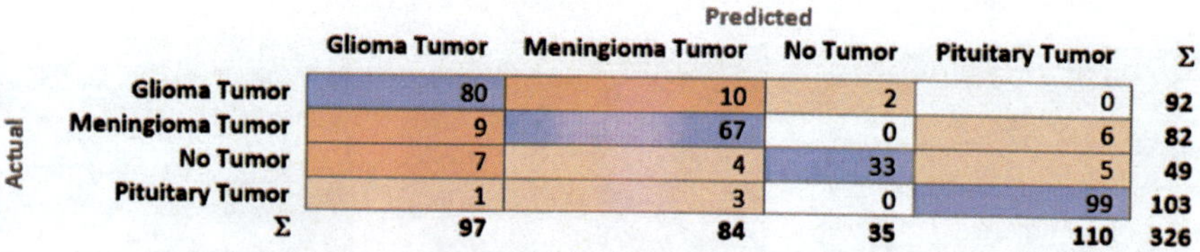

Fig. 4. Confusion matrix for K-NN based multiclass tumor classifier

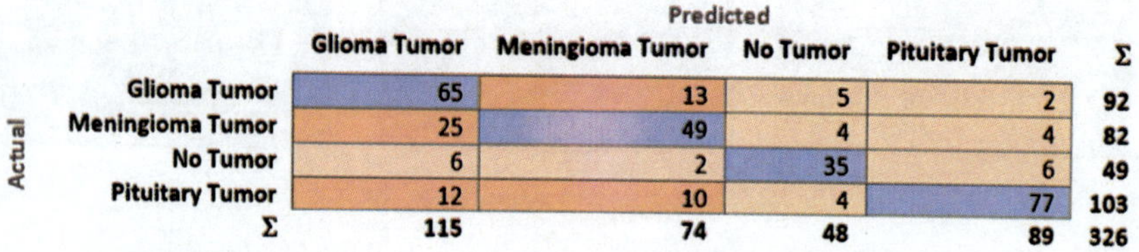

Fig. 5. Confusion matrix for DT based multiclass tumor classifier

Table 2. Average performance of classifiers for all tumors classes of testing dataset

Classifier	AUC	Accuracy	F1 score	Precision	Recall
k-NN	0.951	0.856	0.854	0.859	0.856
DT	0.806	0.715	0.716	0.726	0.715
SVM	0.978	0.850	0.850	0.850	0.850
NN	**0.989**	**0.929**	**0.929**	**0.931**	**0.929**
NB	0.861	0.632	0.625	0.624	0.632
LR	0.964	0.887	0.886	0.886	0.887

Actual \ Predicted	Glioma Tumor	Meningioma Tumor	No Tumor	Pituitary Tumor	Σ
Glioma Tumor	65	26	0	1	92
Meningioma Tumor	6	70	0	6	82
No Tumor	2	0	45	2	49
Pituitary Tumor	2	4	0	97	103
Σ	75	100	45	106	326

Fig. 6. Confusion matrix for SVM based multiclass tumor classifier

Actual \ Predicted	Glioma Tumor	Meningioma Tumor	No Tumor	Pituitary Tumor	Σ
Glioma Tumor	86	5	0	1	92
Meningioma Tumor	3	75	1	3	82
No Tumor	2	3	42	2	49
Pituitary Tumor	1	2	0	100	103
Σ	92	85	43	106	326

Fig. 7. Confusion matrix for NN based multiclass tumor classifier

Actual \ Predicted	Glioma Tumor	Meningioma Tumor	No Tumor	Pituitary Tumor	Σ
Glioma Tumor	52	23	8	9	92
Meningioma Tumor	22	35	10	15	82
No Tumor	6	6	30	7	49
Pituitary Tumor	0	6	8	89	103
Σ	80	70	56	120	326

Fig. 8. Confusion matrix for NB based multiclass tumor classifier

Actual \ Predicted	Glioma Tumor	Meningioma Tumor	No Tumor	Pituitary Tumor	Σ
Glioma Tumor	80	9	2	1	92
Meningioma Tumor	9	69	1	3	82
No Tumor	3	2	42	2	49
Pituitary Tumor	2	1	2	98	103
Σ	94	81	47	104	326

Fig. 9. Confusion matrix for LR based multiclass tumor classifier

Table 3. Performance of classifiers for glioma tumor class of testing dataset

Classifier	AUC	Accuracy	F1 score	Precision	Recall
k-NN	0.943	0.911	0.847	0.825	0.870
DT	0.812	0.807	0.696	0.626	0.783
SVM	0.971	0.887	0.778	0.867	0.707
NN	**0.984**	**0.963**	**0.935**	**0.935**	**0.935**
NB	0.844	0.791	0.605	0.650	0.565
LR	0.948	0.920	0.860	0.851	0.870

Table 4. Performance of classifiers for meningioma tumor class of testing dataset

Classifier	AUC	Accuracy	F1 score	Precision	Recall
k-NN	0.947	0.902	0.807	0.798	0.817
DT	0.732	0.822	0.628	0.662	0.598
SVM	0.953	0.871	0.769	0.700	0.854
NN	**0.983**	**0.948**	**0.898**	**0.882**	**0.915**
NB	0.809	0.748	0.461	0.500	0.427
LR	0.951	0.923	0.847	0.852	0.841

Table 5. Performance of classifiers for no tumor class of testing dataset

Classifier	AUC	Accuracy	F1 score	Precision	Recall
k-NN	0.932	0.945	0.786	0.943	0.673
DT	0.823	0.917	0.722	0.729	0.714
SVM	0.997	0.988	0.957	1.000	0.918
NN	**0.990**	**0.975**	**0.913**	**0.977**	**0.857**
NB	0.846	0.862	0.571	0.536	0.612
LR	0.972	0.963	0.875	0.894	0.857

Table 6. Performance of classifiers for pituitary tumor class of testing dataset

Classifier	AUC	Accuracy	F1 score	Precision	Recall
k-NN	0.975	0.954	0.930	0.900	0.961
DT	0.857	0.883	0.802	0.865	0.748
SVM	0.995	0.954	0.928	0.915	0.942
NN	**0.996**	**0.972**	**0.957**	**0.943**	**0.971**
NB	0.933	0.862	0.798	0.742	0.864
LR	0.987	0.966	0.947	0.942	0.951

4 Discussion and Comparison

4.1 Discussion

Because image classification using machine learning techniques is promising for earlier brain tumor detection and prediction, it is essential to examine the results obtained in this study. Therefore, based on the experimental results captured in Sect. 3.3, it is shown that, for overall performance of all classifiers for testing dataset, the NN based classifier outperforms all other classifiers with an F1-score of 92.9% value rate. The performance of the NB based classifier is the least among all the classifiers, with an F1 score of 63.2% value rate. Similarly, AUC for the NN based classifier is the highest among all classifiers, with a value of 0.989. However, the DT based classifiers comes in the last place with an AUC value of 0.806. For glioma tumor class, the NN based classifier has F1 score and AUC values of 93.5% and 0.984, respectively. The NB based classifier has the lowest F1 score value (60.5%), and the DT based classifier has the lowest AUC value (0.812).

For meningioma tumor class, the NN based classifier has F1 score and AUC values of 89.8% and 0.983, respectively. The NB based classifier has the lowest F1 score value (50%), and the DT based classifier has the lowest AUC value (0.732). For no tumor class, the SVM based classifier has an F1 score and AUC values of 95.6% and 0.997, respectively. The NB based classifier has the lowest F1 score value (57.1%), and the DT based classifier has the lowest AUC value (0.823). For pituitary tumor class, the NN based classifier has F1 score and AUC values of 95.7% and 0.996, respectively. The NB based classifier has the lowest F1 score value (79.8%), and the DT based classifier has the lowest AUC value (0.857).

4.2 Comparison with Classification of MRI Images Related Studies

After looking at the literature [23–28], some researchers studied the classification of brain tumor MRI images using machine learning and deep learning. In most studies, researchers measured accuracy for classifying various types of brain tumors. Therefore, we have used classification accuracy as a comparison parameter to compare the performance of different existing approaches in the literature with the proposed approach. However, researchers used their datasets for most current approaches, so it's challenging to get every dataset for testing purposes. Therefore, for performance comparison, we used similar existing approaches which are near to our proposed work are used in this paper.

For instance, to classify glioma, meningioma, and pituitary tumors, the authors in [23] have a classification accuracy of 90.9%; the authors in [24] has a classification accuracy of 90.67%; the authors in [25] has a classification accuracy of 91.3%. However, in this paper, the classification accuracy for classifying different types of brain tumor has achieved an average value of 92.9% for classification of all tumor classes (see Table 2), 96.3% for glioma class (see Table 3), 94.9% for meningioma class (see Table 4), 98.8% for no tumor (see Table 5), and 97.2% for pituitary class (see Table 6). This comparison ensures that the proposed model has classification accuracy compared to accuracy of existing classifiers [23–25] in the literature.

Some other researchers studied the classification of a single brain tumor class (glioma tumor). For instance, the experimental results in [26] and [27] showed classification accuracy values of 96% and 92%, respectively. However, in this study, glioma brain tumor classification shows an accuracy value of 96.3% (see Table 3). Moreover, using SVM based classifier, it is shown in this paper that the accuracy for no tumor class is 98.8%. However, in [28], the classification accuracy for no tumor class showed performances of 95.6% (see Table 5).

5 Conclusion

This paper discusses the classification of multi class brain tumor using MRI images. The new dual learning-based model is proposed in this paper which classified input MRI images as image has tumor (glioma, meningioma, and pituitary) or normal image. This model developed using concept of deep learning and machine learning. The convolutional layers of SqueezeNet used for feature extraction from the MRI image while machine learning based classifier used for classification of these features as image has tumor or not. The various classifiers such as k-NN, DT, SVM, NN, NB and LR are used in the paper for multiclass classification of brain tumor in MRI image. The experimental results show a resilient performance of the NN based classifier with a 92.9% prediction F1 score and 0.989 AUC value.

References

1. Litin, S.C.: Mayo clinic family health book. In: Nanda, S. (ed.) Time Incorporated Home Entertainment, Time Inc.(2009)
2. Brain Tumor Types: Health Hopkins Medicine (2021). https://www.hopkinsmedicine.org/health/conditions-and-diseases/brain-tumor/brain-tumor-types. Accessed 1 Oct 2021
3. Mukambika, P.S., Uma Rani, K.: Segmentation and classification of MRI brain tumor. Int. Res. J. Eng. Technol. **1**, 683–688 (2017)
4. Iandola, F.N., Han, S., Moskewicz, M.W., Ashraf, K., Dally, W.J., Keutzer, K.: SqueezeNet: AlexNet-level accuracy with 50x fewer parameters and 0.5 MB model size (2016). arXiv preprint arXiv:1602.07360
5. Díaz-Pernas, F.J., Martínez-Zarzuela, M., Antón-Rodríguez, M., González-Ortega, D.: A deep learning approach for brain tumor classification and segmentation using a multiscale Convolutional Neural Network. Healthcare **9**(2), 153. Multidisciplinary Digital Publishing Institute, February 2021

6. Sharif, M., Tanvir, U., Munir, E.U., Khan, M.A., Yasmin, M.: Brain tumor segmentation and classification by improved binomial thresholding and multi-features selection. J. Ambient. Intell. Humaniz. Comput. **10**, 1–20 (2018). https://doi.org/10.1007/s12652-018-1075-x
7. Hussain, U.N., et al.: A unified design of ACO and skewness-based brain tumor segmentation and classification from MRI scans. J. Control Eng. Appl. Inform. **22**(2), 43–55 (2020)
8. Khan, M.A., et al.: Brain tumor detection and classification: a frame- work of marker-based watershed algorithm and multilevel priority features selection. Microsc. Res. Tech. **82**(6), 909–922 (2019)
9. Havaei, M., et al.: Brain tumor segmentation with deep neural networks. Med. Image Anal. **35**, 18–31 (2017)
10. Moeskops, P., et al.: Evaluation of a deep learning approach for the segmentation of brain tissues and white matter hyperintensities of presumed vascular origin in MRI. NeuroImage Clin. **17**, 251–262 (2018)
11. Ronneberger, O., Fischer, P., Brox, T.: U-net: Convolutional networks for biomedical image segmentation. In: International Conference on Medical Image Computing and Computer-assisted Intervention, pp 234–241 (2015)
12. Pashaei, A., Sajedi, H., Jazayeri, N.: Brain tumor classification via convolutional neural network and extreme learning machines. In: 2018 8th International Conference on Computer and Knowledge Engineering, pp 314–319 (2018)
13. Abiwinanda, N., Hanif, M., Hesaputra, S.T., Handayani, A., Mengko, T.R.: Brain tumor classification using convolutional neural network. In: World Congress on Medical Physics and Biomedical Engineering, pp 183–189 (2019)
14. Sultan, H.H., Salem, N.M., Al-Atabany, W.: Multi-classification of brain tumor images using deep neural network. IEEE Access **7**, 69215–69225 (2019)
15. Anaraki, A.K., Ayati, M., Kazemi, F.: Magnetic resonance imaging-based brain tumor grades classification and grading via convolutional neural networks and genetic algorithms. Biocybernet. Biomed. Eng. **39**(1), 63–74 (2019)
16. Işın, A., Direkoğlu, C., Şah, M.: Review of MRI-based brain tumor image segmentation using deep learning methods. Procedia Comput. Sci. **102**, 317–324 (2016)
17. Sharif, M.I., Li, J.P., Khan, M.A., Saleem, M.A.: Active deep neural network features selection for segmentation and recognition of brain tumors using MRI images. Pattern Recogn. Lett. **129**, 181–189 (2020)
18. Narmatha, C., Eljack, S.M., Tuka, A.A.R.M., Manimurugan, S., Mustafa, M.: A hybrid fuzzy brain-storm optimization algorithm for the classification of brain tumor MRI images. J. Ambient. Intell. Humaniz. Comput. **1**, 1–9 (2020). https://doi.org/10.1007/s12652-020-02470-5
19. Rehman, A., Khan, M.A., Saba, T., Mehmood, Z., Tariq, U., Ayesha, N.: Microscopic brain tumor detection and classification using 3D CNN and feature selection architecture. Microsc. Res. Tech. **84**(1), 133–149 (2021)
20. Mzoughi, H., et al.: Deep multi-scale 3D convolutional neural network (CNN) for MRI gliomas brain tumor classification. J. Digit. Imaging **33**, 903–915 (2020)
21. Rehman, A., Naz, S., Razzak, M.I., Akram, F., Imran, M.: A deep learning- based framework for automatic brain tumors classification using transfer learning. Circ. Syst. Sig. Process. **39**(2), 757–775 (2020)
22. Brain Tumor Classification (Multi-label) – CNN (2020). https://www.kaggle.com/dhruvanurag20/brain-tumor-classification-multi-label-cnn/data; https://github.com/SartajBhuvaji/Brain-Tumor-Classification-DataSet. Accessed Dec 2021
23. Sajjad, M., Khan, S., Muhammad, K., Wu, W., Ullah, A., Baik, S.W.: Multi-grade brain tumor classification using deep CNN with extensive data augmentation. J. Comput. Sci. **30**, 174–182 (2019)

24. Cheng, J., et al.: Enhanced performance of brain tumor classification via tumor region augmentation and partition. PLoS ONE **10**(10), 1–13 (2015)
25. Ertosun, M.G., Rubin, D.L.: Automated grading of gliomas using deep learning in digital pathology images: a modular approach with ensemble of convolutional neural networks. In: Annual Symposium Proceeding AMIA Symposuim, pp. 1899–1908 (2015)
26. Papageorgiou, E.I., et al.: Brain tumor characterization using the soft computing technique of fuzzy cognitive maps. Appl. Softw. Comput. J. **8**(1), 820–828 (2008)
27. Özyurt, F., Sert, E., Avci, E., Dogantekin, E.: Brain tumor detection based on Convolutional Neural Network with neutrosophic expert maximum fuzzy sure entropy. Measurement **147**(106803), 1–7 (2019)
28. Seetha, J., Raja, S.S.: Brain tumor classification using convolutional neural networks. Biomed. Pharmacol. J. **11**(3), 1457–1461 (2018)

Feature Transformations for Outlier Detection in Classification of Text Documents

Tomasz Walkowiak(✉)

Faculty of Information and Communication Technology,
Wroclaw University of Science and Technology, Wroclaw, Poland
tomasz.walkowiak@pwr.edu.pl

Abstract. In this paper, we investigate the influence of feature transformation on the results of outlier detection of text documents. We tested four outlier detection methods: Local Outlier Factor, Extreme Value Machine, Weibull-calibrated SVM, and the Mahalanobis distance. The analyzed text documents are represented by different feature vectors ranging from TF-IDF, through averaged word embedding (two types), to document embedding generated by the BERT network. Experimenting on two different text corpora, we show how a transformation of the feature space (vector representation of documents) influences the outlier detection results.

1 Introduction

The classification based on machine learning of texts, that is, the automatic assignment of a text to one of the predefined groups, becomes a practical tool useful in many areas [21]. However, as most of the methods are based on supervised learning, it has an important drawback, i.e. it works as a closed set classifier. Due to the learning process, a classifier is trained to assign any input data to one of the predefined classes seen during training. In real tasks, when input data could come from any source, the results achieved for data (a text in natural language in the analyzed case), which are far away from data used in training, are far from reality. The input text is assigned to one of the training categories, which could be false. The solution to this problem is to use one of the available outlier detection methods. A comprehensive overview of such methods can be found in [4]. In general, outlier detection methods try to give information on how far the input data are from the training data. They build a model on training data and calculate a kind of metric (distance or probabilistic one) that shows how far the input data are. Since outlier detection methods work on any type of data, it raises the question whether we should use original feature vectors or should we transform them in any way. The results of [21] suggest that some transformation of feature vectors could increase the results achieved by an outlier detection method.

W. Zamojski et al. (Eds.): DepCoS-RELCOMEX 2022, LNNS 484, pp. 361–370, 2022.
https://doi.org/10.1007/978-3-031-06746-4_35

The main objective of this paper is to analyze the influence of feature vector transformations on outlier detection results. We have tested four different outlier detection methods: LOF [1], EVM [17], W-SVM [19] and Mahalanobis distance [8], five different methods of feature vector transformation: no transformation, the MiniMax scaler, standardization, the Yeo–Johnson transformation, and the PCA transformation with whitening and L2-normalization. They were tested on two corpora of texts in Polish. Since there are a large number of techniques for representing text documents by feature vectors, with have taken into consideration the traditional one TF-IDF [18], two variants of word2vec methods [7] (fastText in the supervised mode and doc2vec trained on a huge corpus), and the most recent approach - BERT [3].

The paper is organized as follows. Section 2 describes the transformations used from the feature vector and outlier detection methods. Section 3 presents the corpora and methods used for the generation of feature vectors. Section 4 introduces the evaluation metrics for outlier detection and presents the results of the comparative study.

2 Methods

2.1 Feature Vector Transformations

Transformation o feature vectors is a commonly used technique in machine learning. In the experiments, we have tested the following methods: MiniMax scaler, standardization, Yeo–Johnson transformation, and PCA transformation with whitening and L2-normalization. The MinMax scaler performs a linear transformation on the original feature vector to the specified interval. In the experiments reported, we used the interval $(-1, 1)$. MiniMax with interval $(-1, 1)$ is suggested as the default transformation of the SVM feature vectors in the libSVM package [2]. It was shown in [12] to be the best method for the detection of intrusion data, a task similar to the detection of outliers. Standardization (*stand.*) removes the mean and scales the feature vectors to the unit variance. It is a widely used transformation of data in machine learning and is known in statistics as the z-score. The mean and standard deviation are calculated from the training data and later used for the test and outlier set. Standardization was inspired by the results of [21] suggesting that it improves the detection of outliers in the case of the LOF method. The Yeo–Johnson (YJ) transformation [24] aims to make the feature vectors more Gaussian-like. This transformation was successfully used for word2vec-like data, as reported in [22]. And finally, we used the PCA-based transformation of the feature vectors. It was inspired by the work on image retrieval [20] in which feature vectors from the deep network are L2-normalized, then PCA-whitened [6] and again L2-normalized. Within the experiments carried out, we used the same transformation chain and kept 128 components of PCA, so the transformed feature vector was reduced to 128 dimensions. Similarly to the standardization, the Yeo–Johnson and PCA transformation models are estimated on the training data set and used for testing and an outlier one.

2.2 Outlier Detection Methods

The Local Outlier Factor (LOF) [1] is based on an analysis of the local density of points. It works by calculating the local reachability distance, defined as the average distance between a given point, its neighbors, and their neighbors. The relative density of a point against its neighbors is used to indicate the degree of the object being an outlier. The local outlier factor is formally defined as the average of the ratio of the local reachability of an object to its k-nearest neighbors. If the LOF value for a given point is larger than some threshold, the point is assumed to be an outlier. The original LOF [1] is based on the Euclidean distance. It could be easily extended to any other distance metric. We have also tested the cosine metric as suggested in [21].

Extreme Value Machine (EVM) [17] constructs a probability model that a given feature vector belongs to the class of the train set. This model is justified by the Extreme Value Theory. In contrast to LOF, it uses the training set labels. EVM builds a parametric model (assuming the Weibull distribution) of the margin distance. The margin distance is defined as half of the distance between a given class point and its nearest neighbor from any other class. Similarly to LOF it was originally based on the Euclidean distance but could be easily extended to the cosine one.

The Weibull-calibrated Support Vector Machine (W-SVM) was proposed in [19]. It combines a 1-Class SVM with a binary SVM specifically for outlier detection problems. The usage of the Weibull distribution in 1-Class SVM results in a decrease in the probability of class membership as feature vectors move from the training set toward outliers. Moreover, the Weibull distribution provides better modeling at the decision boundaries for a binary SVM. It gives a good generalization even in the presence of outliers [15].

The Mahalanobis distance is commonly used in outlier detection for pretrained neural classifiers [8]. It is based on the construction of a parametric model and represents each class by the mean of the class vector and the covariance vector. The outlier score for unseen data is given as the negative of the minimum Mahalanobis distance among available classes. In the experiments reported, we used a robust estimator of covariance [16].

3 Data Sets

To evaluate the pre-processing algorithms in the outlier detection task, we used two text corpora. The first, *wiki*, consists of articles extracted from Wikipedia in the Polish language. It consists of ca. 10,000 documents assigned to 34 categories (classes) [14]. The second corpus *qaul* [13] covers approximately 3,000 documents containing descriptions of qualifications from a Polish public register of the Integrated Qualifications System and descriptions of degrees from Polish universities. Data were manually divided into 36 classes (the sectors to which the qualifications belong, i.e., IT, culture, medicine). The corpus was randomly divided into training and testing sets in the same proportion as the *wiki* corpus (7:3). The features of these data sets are summarized in Table 1. As outliers, we randomly selected articles from the Polish press news [23] data set. The size of the outlier set is equal to the size of the training set.

Table 1. Data sets used; number of classes (labels), number of documents, and performance of closed set recognition (accuracy).

Dataset	Classes	Size of train set	Size of test/Outlier set	Close set accuracy			
				TF-IDF	fastText	doc2vec	BERT
wiki	34	6885	2952	0.771003	0.875678	0.916328	0.942073
qual	36	2108	904	0.795354	0.805310	0.823009	0.806416

There are many different feature generation methods for text documents. For our study, we have selected four of them. TF-IDF uses a bag-of-word model [18] to build feature vectors, i.e. it counts the occurrences of most common words (or their n-grams) in corpora and weights these frequencies by maximum term frequency in a document and by IDF (inverse document frequencies). We have used words from a text. The most frequent words, 2 g and 3 g (1000 in the case analyzed), and the corresponding IDFs were set up in the training set and used for the calculation of the features for the training, testing, and outlier data set. The great step in the area of text classification was the introduction of word2vec methods [11]. In these approaches, individual words are represented by high-dimensional feature vectors (word embedding) trained on large text corpora. Since classical statistical classifiers require constant-length input vectors, the most common solution is to average the vector representations of individual words. This approach is known as doc2vec [7]. Within the experiments reported, we used pre-trained vectors for the Polish language [9][1].

The next method analyzed, fastText [7] is similar to the previous one, since it uses the doc2vec representation of documents (an average of word embeddings). However, word embeddings are learned supervisedly in the analyzed data set, not in the large external corpora. The main idea behind fastText is to perform word embedding and classifier learning simultaneously.

The newest approaches to language modeling are inspired by deep learning algorithms and context-sensitive methods. The state of the art is BERT [3] based on the transformer architecture. In this study, we used the Polbert[2], a pre-trained BERT model for Polish [10]. We have extended the Polbert network by additional classification layers (fully connected) and tuned the whole model (BERT part and a classifier) on each dataset. As a feature vector (768 dimensions), we used the first token (with index zero) from the last Transformer layer (i.e., the one before the classification layers).

[1] http://hdl.handle.net/11321/606.

[2] https://huggingface.co/dkleczek/bert-base-polish-cased-v1.

4 Numerical Experiments

4.1 Evaluation Metrics

In evaluating the feature transformation method in the outlier detection task, we follow the approach used in [5] where the outlier detection is seen as a binary classification. Outliers are defined as the positive class, and closed set examples (testing and training set) as the negative class. All analyzed outlier methods return a value that after thresholding allows for binary classification. As an outlier detection quality metric, we used the area under the receiver operating characteristic curve (AUC). The receiver operating characteristic curve shows the False Positive Rate (FPR) on the X axis and the True Positive Rate (TPR) on the Y axis across multiple thresholds. The higher the AUC (the maximum is 1.0), the better, and an uninformative detector has an AUC of 0.5. The AUC metric represents the outlier detection performance on various thresholds. Therefore, we have also used the FPR95 metric, which represents performance at a strict threshold, that is, when the TPR is equal to 0.95. Lower FPR95 is better. In the experiments carried out, the data was divided into three data sets: training, testing, and outlier. The training data set was used to build feature vector generation and outlier models, where test and outlier sets (as a negative and positive class) were used to calculate the AUC and FPR95 metric.

4.2 Results of Experiments

Table 2. Outlier detection for the LOF method and the *wiki* corpus. Values in bold mark the best results among feature extraction methods and quality metrics.

Dist.	Transform.	AUC				FPR95			
		TF-IDF	fastText	doc2vec	BERT	TF-IDF	fastText	doc2vec	BERT
euc.	-	**0.8278**	0.7273	0.7614	0.9042	**0.4106**	0.6372	0.6104	**0.2642**
	MinMax	0.8064	0.71	0.7599	**0.9044**	0.5254	0.6721	0.6175	**0.2642**
	Standard	0.7935	0.7157	0.7618	0.904	0.4956	0.667	0.6077	**0.2642**
	YJ	0.5934	0.7069	**0.7647**	0.9039	0.9316	0.664	**0.6026**	0.2649
	PCA	0.707	**0.7663**	0.697	0.7623	0.6758	**0.5383**	0.6446	0.4658
cos.	-	**0.8072**	0.8547	0.7646	0.9062	**0.4749**	0.4194	0.6213	0.2534
	MinMax	0.8052	0.7902	0.7715	**0.9086**	0.518	0.5528	0.583	**0.2503**
	Standard	0.6263	0.8791	**0.8132**	0.9085	0.7073	**0.3709**	**0.519**	0.2486
	YJ	0.681	**0.8825**	0.8122	0.9085	0.6697	0.3675	0.5251	0.251
	PCA	0.7033	0.7685	0.6943	0.7582	0.6799	0.5285	0.6413	0.4688

Tables 2, 3, 4, and 5 show the results for the *wiki* corpus. The results for LOF do not suggest any best method of feature vector transformation. Each feature generation method has a different transformation method, achieving the best results. Similar cases are results for the EVM method with Euclidean distance.

Table 3. Outlier detection for the EVM method and the *wiki* corpus. The results for the cosine distance and the TF-IDF method are missing because of the EVM implementation bug that causes the EVM model reduction process to never end.

Dist.	Transform.	AUC				FPR95			
		TF-IDF	fastText	doc2vec	BERT	TF-IDF	fastText	doc2vec	BERT
euclid.	-	**0.7948**	0.8814	**0.8275**	0.898	**0.5058**	0.3357	0.4482	0.6829
	MinMax	0.5498	0.8713	0.8268	0.8986	0.9004	0.3554	**0.3943**	0.6839
	Standard	0.5202	0.8366	0.6414	0.9019	0.9597	0.394	0.7171	0.6402
	YJ	0.5198	0.8366	0.6392	0.8964	0.9604	0.3886	0.7215	0.7246
	PCA	0.718	**0.8889**	0.7869	**0.9196**	0.54	**0.3347**	0.519	**0.268**
cosine	-		0.9165	0.8691	0.8265		0.2862	0.4014	0.5474
	MinMax	0.8162	0.9116	0.879	0.8419	0.4898	0.2825	0.3777	0.5464
	Standard	0.6933	0.9254	0.8916	0.8399	0.7195	0.2683	0.3689	0.5488
	YJ	0.7097	**0.93**	**0.8921**	0.836	0.7575	**0.2575**	**0.3631**	0.5583
	PCA	**0.846**	0.8976	0.8539	**0.9429**	**0.4638**	0.3218	0.4173	**0.1897**

Table 4. Outlier detection for the W-SVM method and the *wiki* corpus.

Transformation	AUC				FPR95			
	TF-IDF	fastText	doc2vec	BERT	TF-IDF	fastText	doc2vec	BERT
-	0.6975	0.8785	0.7324	0.9028	0.8394	0.5437	0.8025	0.3083
MinMax	0.5057	0.7793	0.7137	0.912	1.0	0.6487	0.7371	0.2561
Standard	**0.8753**	0.9007	0.84	0.908	**0.3736**	0.3767	0.6612	0.2818
YJ	**0.8753**	0.8997	0.838	0.9085	0.4292	0.4007	0.6596	0.2801
PCA	0.8353	**0.9268**	**0.8761**	**0.9265**	0.5518	**0.2571**	**0.4875**	**0.2161**

Table 5. Outlier detection for the Mahalanobis method and the *wiki* corpus.

Transformation	AUC				FPR95			
	TF-IDF	fastText	doc2vec	BERT	TF-IDF	fastText	doc2vec	BERT
-	0.5683	0.4916	0.5457	**0.9245**	0.876	0.9126	0.8699	**0.2236**
MinMax	0.6423	0.4954	0.5509	0.9226	0.8862	0.9024	0.8672	0.2314
Standard	0.6232	0.503	0.5531	0.9229	0.8974	0.9082	0.8659	0.231
YJ	0.5191	0.6163	0.5652	0.9202	0.9451	0.7412	0.8543	0.2334
PCA	**0.7247**	**0.7645**	**0.6946**	0.9088	**0.6792**	**0.4675**	**0.643**	0.2761

However, for the cosine distance, which results outperform the Euclidean distance one, TD-IDF and BERT give the best results for the PCA transformation, and fastText and doc2vec for the Yeo–Johnson transformation. The results for W-SVM suggest that PCA is the best, except for the TF-IDF feature generation method. In the case of Mahalonibis-based outlier detection again, PCA gives the best results, in this case except for the BERT method.

Table 6. Outlier detection for the LOF method and the *qual* corpus.

Dist.	Transform	AUC				FPR95			
		TF-IDF	fastText	doc2vec	BERT	TF-IDF	fastText	doc2vec	BERT
euc.	-	**0.9207**	**0.9255**	0.9902	0.9257	**0.1914**	**0.2743**	0.0288	0.3794
	MinMax	0.8618	0.9116	**0.9915**	0.9244	0.4934	0.3131	0.0221	0.3728
	Standard	0.8856	0.9104	0.9908	0.9285	0.302	0.3119	0.0232	0.3673
	YJ	0.2191	0.9093	**0.9915**	**0.9325**	0.9845	0.3252	**0.0199**	0.3673
	PCA	0.8044	0.8628	0.9285	0.8721	0.2998	0.2666	0.2356	**0.2965**
cos.	-	**0.924**	**0.8977**	**0.9911**	**0.924**	**0.1914**	0.3131	**0.0254**	0.3993
	MinMax	0.8514	0.8868	0.9174	0.9134	0.5487	0.3606	0.1836	0.4314
	Standard	0.6448	0.8667	0.8647	0.8995	0.7489	0.3086	0.2732	0.4325
	YJ	0.2961	0.8759	0.8831	0.9026	0.9801	0.3219	0.2201	0.4181
	PCA	0.828	0.8583	0.9317	0.8697	0.2489	**0.2754**	0.2257	**0.3374**

Table 7. Outlier detection results for the EVM method and the *qual* corpus.

Dist.	Transform	AUC				FPR95			
		TF-IDF	fastText	doc2vec	BERT	TF-IDF	fastText	doc2vec	BERT
euc.	-	**0.9433**	**0.9779**	**0.9966**	**0.9255**	**0.1327**	0.0907	**0.0111**	0.1692
	MinMax	0.7345	0.9742	0.9336	0.9457	0.531	**0.0896**	0.1327	**0.1162**
	Standard	0.6908	0.936	0.7959	0.9257	0.6184	0.1195	0.4082	0.1925
	YJ	0.6786	0.9339	0.7926	0.9195	0.6427	0.1283	0.4148	0.1803
	PCA	0.8991	0.9336	0.7373	0.8934	0.344	0.1748	0.4226	0.2179
cos.	-	**0.9921**	0.9468	0.9965	**0.9736**	**0.0199**	0.281	**0.0111**	0.1438
	MinMax	0.9332	0.9273	**0.9967**	0.9682	0.2633	0.2434	0.0111	0.1847
	Standard	0.9576	0.9579	0.9862	0.9696	0.1272	0.1659	0.042	0.1648
	YJ	0.7958	**0.9654**	0.986	0.9714	0.4757	**0.156**	0.0409	0.1361
	PCA	0.9499	0.9522	0.8998	0.9705	0.2179	0.1471	0.2279	**0.1117**

Tables 6, 7, 8, and 9 show the results for the *qual* corpus. It could be noticed that for some cases the outlier detection does not work since the AUC is equal to or below 0.5, for example, in the case of Mahalanobis and BERT. However, for many methods, the AUC is very high, close to 1. In the case of LOF and EVM, the raw feature vectors give the best or almost the best results. This is not true for W-SVM where fastText and BERT give the best results for PCA-based

Table 8. Outlier detection for the W-SVM method and the *qual* corpus.

Transform	AUC				FPR95			
	TF-IDF	fastText	doc2vec	BERT	TF-IDF	fastText	doc2vec	BERT
-	**0.9957**	0.8405	**0.9988**	0.9568	**0.0188**	0.8219	**0.0022**	0.1084
MinMax	0.2511	0.8046	0.9908	0.9172	1.0	0.771	0.0199	0.1416
Standard	0.9358	0.7698	0.5247	0.9435	0.0896	1.0	1.0	0.135
YJ	0.9122	0.7753	0.6716	0.9516	0.0985	1.0	1.0	0.135
PCA	0.8058	**0.9682**	0.9892	**0.9721**	0.792	**0.0973**	0.0387	**0.0785**

Table 9. Outlier detection for the Mahalanobis method and the *qual* corpus.

Transformation	AUC				FPR95			
	TF-IDF	fastText	doc2vec	BERT	TF-IDF	fastText	doc2vec	BERT
-	0.3807	0.5472	0.8403	0.5	**0.9624**	**0.7765**	0.8496	1.0
MinMax	0.4769	0.5375	0.8771	0.5	0.8606	0.844	**0.7389**	1.0
Standard	0.4199	**0.5636**	**0.8818**	0.5	0.948	0.8208	0.7445	1.0
YJ	**0.8826**	0.5422	0.8433	0.5	0.3319	0.8252	0.8241	1.0
PCA	0.3018	0.5866	0.5971	0.5	0.9558	0.9115	0.9723	1.0

transformation. It is hard to analyze Mahalanobis results, since the method is not working at all or gives very poor results. It is probably caused by the *qua* corpora structure, that is, a large number of not equally balanced classes. Moreover, the closed-set classification accuracy close to 0.8 may suggest that classes are not well separated, which makes Mahalanobis wrongly model the training data with its simple model.

Table 10 shows the best AUC for each method of generating vectors of characteristics. In the case of the *wiki* class, EVM with cosine distance is the winner among outlier detection methods and similarly with the Yeo–Johnson transformation. Moreover, it can be seen that for *qual* the most successful case is W-SVM without any transformation. However, the best results are for the doc2vec method with W-SVM and PCA-based normalization.

Table 10. The best outlier detection and feature vector transformation for both corpora and all feature vector generation methods.

Dataset	Best results	TF-IDF	fastText	doc2vec	BERT
wiki	Method	W-SVM	EVM	EVM	EVM
	Distance	-	Cosine	Cosine	Cosine
	Transformation	YJ, Standard	YJ	YJ	PCA
	AUC	0.8753	0.93	0.8921	0.9429
qual	Method	W-SVM	EVM	W-SVM	W-SVM
	Distance	-	Euclidean	-	-
	Transformation	-	-	PCA	
	AUC	0.9957	0.9779	0.9988	0.9721

5 Conclusions

We have performed a large number of numerical experiments analyzing the influence of the feature vector transformation on outlier detection for two corpora and four different methods of feature generation. The conclusions of the analysis are not straightforward. There is no best transformation method. However, results show that feature vector transformation has to be taken into account in outlier detection tasks, as it could significantly improve the results in some cases. For example, the best AUC result for the *wiki* corpus is 0.9429. It was achieved for the EVM method with a cosine distance and PCA-based transformation. Raw features give AUC equal only to 0.8265. Moreover, the results achieved could suggest that outliers far away from the closed set data (like in the case of *qual* corpus) could require the usage of raw features. Whereas in the case of outliers closer to the training data, the feature transformation like Yeo–Johnson gives better results. However, this thesis requires more investigations. Therefore, we plan to perform the next experiments in which we will have more control over the outlierness of the analyzed data sets.

References

1. Breunig, M.M., Kriegel, H.P., Ng, R.T., Sander, J.: LOF: identifying density-based local outliers. In: ACM SIGMOD Record, vol. 29, pp. 93–104. ACM (2000)
2. Chang, C.C., Lin, C.J.: LIBSVM: a library for support vector machines. ACM Trans. Intell. Syst. Technol. **2**, 27:1–27:27 (2011). http://www.csie.ntu.edu.tw/~cjlin/libsvm
3. Devlin, J., Chang, M.W., Lee, K., Toutanova, K.: BERT: pre-training of deep bidirectional transformers for language understanding. arXiv:1810.04805 (2018)
4. Geng, C., Huang, S.j., Chen, S.: Recent advances in open set recognition: a survey. IEEE Trans. Pattern Anal. Mach. Intell. **43**, 3614–3637 (2020)
5. Hendrycks, D., Mazeika, M., Dietterich, T.: Deep anomaly detection with outlier exposure. In: Proceedings of the International Conference on Learning Representations (2019)
6. Jégou, H., Chum, O.: Negative evidences and co-occurrences in image retrieval: the benefit of PCA and whitening. In: ECCV - European Conference on Computer Vision. Firenze, Italy, October 2012. https://hal.inria.fr/hal-00722622
7. Joulin, A., Grave, E., Bojanowski, P., Mikolov, T.: Bag of tricks for efficient text classification. In: Proceedings of the 15th Conference of the European Chapter of the Association for Computational Linguistics, vol. 2, Short Papers, pp. 427–431. Association for Computational Linguistics (2017)
8. Kamoi, R., Kobayashi, K.: Why is the mahalanobis distance effective for anomaly detection? arXiv preprint arXiv:2003.00402 (2020)
9. Kocon, J., Gawor, M.: Evaluating KGR10 polish word embeddings in the recognition of temporal expressions using BILSTM-CRF. CoRR abs/1904.04055 (2019). http://arxiv.org/abs/1904.04055
10. Kłeczek, D.: PolBERT: attacking polish NLP tasks with transformers. In: Ogrodniczuk, M., Kobyliński, Ł. (eds.) Proceedings of the PolEval 2020 Workshop, pp. 79–88. Institute of Computer Science, Polish Academy of Sciences (2020)
11. Le, Q., Mikolov, T.: Distributed representations of sentences and documents. In: International Conference on Machine Learning, pp. 1188–1196 (2014)
12. Li, W., Liu, Z.: A method of SVM with normalization in intrusion detection. Proc. Environ. Sci. **11**, 256–262 (2011)
13. Marcińczuk, M., Gniewkowski, M., Walkowiak, T., Będkowski, M.: Text document clustering: WordNet vs. TF-IDF vs. word embeddings. In: Proceedings of the 11th Global Wordnet Conference, pp. 207–214. Global Wordnet Association, University of South Africa (UNISA), January 2021

14. Młynarczyk, K., Piasecki, M.: Wiki train - 34 categories (2015). http://hdl.handle.net/11321/222. CLARIN-PL digital repository
15. Rattani, A., Scheirer, W.J., Ross, A.: Open set fingerprint spoof detection across novel fabrication materials. IEEE Trans. Inf. Foren. Secur. **10**(11), 2447–2460 (2015)
16. Rousseeuw, P.J., van Driessen, K.: A fast algorithm for the minimum covariance determinant estimator. Technometrics **41**(3), 212–223 (1999)
17. Rudd, E.M., Jain, L.P., Scheirer, W.J., Boult, T.E.: The extreme value machine. IEEE Trans. Pattern Anal. Mach. Intell. **40**(3), 762–768 (2017)
18. Salton G, B.C.: Term-weighting approaches in automatic text retrieval. Inf. Process. Manag. **24**(5), 513-523 (1988)
19. Scheirer, W.J., Jain, L.P., Boult, T.E.: Probability models for open set recognition. IEEE Trans. Pattern Anal. Mach. Intell. **36**(11), 2317–2324 (2014)
20. Tolias, G., Sicre, R., Jégou, H.: Particular object retrieval with integral max-pooling of CNN activations. In: ICLR 2016 - International Conference on Learning Representations, pp. 1–12. International Conference on Learning Representations, San Juan, Puerto Rico, May 2016. https://hal.inria.fr/hal-01842218
21. Walkowiak, T., Datko, S., Maciejewski, H.: Distance metrics in open-set classification of text documents by local outlier factor and Doc2Vec. In: Wotawa, F., Friedrich, G., Pill, I., Koitz-Hristov, R., Ali, M. (eds.) IEA/AIE 2019. LNCS (LNAI), vol. 11606, pp. 102–109. Springer, Cham (2019). https://doi.org/10.1007/978-3-030-22999-3_10
22. Walkowiak, T., Gniewkowski, M.: Evaluation of vector embedding models in clustering of text documents. In: Proceedings of the International Conference on Recent Advances in Natural Language Processing (RANLP 2019), pp. 1304–1311. INCOMA Ltd., Varna, Bulgaria, September 2019
23. Walkowiak, T., Malak, P.: Polish texts topic classification evaluation. In: Proceedings of the 10th International Conference on Agents and Artificial Intelligence, vol. 2, ICAART, pp. 515–522. INSTICC, SciTePress (2018)
24. Yeo, I.K., Johnson, R.A.: A new family of power transformations to improve normality or symmetry. Biometrika **87**(4), 954–959 (2000)

Automatic Classification of Complaint Reports in Waste Management Systems Using TF-IDF, fastText, and BERT

Tomasz Walkowiak[1(✉)], Alicja Dąbrowska[2], Robert Giel[2], and Sylwia Werbińska-Wojciechowska[2]

[1] Faculty of Information and Communication Technology, Wroclaw University of Science and Technology, Wroclaw, Poland
tomasz.walkowiak@pwr.edu.pl

[2] Faculty of Mechanical Engineering, Wroclaw University of Science and Technology, Wroclaw, Poland

Abstract. The paper concerns the issue of automatic text classification of complaint letters written in Polish that were sent to the municipal waste management system operating in one of the largest Polish cities. The problem analyzed regards a multi-class classification task with information source separation. The authors compare five approaches, starting from TF-IDF, through word2vec methods, and to transformer-based BERT models. The article includes a detailed analysis of the experiments performed and the data set used. The analysis was performed according to the stratified k-fold cross-validation with 10 folds. The classification results were analyzed using three measures: precision, average F1 score, and weighted F1 score. The results obtained confirm that the BERT-based approach outperforms the other approaches. Indeed, the HerBert large model is recommended for use in similar downstream tasks in Polish.

Keywords: Text classification · Complaint reports · Waste management · Word embedding · Language model · fastText · BERT

1 Introduction

In this paper, the authors focus on the problem of ensuring effective waste management in the context of the assessment of information management systems. A conducted investigation into the operation of the municipal waste management system in a selected Polish city clearly indicates the problem of long delays in responding to residents' complaints resulting from inadequate categorization of complaints. Many unresolved resident problems have been reported due to inefficiency of the information flow [3]. Therefore, it is reasonable to introduce an automatic classification of residents' complaints according to different criteria adapted to the requirements of municipal waste management systems.

A good summary of the implementation of the machine learning (ML) technique for clustering and categorization of Polish text is given, for example, in [17].

W. Zamojski et al. (Eds.): DepCoS-RELCOMEX 2022, LNNS 484, pp. 371–378, 2022.
https://doi.org/10.1007/978-3-031-06746-4_36

The specificity of the morphology of the Polish language is given in [16]. Furthermore, the problem of a classification of texts in Polish was investigated, e.g., in [11,15]. However, the types of databases already analyzed (e.g. Wikipedia, Twitter) do not allow direct implementation of the obtained experiment results to municipal waste management system problems. Furthermore, despite the growing interest in the use of machine learning algorithms in the area of text classification, there are very few works dedicated to waste information management in the context of text classification implementation. The previous work of the authors (see [2,4]) implemented the classical algorithms for supervised learning and active learning approaches. The proposed study considered specific requirements related, for example, to the necessary reaction speed, the occurrence of different types of reporting users with different levels of knowledge related to waste or the specificity of the waste language. The main limitation of that study was the structured data set taken for the investigation.

Therefore, the main objective of this document is to expand knowledge about waste information management in the context of the implementation of automatic text classification. The authors' primary intention is to compare different approaches to automatic text classification for the analysis of complaints about municipal waste management written in Polish and the analysis of the results obtained taking into account the unstructured data set. The article is a continuation of the research work of the authors on ML algorithms used in waste management areas (see, e.g. [2,4]) and automatic text classification for Polish (see, e.g. [14]). The paper is structured as follows; first, we describe the data sets used in the analysis. In Sect. 3, we provide an overview of the classification methods used. After that, the experiments and the results are presented. Finally, Sect. 6 concludes the investigation and presents recommendations for future research work directions in the given area.

2 Complaint Reporting System

The data set used for the classification includes data on reports of irregularities in the execution of the waste collection process and cleanliness in the city of Wroclaw in Poland. Users reported irregularities in 7 classes. The most numerous group was waste collection (32%), followed by the administrator zone (24%), container/waste bags (21%), city cleaning (14%), others (7%), segregation (1%) and winter related problems (1%). Data were collected from August 2017 to April 2019 and included 24,539 complaint reports (about 1,169 reports per month). Anyone can report the maintenance and cleanliness of the special purpose vehicle of the city of Wroclaw through the website. The reports were made by three groups of users: residents, employees, and administrators. Each report should consist of the details of the reporting person and a description of the irregularity. The description contains an average of 38 words, with a standard deviation of 31. The maximum number of words per complaint report is 667 [2].

3 Methods

There are a large number of different classification methods for text documents. In the experiments carried out we have six of them following the schema used in [14]. The most basic one is a bag-of-word based TF-IDF. The method is based on counting the occurrences of words from texts and weighting these frequencies by the maximum word frequency in a document and by IDF (inverse document frequency). The weighted frequencies were standardized, that is, we removed the mean and scaled the feature vectors to the unit variance. Standardization is a widely used transformation of data in machine learning and is known in statistics as the z-score. As a classifier, we have used the Support Vector Machine with Sigmoid Kernel [6] following the results from [2] (where it gave the best results). The list of words, the corresponding IDF, the mean and variance of each feature were set on the train set and used for the calculation of features in the test set. The big step in the area of text classification was the introduction of the word2vec method [10]. In these approaches, individual words are represented by high-dimensional feature vectors (word embeddings) trained on large text corpora, that is, why such methods are called language models (LM). Since classical statistical classifiers require constant-length input vectors, the most common solution is to average vector representations of individual words. This approach is known as doc2vec and is used in the fastText package [7]. The original word2vec method was extended [5] with position weights and subword information (character n-grams) that allow the generation of embedding for unseen words and behave much better for morphologically rich languages such as Polish. Within the reported experiments, we used pre-trained vectors for the Polish language [8][1]. Similarly to TF-IDF, fastText LM embeddings were normalized by z-score. As a classifier, we have used the multilayer perceptron (MLP) [6] trained using the stochastic gradient descent method (SGD) [6].

The next method analyzed, fastText supervised [7] is similar to the previous one, since it uses the doc2vec representation of documents (an average of word embeddings). However, word embeddings are learned in a supervised way on the analyzed data set, not on the external large corpora. The main idea behind fastText supervised is to perform word embedding and classifier learning simultaneously.

The newest approaches to language modeling are inspired by deep learning algorithms and are context-aware methods. The state of the art is BERT [1] based on the transformers [13] architecture. We have extended the BERT network by the classification layer (fully connected) and tuned the whole model (BERT part and classifier) on the train corpus. In this study, we used the following pre-trained BERT models for Polish:

- PolBert[2] [9],
- HerBert (base)[3] [12],

[1] http://hdl.handle.net/11321/606.
[2] https://huggingface.co/dkleczek/bert-base-polish-cased-v1.
[3] https://huggingface.co/allegro/herbert-base-cased.

- HerBert (large)[4] [12].

The first are the BERT base models, consisting of 12 layers, 768 hidden unit embeddings, 12 attention heads, and 110M parameters in total. The last one is the BERT large language model, consisting of 24 layers, 1024 hidden unit embeddings, 16 attention heads, and 335M parameters in total. It is three times larger than the Bert base model. The main drawback of the BERT method is the requirement of using GPU cards, whereas other methods (TF-IDF and fastText in both versions) could work on CPU. For a time performance analysis, see [14].

4 Full Data Set

4.1 Data Cleaning

The data obtained from the special purpose vehicle of the city of Wroclaw for maintenance and cleanliness included 24,539. However, due to the nature of the study carried out, the data were subject to certain modifications. Entries of a test nature (made by IT specialists) and entries shorter than 24 characters were removed. Due to the fact that some entries had a structure, fields such as subject, place, and author have been removed from the description, leaving only text descriptions of complains. After the changes, the database contained 23,639 unique entries.

4.2 Experiment Organization

The study was carried out according to the stratified k-fold cross-validation [6] (with 10 folds). Each data set was divided randomly into 10 folds, and 9 folds were used for training and renaming one for testing. We use three metrics calculated on the test dataset to report method performance: accuracy, average f1 score, and weighted f1 score. Accuracy is defined as all correctly classified objects over all objects. The accuracy is not well suited for non-equally sized labels (and this is our case). Therefore, we have also used the f1 measure. However, it is defined (as a harmonic mean of precision and recall) for binary classification. For multiclass problems, we can average the f1 scores for each label or take into account the support (the number of examples in each class) and make a weighted average that takes into account the label imbalance.

4.3 Results

The results, that is, the mean values and standard deviations of accuracy, the average f1 and the weighted f1, are presented in Table 1. It could be seen that the best method is the BERT classifier built on the HerBert language model. It outperforms other methods, confirming the thesis about BERT as the state-of-the-art in text classification. The sequence of methods ordered by effectiveness,

[4] https://huggingface.co/allegro/herbert-large-cased.

that is, TF-IDF, fastText LM, fastText supervised, Bert (base), Bert (large), follows the results for other data sets in Polish presented in [14]. The standard deviation of accuracy and weighted f1 is relatively small (0.3–0.6% points), confirming that the method performance is not based on the train/test split. The averaged f1 is smaller than the weighted f1 for all methods, as the labels are unbalanced and the classifier is less accurate for rarer labels. It also results in higher standard deviations for averaged f1.

Table 1. Classification results for the full data set

Method	Accuracy		Average f1		Weighted f1	
	Mean	Std	Mean	Std	Mean	Std
TF-IDF	81.53	0.3	65.81	1.07	80.44	0.3
fastText LM	84.18	0.36	76.15	0.64	83.70	0.30
fastText supervised	86.31	0.27	78.65	0.98	85.91	0.29
PolBert	87.70	0.43	81.26	1.15	87.43	0.50
HerBert (base)	88.6	0.63	**82.47**	1.05	88.32	0.59
HerBert (large)	**89.08**	0.56	82.33	1.19	**88.50**	0.61

5 Annotated Data Set

5.1 Problem Statement

Based on the analysis of the irregularity reports, it was observed that the content of the description does not always match the label given by the user. For this purpose, 2,400 reports from the entire database and manually labeled by two experts [2]. After it, the expert's assessment was cross-checked (it was checked to what extent the experts agreed). Only submissions containing several problems at the same time, which could be classified into more than one class, raised doubts. In such a situation, the submission was classified as a label other. Reporting persons can be divided into three groups: residents, company employees, and administrators. Our analysis of 2,400 reports indicates that the classification accuracy is the lowest for residents (84.37%). The company's employees and administrators, in contrast, have achieved accuracy of 94.30% and 98.05% respectively. Therefore, there is a need to improve the accuracy of classification.

5.2 Experiment Organization and Results

The annotated data set was cleaned as described in Sect. 4.1. It results in 2,365 complaints. The corpus was analyzed in the same way as described in 4. The results are presented in Table 2. We can notice that again the HerBert (large) outperforms other methods. However, the achieved values raise important questions:

- Why the values of performance metrics are lower than those in Table 1?
- Why the accuracy of TF-IDF is much smaller than that reported in [2] (76.68% compared to 86.0%), if almost the same method is used?

The answer to the first question is suggested by the high values of standard deviation. For example, 2.61 percentage of points for the accuracy in the case of PolBert, and 7.61 percentage of points for the average f1 accuracy in the case of HerBert (large). It suggests that there is great randomness in the data split of the train test. And when we look at the number of examples for the rarest labels (1 percentage of 2,365), it becomes clear that ca. 20 examples divided in a proportion of 9:1 for train and test sets results in only 2–3 examples for testing. Therefore, there are too few examples in the corpus. Moreover, guidelines for 'other' label annotation, i.e., submissions containing several problems at the same time, with a very low representation in the train set (ca. 20) makes this class very hard to distinguish by a machine learning-based classifier.

Table 2. Results for the data set annotated by experts and cleaned as described in Sect. 4.1. 2,365 complaints randomly divided into training (90%) and testing 10 times.

Method	Accuracy		Average f1		Weighted f1	
	Mean	Std	Mean	Std	Mean	Std
TF-IDF	76.66	1.0	50.72	2.19	76.13	1.08
fastText LM	80.92	1.76	55.96	4.34	80.27	1.91
fastText supervised	83.58	1.05	56.07	4.16	82.42	1.13
PolBert	86.53	2.43	68.35	6.2	86.06	2.4
HerBert (base)	88.06	2.14	74.02	6.38	87.62	2.06
HerBert (large)	**88.86**	1.6	**76.54**	7.61	**88.5**	1.7

To find the answer to the second question, we performed the next set of experiments. This time data set was not cleaned; we left all text data, for example, the subject of complaint (i.e., the class label selected by system users). We have only lower-cased text (as in [2]), since some of the reports were written only in capital letters. The results are shown in Table 3. We can see that the results are better than in previous experiments (Table 2), and the results for TF-IDF are very similar to those reported in [2]. It shows that report subjects (existing in about 60% of reports), containing text very close to the label name, are detected by classifiers and used for the final classification. It raises doubts as to whether the subject element of the reports should be used as input for the classifier. On the other hand, again HerBert (large) is the best solution for the analyzed data.

Table 3. Results for data set annotated by experts and not cleaned (text was only lowercased). 2400 complaints randomly divided into training (90%) and testing

Method	Accuracy		Average f1		Weighted f1	
	Mean	Std	Mean	Std	Mean	Std
TF-IDF	87.37	1.11	65.08	6.26	86.93	1.13
fastText LM	87.23	1.22	63.06	5.28	86.66	1.29
fastText supervised	89.78	0.91	65.35	4.78	88.99	0.99
PolBert	90.62	1.68	77.58	10.82	90.33	1.76
HerBert (base)	91.25	1.5	77.36	6.74	90.84	1.62
HerBert (large)	**91.58**	1.29	**79.86**	6.15	**91.29**	1.2

6 Conclusion

In this paper, the authors have analyzed six approaches for automatic text classification of complaint letters written in Polish. The investigated models were based on TF-IDF, through word2vec methods up to transformer-based BERT models.

The obtained results confirm that BERT based approach is the SOTA for subject classification of texts in Polish. The results show how important is transfer learning in modern text analysis. The usage of the language model (BERT trained on a large corpus of texts) allows to beat other approaches. We recommend to use HerBert large model for fine tuning in similar downstream tasks in Polish. Moreover, we have shown how important is a preprocessing of data. Data scientists have to carefully analyse the input data from the point of content (remove 'unfair' information, like subjects in the analysed case) and the used labels ('other' label used in the annotation process is very doubtful and have too few examples). An interesting area of further investigation is to use the full data set information in the process of building the classifier for the hand annotated data set (ten times smaller than the full data set). It could be achieved by fine-tuning the BERT model learned on the full data set or a use of Graph Neural Networks that will allow few-shot learning.

References

1. Devlin, J., Chang, M.W., Lee, K., Toutanova, K.: BERT: pre-training of deep bidirectional transformers for language understanding. arXiv preprint arXiv:1810.04805 (2018)
2. Dąbrowska, A., Giel, R., Werbińska-Wojciechowska, S.: Automatic multi-class classification of polish complaint reports about municipal waste management. In: Zamojski, W., Mazurkiewicz, J., Sugier, J., Walkowiak, T., Kacprzyk, J. (eds.) DepCoS-RELCOMEX 2021. AISC, vol. 1389, pp. 44–52. Springer, Cham (2021). https://doi.org/10.1007/978-3-030-76773-0_5

3. Giel, R.: A multi-criteria method for assessing the operation of a waste processing system (in Polish). Ph.D. thesis, Publ. House of Wroclaw University of Science and Technology, Wroclaw (2017)
4. Giel, R., Dąbrowska, A., Werbińska-Wojciechowska, S.: Active learning for automatic classification of complaints about municipal waste management. Environ. Protect. Eng. **47**, 53–66 (2021)
5. Grave, E., Bojanowski, P., Gupta, P., Joulin, A., Mikolov, T.: Learning word vectors for 157 languages. In: Proceedings of the International Conference on Language Resources and Evaluation (LREC 2018), pp. 3483–3487 (2018)
6. Hastie, T., Tibshirani, R., Friedman, J.: The Elements of Statistical Learning. SSS, Springer, New York (2009). https://doi.org/10.1007/978-0-387-84858-7
7. Joulin, A., Grave, E., Bojanowski, P., Mikolov, T.: Bag of tricks for efficient text classification. In: Proceedings of the 15th Conference of the European Chapter of the Association for Computational Linguistics: Volume 2, Short Papers, pp. 427–431. Association for Computational Linguistics (2017). http://aclweb.org/anthology/E17-2068
8. Kocon, J., Gawor, M.: Evaluating KGR10 polish word embeddings in the recognition of temporal expressions using BILSTM-CRF. CoRR abs/1904.04055 (2019). http://arxiv.org/abs/1904.04055
9. Kłeczek, D.: PolBERT: Attacking polish NLP tasks with transformers. In: Ogrodniczuk, M., Łukasz Kobyliński (eds.) Proceedings of the PolEval 2020 Workshop, pp. 79–88. Institute of Computer Science, Polish Academy of Sciences (2020)
10. Le, Q., Mikolov, T.: Distributed representations of sentences and documents. In: International Conference on Machine Learning, pp. 1188–1196 (2014)
11. Piskorski, J., Sydow, M.: Experiments on classification of polish newspaper. Arch. Control Sci. **15**, 613–625 (2005)
12. Rybak, P., Mroczkowski, R., Tracz, J., Gawlik, I.: KLEJ: comprehensive benchmark for polish language understanding. In: Proceedings of the 58th Annual Meeting of the Association for Computational Linguistics, pp. 1191–1201. Association for Computational Linguistics, Online, July 2020. https://www.aclweb.org/anthology/2020.acl-main.111
13. Vaswani, A., et al.: Attention is all you need. In: Guyon, I., et al. (eds.) Advances in Neural Information Processing Systems, vol. 30. Curran Associates, Inc. (2017). https://proceedings.neurips.cc/paper/2017/file/3f5ee243547dee91fbd053c1c4a845aa-Paper.pdf
14. Walkowiak, T.: Subject classification of texts in Polish - from TF-IDF to transformers. In: Zamojski, W., Mazurkiewicz, J., Sugier, J., Walkowiak, T., Kacprzyk, J. (eds.) DepCoS-RELCOMEX 2021. AISC, vol. 1389, pp. 457–465. Springer, Cham (2021). https://doi.org/10.1007/978-3-030-76773-0_44
15. Walkowiak, T., Datko, S., Maciejewski, H.: Bag-of-words, bag-of-topics and word-to-vec based subject classification of text documents in polish - a comparative study. In: Zamojski, W., Mazurkiewicz, J., Sugier, J., Walkowiak, T., Kacprzyk, J. (eds.) Contemporary Complex Systems and Their Dependability, pp. 526–535. Springer International Publishing, Cham (2019)
16. Walkowiak, T., Malak, P.: Polish texts topic classification evaluation. In: Proceedings of the 10th International Conference on Agents and Artificial Intelligence - Volume 2: ICAART, pp. 515–522. INSTICC, SciTePress (2018)
17. Wielgosz, M., et al.: Experiment on methods for clustering and categorization of polish text. Comput. Inform. **36**, 186–204 (2017)

Mobile Application for Diagnosing and Correcting Color Vision Deficiencies

Natalia Wcisło[1](✉), Michał Szczepanik[1], and Ireneusz Jóźwiak[2]

[1] Department of Applied Informatics, Faculty of Information and Communication Technology, Wroclaw University of Science and Technology, Wybrzeże Wyspiańskiego 27, 50-370 Wrocław, Poland
{240255,michal.szczepanik}@pwr.edu.pl

[2] Military University of Land Forces, ul. Piotra Czajkowskiego 109, 51-147 Wrocław, Poland
ireneusz.jozwiak@awl.edu.pl
https://kis.pwr.edu.pl/
https://www.wojsko-polskie.pl/awl/

Abstract. The main functionalities of the application are color recognition, applying filters to the camera feed and photos from the user's gallery, and diagnosis of color vision disorders for a green-red defect using the Ishihara test, and testing a specific area of the defect using the 100 hue test. The application includes additional functionalities, such as color descriptions, color name text-to-speech, and password authentication.

Keywords: Mobile solution · Color vision · Color blindness · Daltonism

1 Introduction

This paper presents an application for iOS devices that will allow users to diagnose the degree of their color vision defect and help them in their everyday activities regarding color recognition. The application is divided into two main functions: diagnosing and supporting the user. The diagnostic functionality consists of two screening tests for the patient's CVD. The functionality of the application related to supporting the user is color recognition and applying a filter to the camera. Color recognition can be used for real-time image recognition or for color recognition in an image retrieved from a user's device. When using this functionality, the user will receive information about the current recognized color, such as the name of the color, and its description, along with examples of recognizable objects that may represent this color in everyday life. The application contains four filters: three filters for the most popular varieties of CVD: protanopia, deuteranopia, tritanopia, and monochromatic.

The application is intended for people of all ages. Therefore, before each test, it presents a tutorial as well as a description of the test results. The application contains text-to-speech functionality for color recognition. The user is able to hear the name of the currently recognized color on the screen.

W. Zamojski et al. (Eds.): DepCoS-RELCOMEX 2022, LNNS 484, pp. 379–388, 2022.
https://doi.org/10.1007/978-3-031-06746-4_37

The article shows the results of work on the application and further steps in the application's development.

2 Classification of Color Vision Deficiencies

Color vision deficiency (CVD) is also known as color view defect [3]. CVD is a disease that affects plenty of people of different ages. Color vision deficiencies can be classified into acquired and congenital [4,16]. There are two types of photoreceptor cells: rods and cones [8]. The cause of CVDs is missing or damaged photoreceptor cones [7]. The sensitivity spectra of normal color vission and common types of CVD in Fig. 1. CVDs are classified into three levels of vision anomalies: tetrachrome, dichromization, and monochrome. In case of trichrome, one of the photoreceptor cones is damaged. In the case of dichrome, one of the cones is completely missing. In the case of monochrome, at least two photoreceptor cones are absent. Monochrome is the rarest type of CVD.

Table 1. Prevalence of CVD types (*source:* [13]).

CVD form	Type	Prevalence [%]
2*Monochromacy	Achromatopsia	0.003
	Blue cone monochromacy	0.001
3*Dichromacy	Protanopia	1.01
	Deuteranopia	1.27
	Tritanopia	0.2
3*Anomalous trichromacy	Protanomaly	1.08
	Deuteranomaly	4.63
	Tritanomaly	0.2

Prevalence of these CVD types in the population is presented in Table 1.

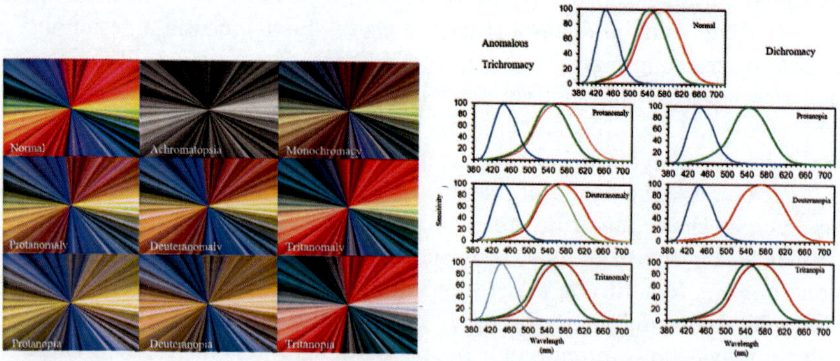

(a) Human color perception with different CVDs perceiving a rainbow of colors.

(b) Sensitivity spectra of normal color vision and common types of CVD.

Fig. 1. CVD color perception (*source:* [13]).

Normal human color vision is called trichromatic—Trichromats possess all three types of cones. Color blindness can be divided into three types: protanopia, deuteranopia, and tritanopia [7]. Protanopia is called red color blindness, and deuteranopia is called green color blindness. A person with protanopia does not recognize the red color or confuses it with the color green. Deuteranopia is the opposite of protanopia. In this case, the person does not recognize the green color or confuses it with the color red. In color blindness, the perception of red and green is varying—either the colors are swapped, or both are recognized as the same color. If the change in perception is mild, protanopia is called protanomaly, and deuteranopia is called deuteranomaly [4]. Human color perception with different CVDs perceiving a rainbow of colors in Fig. 1a.

Correction and treatment of color blindness are long-lasting and laborious or sometimes not possible. There is no cure for inherited color deficiency, but if the cause is an illness or eye injury, treating these conditions may improve color vision. In some cases, surgery is possible but very expensive. Ophthalmologists and surgeons undertake difficult operations that are often unsuccessful.

3 Solution

The purpose of the thesis is to design an application for iOS devices, which will allow users to diagnose CVD and help them in their everyday activities regarding color recognition. The application is divided into two main functions: diagnosing and supporting the user.

3.1 Comparison of the Features of Displays in Various Mobile Devices

A common problem with computerized diagnostic tests is their reliability and precision due to displaying or other adversities that may affect the test result. The selection of screen parameters has a significant impact on the patient's test results. The application was designed for iPhones, as their displays do not differ significantly from each other when it comes to displaying colors. The display parameters are presented in Table 2.

Table 2. Comparison of different models of telephone displays (*source: own*).

Parameters	iPhone 12	iPhone X	iPhone 8
Year of the premiere	2021	2018	2017
Display	Super Retina XDR	Super Retina HD	Retina HD
Maximum brightness of	625 nits	625 nits	625 nits
Wide color display (P3)	Yes	Yes	Yes
True Tone display	Yes	Yes	Yes
Contrast ratio	2,000,000:1	1,000,000:1	400:1

According to the Table 3, the difference in PPI and max screen brightness in different iPhone models is less widespread than in other producers mobile devices. That is why the differences during the 100 Hue test on Apple devices would be less visible and the test more accurate [10,11].

Table 3. List of parameters of various phone companies (*source: own*).

Model	PPI	Max brightness
Xiaomi Civi Pro	673	950 nits
Xiaomi Redmi Note 8 2021	409	559 nits
Google Pixel 6 Pro	512	497 nits
Huawei Mate 40 Pro 4G	456	476 nits
Samsung Galaxy Z Flip3 5G	426	935 nits
Motorola Moto G31	411	700 nits
Vivo S10	409	646 nits
iPhone 12	460	625 nits
iPhone 11	326	625 nits
iPhone X	458	634 nits
iPhone 8	326	625 nits

3.2 User Interface

The interface was built using SwiftUI, which is a relatively new technology. Moreover, it is a declarative framework for building Apple applications that comes with a state-based declarative approach. SwiftUI simplifies the interface implementation because the user no longer uses the storyboard but the declarative UI structure. Exits and actions are effectively checked at compile-time, reducing the risk of UI failure at runtime [5]. The visual layer, which is what the user sees on the screen, is new-fashioned, intuitive, and simple in use. In addition, interfaces follow Apple's Human Interface Guidelines guidelines.

3.3 Style Guide

The colors of the application should be matched based [6], which presents a list of the best to worst color options. The application should implement a combination of *better* and *best*, as seen on Fig. 2. The important feature is high contrast between the accent and the background color of the application, the ability to adjust the screen according to user's preferences. The colors should be matched to the type of color vision defect of the user. [1]. an example of adjusting the appearance of the application to the user's preferences is shown in Fig. 5c

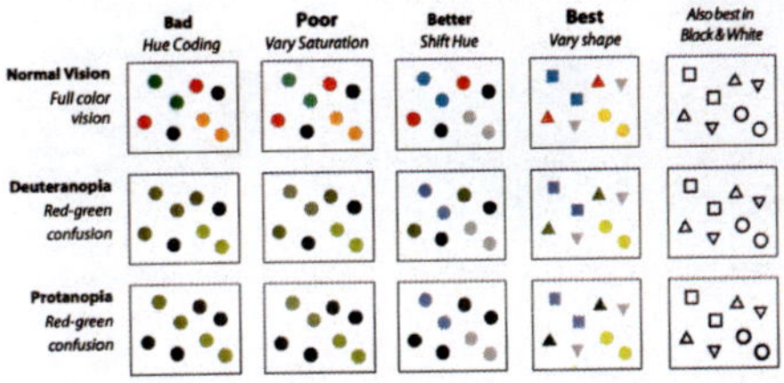

(a) Point classes typical of a dot map distinguished by saturation, hue and shape.

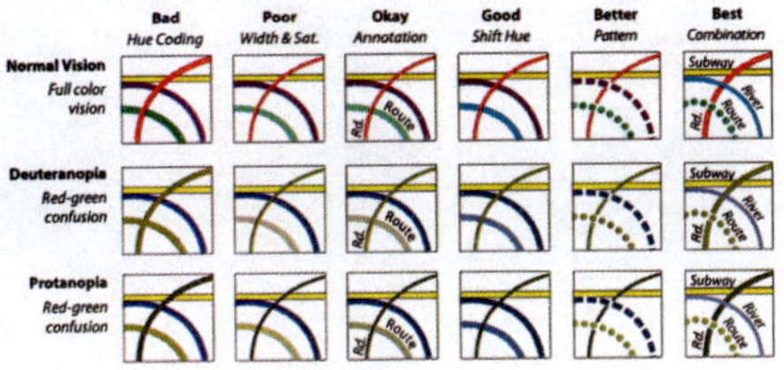

(b) Line classes distinguished by width and saturation, annotation, hue and line pattern.

Fig. 2. Color design for the Color Vision Impaired (*source:* [6]).

3.4 Supporting for the CVD Users

The functionality of the application related to supporting the user is color recognition and applying a filter to the camera.

Color recognition can be used for real-time image recognition (for instance when using a camera) or used for color recognition in an image retrieved from a user's device. When using this functionality, the user will receive information about the current recognized color, such as the name of the color, and its description, with examples of recognizable objects that may represent this color in everyday life. The application aims to help associate specific colors with objects, for example, grass → green. It utilizes a method used for people who have never seen certain colors so that they can associate colors with things, as descriptions are tailored to people with all color vision defects. The user also has the option to save the photo along with information about the color (the hue, as well as the name and description of the color).

The application contains a text-to-speech functionality for color recognition. The user is able to hear the name of the currently recognized color on the screen. It is great for users who have accompanying diseases such as blindness or severe vision impairment which are very common among people with CVDs [12]. The application's database contains 32 of the most basic colors, it has 20 different varieties based on primary colors along with silver and gold and 4 colors combining black and white.

The second functionality related to supporting the user is the application of filters to the camera or photos, which increase the contrast between colors. The application contains four filters:

- three filters for the most popular varieties of CVD: protanopia, deuteranopia, and tritanopia,
- monochromatic filter intended for people who are completely deprived of color vision, and only see shades of white and black,

These filters address color vision deficiencies that are visualized in Fig. 3.

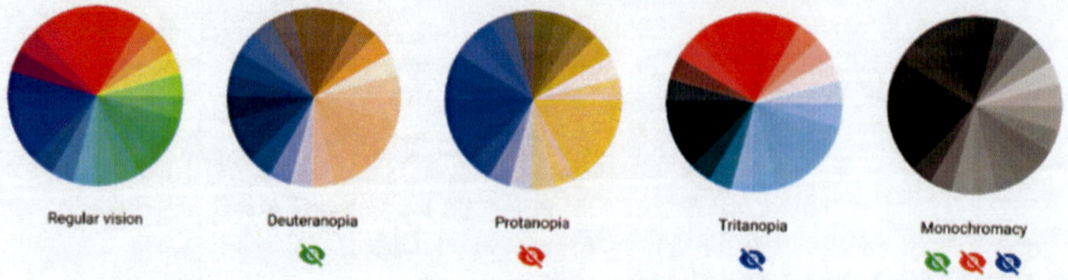

Fig. 3. Color circles as seen by people with various color defects compared to normal vision (*source:* [15]).

Filters. The filtering functionality of the application transforms the color space of a given image into a range more recognizable by people with color vision defects. This filter uses a tree-dimensional color table to transform pixels in the source image. The roots of lookup tables come from computer graphics and video games in order to quickly alter the appearance of graphics without a heavy cost.

To actually compute the lookup table the application uses the *LMS Daltonization algorithm* described in [14]. LMS Daltonization is one of the most famous algorithms when it comes to correction of colors for color-blindness. The idea of the algorithm is using the information lost in the simulation of color blindness and use LMS color space to compensate colors missing in each group/type of cones, long (L), medium (M), and short (S) in order to transform the colors into ones that are distinguishable to the viewer with a color vision defect. LMS color space is a color space which represents the response of the three types of cones of the human eye. The filter function screen is shown in Fig. 4c.

Color Recognition. The color recognition functionality allows the user to pick a color from the screen and find a color in a given set of predefined colors that most closely represents the picked color.

The process of picking a color is done by passing the XY coordinates of the pixel to the algorithm, translated by specified coordinates, effectively rendering a single pixel, which is then interpreted as the color picked by the user.

The next step is the creation of set of colors, which the algorithm will eventually have to look through. Color is represented as three separate floating-point numbers for red, green, and blue values. Each color is then assigned with a unique ID, which is computed the same way an RGB color would be represented as a single number:

$$V = *R \cdot 255 * 255^2 + *G \cdot 255 * 255 + *B \cdot 255$$

which essentially first transforms the ranges of all values from $[0, 1]$ to $[0, 255]$, then floors each value, and finally multiplies each value by 255^i, where i is the "position" of the value. Along with the ID, the algorithm also computes the HSL representation of the color from the default RGB representation. This results in a list of entries, each one defined by an ID, an RGB value, a HSL value, a color's name, and color's description.

To find the color x in the list *nearest* to the given color a, the algorithm "scores" each color x. After associating each color with a score, the algorithm picks the color with a highest score value and returns it as a color *nearest* to the provided color a.

The function computes two euclidean distances from a and x: one in terms of the RGB color space, and one in terms of the HSL color space. The distances are then added together to form a single score value, then returned by the function.

The algorithm results in a single list entry, which contains the color's name and description. These are then displayed to the user in the front-end, along with a colored dot. The color recognition function screen is shown in Fig. 4a and color description screen is shown in Fig. 4b.

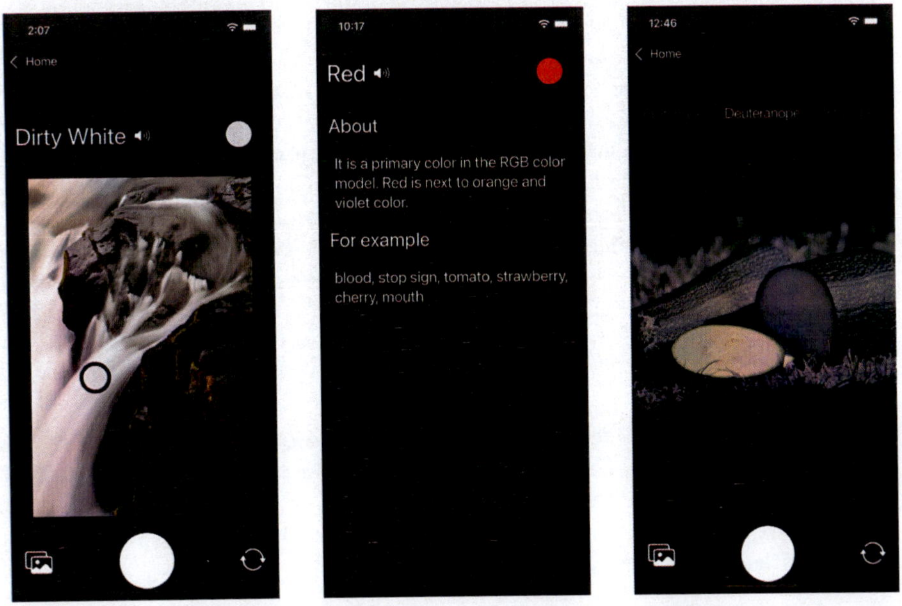

(a) Recognizing colors on a static photo.
(b) The details of the recognized color.
(c) The filter for the deuteranope CVD.

Fig. 4. Screenshots of views related to the supporting functions (*source: own work*).

3.5 Diagnosis of CVD Users

The diagnosing functionality consists of 2 screening tests for the patient's CVD.

The first is the Ishihara test, which is the most popular topic among the public, as it is widely used for the diagnosis of CVD when visiting an ophthalmologist. The test contains 38 plates, based on which the user will be diagnosed regarding the recognition of green and red color. In order for the test to be carried out reliably, the application uses a timer—the patient sees the test only for a few seconds so that the ophthalmologist can give an appropriate result.

The second test is the 100 hue test, also called an arrangement test. Its purpose is to diagnose and detect a specific wavelength that the user is unable to see. The test contains a set of 44 colors, based on which the application can determine the appropriate range of colors. This is a practical test when the disease is acquired or is an uncommon form of CVD [4,16]. The results of the test are presented in the form of a radial bar which is intended to visualize the color palette along with the error of the user. Its purpose is to diagnose and detect a specific wavelength that the user is unable to see.

The application is intended for people of all ages, which is why, before each test, a tutorial is presented.

The Ishihara Test. Ishihara plates displayed in the Ishihara tests are not regular raster images. Instead, each plate is defined as a set of circles, each with a position, radius, and a color. To draw the Ishihara plate, the application first has to load the CSV file of an appropriate plate, parse it, and draw each circle on a canvas. Images of Ishihara plates retrieved from materials available on the Internet [2] proved to have too low resolution for the purpose of this project. Additionally, because of the JPG format, they contain a lot of noise and have a white background, which would be unacceptable, given that the application provides a non-white background to users. For these reasons, the plates have been created by hand with the use of a specialized editor created just for this purpose. The editor was implemented as a simple web application with the use of the p5.js library and JavaScript. The application allows the user to set the background image of the plate to serve as a reference. The user then traces out the circles on the plate with the mouse by clicking on origins of circles and dragging the mouse to their circumferences. The colors are automatically picked from the original plate. After tracing all circles on the image, the user has the option to export all circles in JSON and CSV data formats. The test screen is shown in Fig. 5b.

The Hue Test. The hue test is divided into four color groups. These groups are defined as lists of colors, where each color c gets associated with an index i—the "correct" position in the color gradient. At the start of the test, the list of (c, i) pairs gets shuffled randomly. The user then has the ability to swap the pairs on the list by clicking on the colored blocks. After finishing the test, each (c, i) pair is compared to its final position on the list. This way, the application computes a score for each pair, which is essentially the distance from the final position of the pair to its "correct" i value. These scores are then presented to the user in the form of a radar chart. The test screen is shown in Fig. 5a.

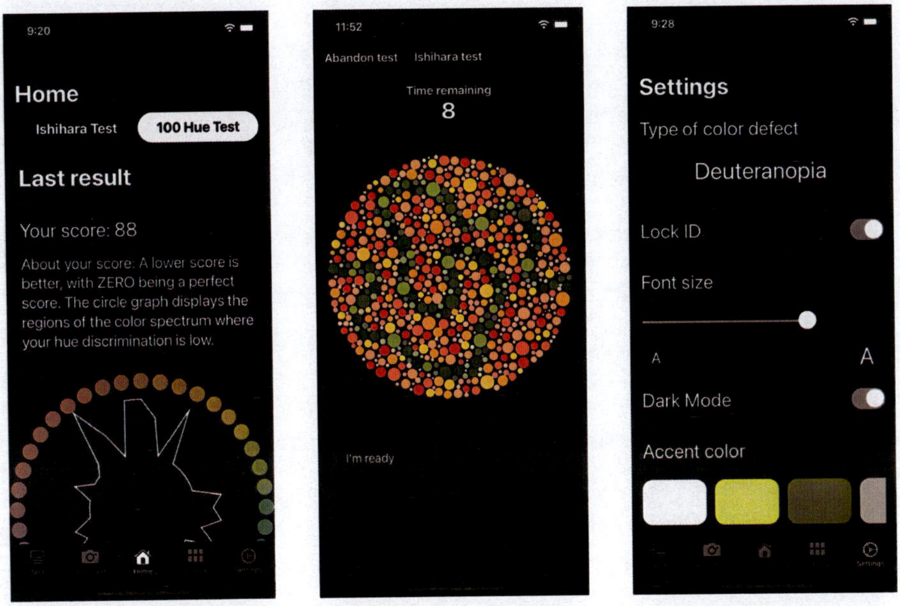

(a) The home view with the 100 hue test results.

(b) The details of the recognized color.

(c) The settings screen for deuteranopia.

Fig. 5. Screenshots of views related to the tests and settings (*source: own work*).

4 Conclusion and Future Work

Mobile devices have numerous capabilities that could aid in the improvement of color blindness diagnostics. Well-chosen devices allow you to perform an accurate test. This application could facilitate the work of optometrists and ophthalmologists. Based on research in a registry of clinical trials (ClinicalTrials.gov) and National Institutes of Health (nih.gov) [9], 1 in 12 men and 1 in 200 women suffer from the red-green recognition defect. Furthermore, 1 in 10,000 people worldwide cannot differentiate between blue and yellow, and 1 in 100,000 people worldwide has blue cone deficiency. In additional, mobile devices be able to help people in help them in their everyday activities regarding color recognition.

There are future development plans to constantly adapt the appearance of the application to keep up with the latest trends. The main points of the development plan are: a filter generated based on the hue test result. For instance, when the user has a hard time differentiating the red color, the application will increase the contrast between red and other colors to sharpen and improve the user's vision. The addition of new functionality for the application is called the "screen freeze function", in which users can freeze the screen and recognize colors directly on the static camera. The application will add a control test consisting of a set of 15 hues, the purpose of which this is to make an initial diagnosis of the patient's health condition.

References

1. Apple Developer: Human Interface Guidelines. https://developer.apple.com/design/human-interface-guidelines/. Accessed 19 Jan 2022
2. Colblindor: Ishihara's Test for Colour Deficiency: 38 Plates Edition. https://www.color-blindness.com/ishiharas-test-for-colour-deficiency-38-plates-edition/. Accessed 09 May 2022
3. Colour Blind Awareness Inherited colour vision deficiency. http://www.colourblindawareness.org/colour-blindness/inherited-colour-vision-deficiency/. Accessed 20 Jan 2022
4. Hasrod, N., Rubin, A.: Defects of colour vision: a review of congenital and acquired colour vision deficiencies. African Vision Eye Health **75**, 1 (2016)
5. Hudson, P.: SwiftUI lets us build declarative user interfaces in Swift. https://www.hackingwithswift.com/articles/191/swiftui-lets-us-build-declarative-user-interfaces-in-swift. Accessed 11 May 2022
6. Jenny, B., Kelso, N.V.: Color design for the color vision impaired. Cartogr. Perspect. **58**, 61–67 (2007)
7. Michael, K., Charles, L.: Color perception (2011). https://webvision.med.utah.edu/. Accessed 10 May 2022
8. Molday, R.S., Moritz, O.L.: Photoreceptors at a glance. J. Cell Sci. **128**(22), 4039–4045 (2015)
9. National Eye Institute: Color blindness. National Eye Institute, July 2019
10. PhonesData: phonesdata. https://phonesdata.com/en/best/screenppi/2021/. Accessed 22 Jan 2022
11. Pixensity: pixensity. https://pixensity.com/list/phone/. Accessed 22 Jan 2022
12. Pokorny, J., Smith, V.C.: Eye disease and color defects. Vision Res. **26**(9), 1573–1584 (1986)
13. Salih, A.E., Elsherif, M., Ali, M., Vahdati, N., Yetisen, A.K., Butt, H.: Ophthalmic wearable devices for color blindness management. Adv. Mater. Technol. **5**(8), 1901134 (2020)
14. Tecson, G.R., Calanda, F.B., Cayabyab, G.T., Reyes, F.C.: Covisance. In: Proceedings of the 2017 International Conference on Computer Science and Artificial Intelligence - CSAI 2017, ACM Press (2017)
15. Tuchkov, I.: Color blindness: how to design an accessible user interface. Medium, August 2018
16. Turgut, B.: Discriminators for optic neuropathy and maculopathy. Adv. Ophthalmol. Visual Syst. **7**, 7 (2017)

Tool for Monitoring Time Reserves in a Warehouse

Klaudia Winiarska[1](✉), Marina Chmil[2], and Karolina Radzajewska[3]

[1] Wroclaw University of Science and Technology, Wroclaw, Poland
klaudia.winiarska@pwr.edu.pl
[2] Kozminski University, Warszawa, Poland
[3] Bialystok University of Technology, Białystok, Poland

Abstract. Numerous articles describe issues related to optimizing external and internal transport routes. A crucial problem is the occurrence of deadheading, i.e., transporting a vehicle without a load. This phenomenon is the cause of the inefficient management of vehicles and increased logistics expenses.

This article presents an analysis of processes taking place in a warehousing company, locates places for possible improvements, and proposes a tool for monitoring time reserves by combining two processes (unloading process and picking process) into one process. Forklift companies are committed to minimizing deadheading runs, as this increases the efficient use of vehicles and employees and reduces logistics costs.

Keywords: Deadheading · Forklift · Storage process optimization

1 Introduction

Nowadays, planning and optimizing routes is a significant challenge. This problem occurs not only in external transport (see, e.g. [12]) but also in internal transport. Planning optimal routes of the means of transport is aimed at improving efficiency and reducing logistics costs.

An important point in planning routes is to optimize the degree usage of a vehicle's capabilities and minimize the situation in which the conveyance moves partially loaded or completely unloaded, otherwise called deadheading [2]. Deadheading occurs during the trip to collect cargo and returning trips after the cargo has been delivered. Deadheading generates a significant increase in costs.

In the literature, articles describe the problem of deadheading in means of external transport [1–4] and present ways of estimating expenses of deadheading [7]. The solution to minimizing deadheading in external transport is, among others, freight platforms that make it possible for co-partners, which is possible to reduce deadheading by transporting their cargo.

Authors in their publications describe the problem of deadheading in means of internal transport through the context of the analysis of the logistics system supporting production [8] and the impact on the environment through fuel consumption, the emission

W. Zamojski et al. (Eds.): DepCoS-RELCOMEX 2022, LNNS 484, pp. 389–397, 2022.
https://doi.org/10.1007/978-3-031-06746-4_38

of greenhouse gasses [5], and energy consumption [6, 9]. The deadheading problem is usually a side topic of the analyzed publications. The articles present methods of effective warehouse management [10]. One of the most important activities related to warehouse management is to make its products as high as possible in terms of effectively using resources. It is means of internal transport or a team of employees. Authors [10] propose a storage space allocation model that considers the availability of a forklift fleet in a warehouse. Using this model, they are minimized the volume of forklift fleet idle hours. Other authors have developed models that calculate the number of forklifts needed [11]. For companies running forklifts, it is imperative to optimize processes that can positively influence a change in the context of costs acquired. Even the slightest change in storage processes can reduce logistical costs in warehousing companies.

This article presents the problem of a warehousing company in which there are vacant passages in unloading cargo and picking them up. The article's main goal is to present a supporting tool to minimize deadheading in the case company.

The remainder of this paper is organized as follows. Section 2 describes the process of unloading and picking in the warehouse. In Sect. 3, it is approximated problem statement. The solution, which is the proposed tool, is described in Sect. 4. The Simulation results are presented in Sect. 5, and conclusions are drawn in Sect. 6.

2 Unloading and Picking Processes in a Warehouse

The investigated supporting tool was developed for a specific enterprise. Therefore, the processes connected with warehousing are presented first.

2.1 The Unloading Process

The planner receives information from the carrier about the date of arrival of the transport vehicle with the delivery of the cargo. From there, he plans out the work in the warehouse. He chooses which ramp the vehicle should dock to, based on the available spaces. Three forklift operators work one shift during the unloading process, who only perform this. The unloading process is divided into two stages (Fig. 1):

Stage I: Unloading the cargo from the loading space of the vehicle to the storage location next to the ramps.

Stage II: The organizing of cargo on racks. Forklift operators pick up the pallet from the storage place, transport it to the first free place on the rack, and then return to the place of unloading and repeat the process when cargo is to be transported. If the storage place is vacant, forklift operators wait for the subsequent transport with new cargo.

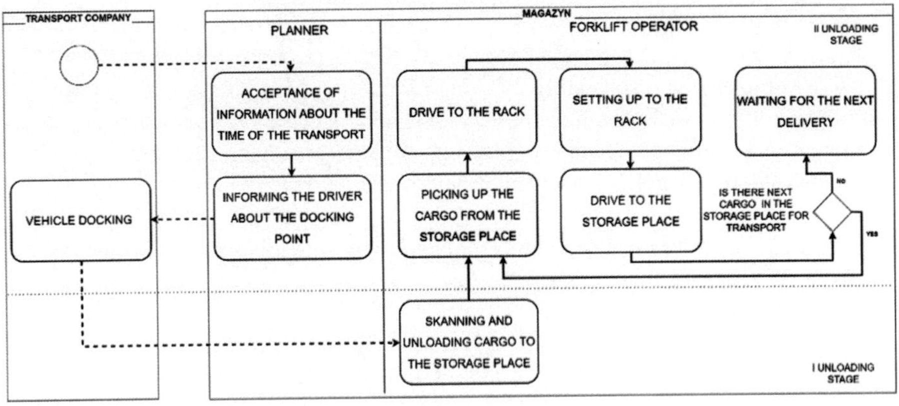

Fig. 1. Operating scheme of the unloading process in an investigated company's warehouse

2.2 Picking Process

In the described warehouse, there is no depalletization of the cargo. It is sold in the same condition as it was brought into the warehouse. Therefore, there is no need to set aside an unpacking zone. The storage area is also a picking zone. During the picking process, two forklift operators work one shift. After accepting the order, the planner assigns employees tasks in the system. The forklift operator receives a list of tasks that must be performed with the location of the cargo to be downloaded. The employee drives to the rack alley, picks up the cargo, then transports them to the storage location at the selected ramp, to which he will dock the vehicle scheduled to arrive for a given order. The process is repeated until the entire order is completed (Fig. 2).

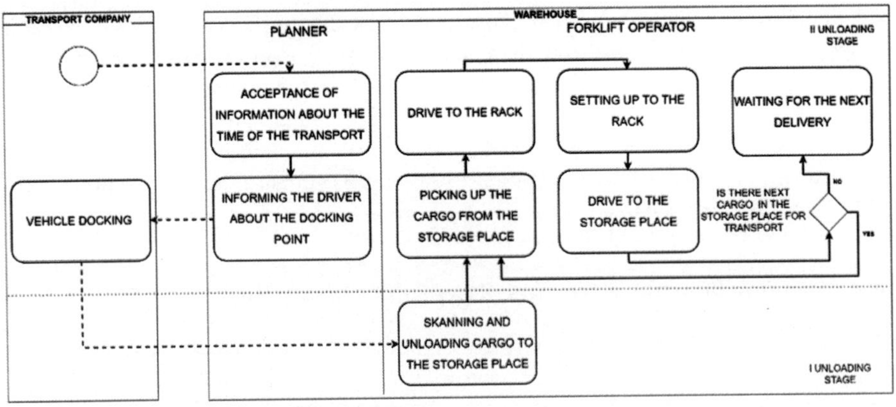

Fig. 2. Operating scheme of the picking process in a selected company's warehouse

3 Problem Statement

Modern companies are constantly focusing on refining processes to satisfy customers, minimize costs, or make the most of personnel and machine resources. For this reason, optimization of storage systems is carried out. After analyzing the processes taking place in the investigated warehouse, several problems can be noted:

a) the occurrence of deadheading crossings in the unloading process (the way back from the rack to the storage site) and in the picking process (the way from the storage site to the rack),
b) assigning a specific number of employees to a given process,
c) inefficient management of the operating time of forklifts and their operators.

In order to minimize the occurrence of these indicated problems, the creation of a tool for monitoring time reserves has been proposed.

4 Time Reserves Monitoring Tool

In the inner workings of the described tool, it was proposed to eliminate the division among employees into two separate tasks. All employees perform the same activities, and the combination of two processes (unloading and picking), currently carried out separately, into one process for employees.

The storage zone and the picking zone are combined into one. This means that the cargo transported during unloading is deposited on the same racks from which the cargo is taken for picking. Thanks to this, it is possible to combine the unloading and picking processes into one process for the employee to carry out. The proposed new process will eliminate deadheading crossings. The new process replaces the vacant passage with the handling of the second order. This means that the employee performs unloading in one direction, and on the way back, he picks up the load and completes the order. In the task list for each employee, there is a change in the unloading task and picking order. This procedure shortens the total time of completing tasks in the warehouse and minimizes the occurrence of deadheading crossings. By combining two processes into one and eliminating the division of labor, all employees unload and pick together. Figure 3 shows an operating scheme of the described process.

4.1 The Functionality of the Time Reserve Monitoring Tool

The tool functions were based on the assignment of importance ranks. The rank of importance determines the most time-sensitive task – task labeled #1 shall be performed first. The tool sorts the data according to a specific order of criteria to assign ranks to the task. Firstly, the time left until the arrival of the CZ transport is analyzed, sorting from the smallest value to the largest takes place. The second step is to sort the K value from smallest to largest. K is an indicator that determines whether it can combine the unloading and picking processes. If the K index is lower than the previously determined value adopted by the company, the tool will perform only one process without adding on

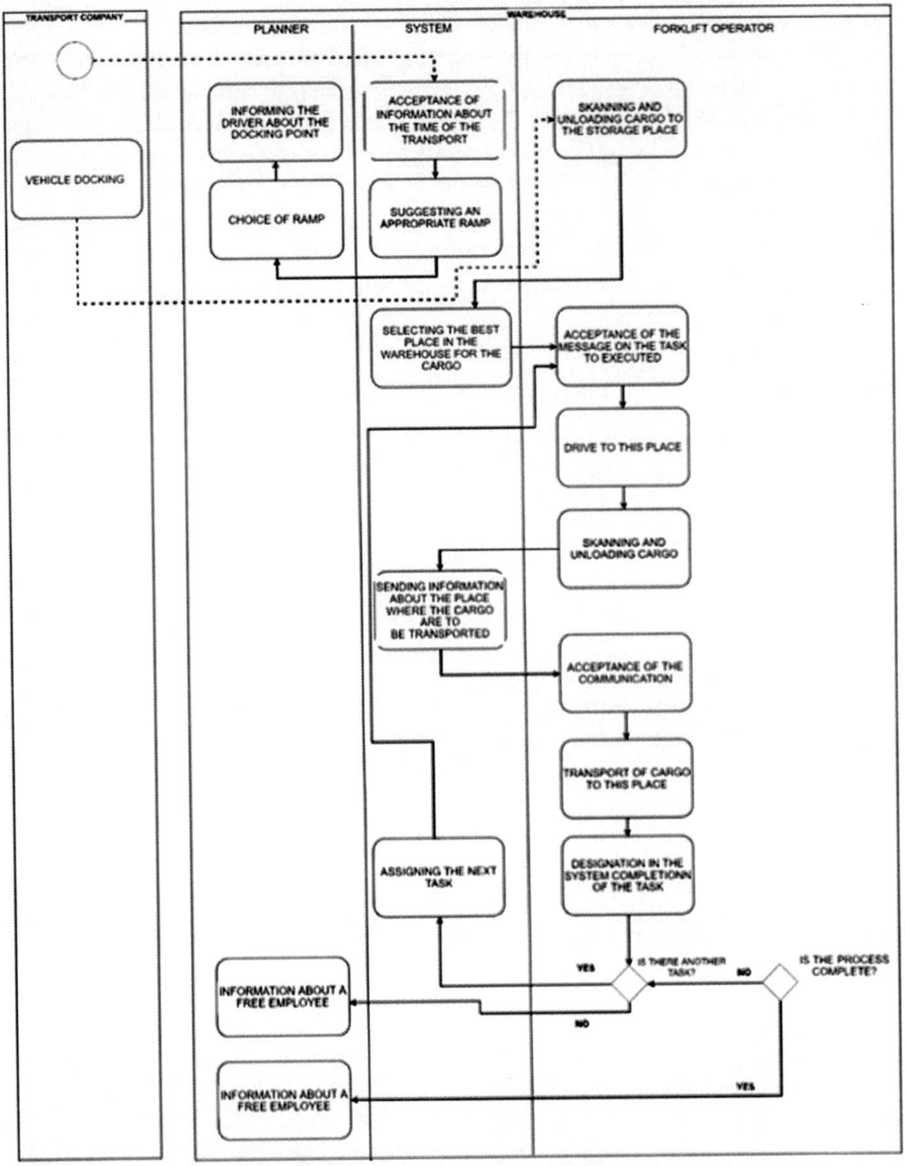

Fig. 3. Operating scheme of the described process.

another. There may be a situation whereby the time for which the cargo arrives is less than the total time of task completion; then, all employees are redirected to the completion of one order so that it is finished on time. The third step is to sort the tasks according to the execution time of a single task CWZ, and this is ranked from the longest task lead time to the shortest. The last step is to sort by the distance between the employee and the location of the cargo O-P. This distance is sorted from the smallest to the largest. This step prioritizes when several tasks have the same execution time in the same order. Table 1 shows the data before sorting, while Table 2 shows the data sorted by all steps.

Table 1. Sample orders before sorting

Rank	ID order	ID cargo	Location address	Destination address	Process	O-P	CWZ	CPT	CZ	Employees number	k	T
1	1	ID-1	AP-1	AD-1	Z	15	10	14:45	40	2	16.33	Yes
2	1	ID-2	AP-2	AD-2	Z	20	12	14:45	40	2	16.33	Yes
3	1	ID-3	AP-3	AD-3	Z	11	17	14:45	40	2	16.33	Yes
4	1	ID-4	AP-4	AD-4	Z	5	19	14:45	40	2	16.33	Yes
5	2	ID-5	AP-5	AD-5	Z	10	10	14:35	30	3	10.11	Yes
6	2	ID-6	AP-6	AD-6	Z	12	1	14:35	30	3	10.11	Yes
7	2	ID-7	AP-7	AD-7	Z	10	16	14:35	30	3	10.11	Yes
8	2	ID-8	AP-8	AD-8	Z	7	18	14:35	30	3	10.11	Yes
9	3	ID-9	AP-9	AD-9	R	9	14	14:40	35	2	15.91	Yes
10	3	ID-10	AP-10	AD-10	R	10	12	14:40	35	2	15.91	Yes
11	3	ID-11	AP-11	AD-11	R	13	16	14:40	35	2	15.91	Yes

Table 2. Sample orders after sorting

Rank	ID order	ID cargo	Location address	Destination address	Process	O-P	CWZ	CPT	CZ	Employees number	k	T
1	2	ID-5	AP-8	AD-8	Z	7	18	14:35	30	3	10.11	Yes
2	2	ID-6	AP-7	AD-7	Z	10	16	14:35	30	3	10.11	Yes
3	2	ID-7	AP-6	AD-6	Z	12	15	14:35	30	3	10.11	Yes
4	2	ID-8	AP-5	AD-5	Z	10	10	14:35	30	3	10.11	Yes
5	3	ID-9	AP-11	AD-11	R	13	16	14:40	35	2	15.91	Yes
6	3	ID-10	AP-9	AD-9	R	9	14	14:40	35	2	15.91	Yes
7	3	ID-11	AP-10	AD-10	R	10	12	14:40	35	2	15.91	Yes
8	1	ID-1	AP-4	AD-4	Z	5	19	14:45	40	2	16.33	Yes
9	1	ID-2	AP-3	AD-3	Z	11	17	14:45	40	2	16.33	Yes
10	1	ID-3	AP-2	AD-2	Z	20	12	14:45	40	2	16.33	Yes
11	1	ID-4	AP-1	AD-1	Z	15	10	14:45	40	2	16.33	Yes

5 Results

The application possibilities of the developed solution were analyzed using the example of a typical workday in the selected company investigation.

A simulation of a workday in the warehouse was made for two cases using:

a) implementation of tasks using two separate processes,
b) implementation of tasks using the proposed tool to monitor time reserves.

With the exact entry times, when using the tool to monitor time reserves, the number of deadheading has been minimized to 0, and the number of completed tasks increased compared to the implementation of tasks using two separate processes. Using two separate processes caused each completed task to be deadheading. Comparing the calculations from both examples, it was noticed that in the case of b), it is 34% more than in the case a) complete deliveries are distributed and 66% more picking up full orders.

FlexSim 19.0 simulates the working time of forklifts by introducing random variables, such as:

a) orders frequency,
b) delivery frequency,
c) time to complete individual tasks.

The purpose of the presented case study is to minimize the deadheading of forklifts in the warehouse, which is the most frequent use of combining the unloading process with the picking process. For this purpose, it was checked which frequencies of the orders and deliveries are the best using five forklifts. The simulation analysis shows that it is best if orders will appear every 37 min. If this time is longer, a deadheading will appear because it will not be picking tasks, and all forklifts will only unload deliveries. At this time, there is the smallest number of unrealized orders. Deliveries as best as appear every 27 min because deadheading is the smallest. Unrealized orders and deliveries result from the appearance of the order at the end of the 24h cycle; therefore, the employees did not make them. This criterion was disregarded when determining the best frequency of deliveries using five forklifts. Figure 4 shows the simulation results for one of the five forklifts.

A survey was also conducted among warehouse employees to check their attitude about the possibility of introducing a combination of unloading and picking processes, what benefits they see and what concerns they have about this type of solution. The respondents were both forklift operators and foremen. Figure 5 shows the results of the surveys. The surveys show that many employees support introducing this type of solution. The concerns of employees who are against the introduced changes are related to the habits of performing specific tasks with separation into two separate processes and a generally negative attitude towards any changes.

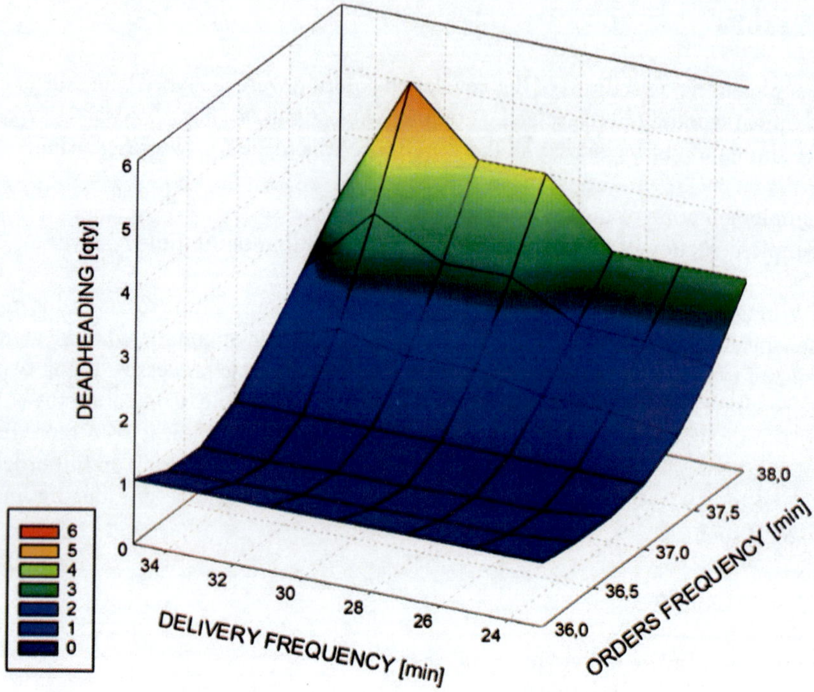

Fig. 4. The simulation results for forklift number 1 of the five forklifts.

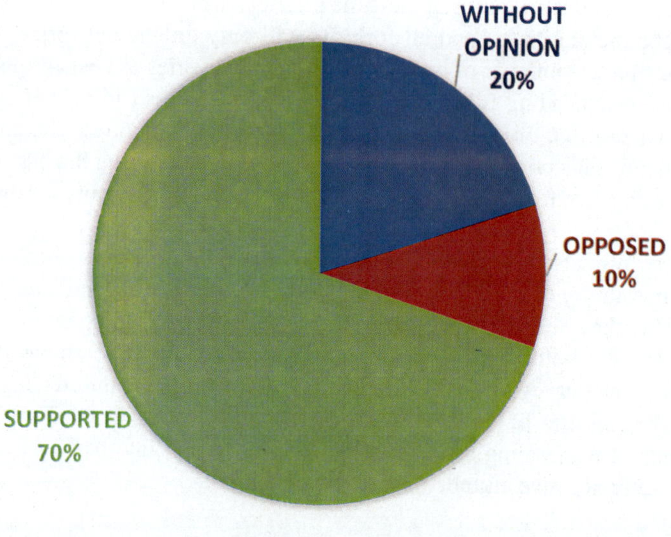

Fig. 5. The results of the surveys conducted among warehouse employees

6 Summary

In the presented article, the authors introduced a supporting tool to monitor time reserves in the warehouse, where there is no division of a picking zone and a storage zone. Optimizing the processes in the warehouse using the proposed tool will significantly shorten the time of order fulfillment and increase the percentage of machines in use – forklifts and personnel. The described tool is aimed at minimizing deadheading done by forklifts in the warehouse, while at the same time, thanks to this procedure, eliminating the idle time of employees and forklifts.

The proposed solution was developed during the Top Young 100 students project focused on creating new challenges to provide logistics students with as broad and up-to-date a range of opportunities to further their knowledge and test their practical skills as possible. The authors' further research may focus on the issues of interleaving tasks optimization to improve warehouse productivity based on various quantitative methods.

References

1. Ziółkowski, J., Lęgas, A.: Minimisation of empty runs in transport. J. KONBiN **48**, 465–491 (2018)
2. Nowotyńska, I.: Issue minimization of empty runs in express delivery company. Mod. Manag. Rev. **XIX**(21), 77–83 (2014). (in Polish)
3. Milewski, D.: Optimisation of full truckloads. Logistyka **3**(2007), 37–40 (2007). (in Polish)
4. Starkowski, D., Zielińska, J.: Exploiting the transport Wtransnet stock exchange in the planning system of the transport operation by the carrier in a transport enterprise and forwarding with using the electronic business. Autobusy Technika Eksploatacja Systemy Transportowe **12**, 1863–1870 (2016)
5. Fuc, P., Kurczewski, P., Lewandowska, A., Nowak, E., Selech, J., Ziolkowski, A.: An environmental life cycle assessment of forklift operation: a well-to-wheel analysis. Int. J. Life Cycle Assess. **21**(10), 1438–1451 (2016). https://doi.org/10.1007/s11367-016-1104-y
6. Zając, P.: Energy in the methods the optimization of warehouse systems summary. Logistyka i Transport **8**, 193–198 (2009). (in Polish)
7. Huka, M.A., Gronalt, M.: Log yard logistics. Silva Fennica **52**(4) (2018). Article ID 7760
8. Wiśniewski, C., Walczak, R., Szawłowski, J., Piegat, J.: The improvement of logistic system in manufacturing as the effect of actions based on the Kaizen idea. Logistyka **6**(2014), 11140–11149 (2014). (in Polish)
9. Saricicek, I., Keser, S.B., Cibi, A., Ozdemin, T., Yazici, A.: Energy-efficient routing and task scheduling for autonomous transfer vehicles in intralogistics. Kuwait J. Sci. **49**(1), 1–11 (2022)
10. Ghalehkhondabi, I., Masel, D.: Storage allocation in a warehouse based on the forklifts fleet availability. J. Algorithms Comput. Technol. **12**(2), 127–135 (2018)
11. McAree, P., Bodin, L., Ball, M., Segars, J.: Design of the federal express large package sort facility. Ann. Oper. Res. **144**, 133–152 (2006)
12. Emde, S., Zehtabian, S.: Scheduling direct deliveries with time windows to minimize truck fleet size and customer waiting times. Int. J. Prod. Res. **57**(5), 1315–1330 (2019)

Author Index

W. Zamojski et al. (Eds.): DepCoS-RELCOMEX 2022, LNNS 484, pp. 399–400, 2022.
https://doi.org/10.1007/978-3-031-06746-4